The Complete Guide to Fiber Optic Cable System Installation

The Complete Guide to Fiber Optic Cable System Installation

Eric R. Pearson
Certified Professional Consultant
President
Pearson Technologies Inc.

Delmar Publishers

I(T)P® International Thomson Publishing

Albany • Bonn • Boston • Cincinnati • Detroit • London • Madrid
Melbourne • Mexico City • New York • Pacific Grove • Paris • San Francisco
Singapore • Tokyo • Toronto • Washington

NOTICE TO THE READER

products described herein or perform any independent analysis in connection with any of the prod-
not assume, and expressly disclaims, any obligation to obtain and include information other than

The reader is expressly warned to consider and adopt all safety precautions that might be indicated by the activities herein and to avoid all potential hazards. By following the instructions contained herein, the reader willingly assumes all risks in connection with such instructions.

The publisher makes no representation or warranties of any kind, including but not limited to, the warranties of fitness for particular purpose or merchantability, nor are any such representations implied with respect to the material set forth herein, and the publisher takes no responsibility with respect to such material. The publisher shall not be liable for any special, consequential, or exemplary damages resulting, in whole or part, from the readers' use of, or reliance upon, this material.

Cover photos courtesy of Charles Feil, Condux International, Inc., Tektronix, and Eric Pearson

Delmar Staff
Publisher: Robert D. Lynch
Acquisitions Editor: Mark W. Huth
Developmental Editor: Michelle Ruelos Cannistraci
Production Coordinator: Toni Bolognino
Art/Design Coordinator: Michael Prinzo

Printed in the United States of America

For more information, contact:

Delmar Publishers
3 Columbia Circle, Box 15015
Albany, New York 12212-5015

International Thomson Publishing Europe
Berkshire House 168–173
High Holborn
London WC1V 7AA
England

Thomas Nelson Australia
102 Dodds Street
South Melbourne, 3205
Victoria, Australia

Nelson Canada
1120 Birchmont Road
Scarborough, Ontario
Canada M1K 5G4

International Thomson Editores
Campos Eliseos 385, Piso 7
Col Polanco
11560 Mexico D F Mexico

International Thomson Publishing GmbH
Konigswinterer Strasse 418
53227 Bonn
Germany

International Thomson Publishing Asia
221 Henderson Road
#05–10 Henderson Building
Singapore 0315

International Thomson Publishing—Japan
Hirakawacho Kyowa Building, 3F
2-2-1 Hirakawacho
Chiyoda-ku, Tokyo 102
Japan

2 3 4 5 6 7 8 9 10 XXX 02 01 00 99 98 97

Library of Congress Cataloging-in-Publication Data

Pearson, Eric R.
The complete guide to fiber optic cable system installation / Eric R. Pearson.
p. cm.
Includes index.
ISBN 0-8273-7318-X
1. Telecommunication cables. 2. Optical wave guides. I. Title.
TK5103.15.P43 1997 96-23138
821.382'75—dc20 CIP

Contents

Foreword

Fiber optics is in a period of transition. When I began working with fiber optics over fifteen years ago, it was installed by the same researchers who designed the components and networks. Since then, installation has been done by engineers, and highly skilled and well trained tradespeople. Today, installation is done by those who routinely install electrical and communications copper cables. This transition has occurred because the components have been developed to cost less, and require considerably less skill and training to install, and the number of installations has grown immensely, requiring more installers to be trained to meet the demand.

With this change, better training and reference materials are essential for the continued growth of the fiber optic market. These materials need to be aimed at a more basic level of learning, teaching technique rather than theory, focusing on component installation rather than design, and providing information gained from practical experience in real-world field installations.

Few people in fiber optics are better qualified to write such a book than Eric Pearson. His experience spans the history of fiber optics and the breadth of the industry. He has developed fiber processes, designed and installed networks, tested components, and then, written and taught widely using this experience. Eric has always questioned "common knowledge," looking for better and less expensive ways of doing things, just what is needed to help reduce the cost and difficulty of fiber optic installations. From his years of teaching, he knows what questions are asked and from his experience, he knows what answers really work.

This book is just what the fiber optic industry needs right now—a cookbook to help make work easier for everyone involved with the technology.

Jim Hayes
President
Fotec, Inc.

Preface

The goal of this book is to guide the reader to successful implementation of fiber optic cable systems. By "successful," I mean achievement of low optical power loss, low installation cost, and high reliability. To achieve this goal, this book provides extensive detail on the complete process of cable system installation. This book includes the steps required to install cable, prepare cable ends, install connectors, inspect connectors, make both fusion and mechanical splices, test cable systems, interpret test results, and certify, commission, and troubleshoot installed cable systems.

This extensive detail is required for two reasons: the wide variety of products the reader may use, and the wide variety of procedures, all of which can work if the reader follows the steps properly. The wide variety of products requires two cable preparation procedures, four connector installation procedures, three test procedures, and two splicing procedures. Access to all these procedures will allow the reader to succeed when working with any products or combination of products.

The wide variety of procedures is reflected in the detail, in the options, and in the notes. I recognize that different procedures will result in equivalent results. Wherever appropriate, I have included options and alternative procedures that I, or my professional associates, have used successfully.

This detail will enable the reader to install products not explicitly covered in this book. For example, some connectors require no epoxy or adhesive and no polishing. While the instructions for these products are not included in this book, the steps required, cleaving and crimping, are. In short, the reader will be able to install any product on the market.

During my 14 years of installation and training activities, I have experimented with and tested alternative procedures to achieve reduced loss, improved yield, and improved reliability. I have also observed what steps and procedures cause problems for the novice during the training of over 2,800 people. Finally, I have included procedures that my professional associates use and have shared with me. From these experiments, observations, and shared information, I have developed procedures that maximize the probability of success.

Many of these procedures are improvements on those procedures recommended by manufacturers. I do not say this out of arrogance, but out of testing. Manufacturers have accepted, and, in some cases, endorsed these improvements.

For instance, the 3M Hot Melt® instructions do not include an air polish. The instructions in this book include this air polish because it increases the success of both novice and professional installers.

In some cases, I've eliminated steps because my experience has shown them to be unnecessary. In other cases, the changes I've made result in greater success in the training environment.

This book is written for two types of readers: those studying the basics of fiber optic communication systems; and those who install or are training to install fiber optic systems. Those studying the basics of fiber systems will be able to understand how fiber systems function and how installation procedures can influence system performance. Those training to become involved in installation will benefit in three ways: they will gain a basic understanding of system functioning, a sensitivity to how their activities influence system performance, and a detailed knowledge of installation procedures. With practice, these procedures will give readers the ability to become professionals in this powerful and rapidly growing industry.

Acknowledgments

Companies and individuals in this industry have been exceedingly generous in their support of both this book and my activities in fiber optics. Without this support, this book would not be as beneficial to the reader as it is. To these companies and individuals, I express my deep appreciation.

These companies include: Alcoa-Fujikura Ltd., AMP Inc., AT&T, Belden Wire and Cable, Chromatic Technologies, Inc., Corning Inc., Fotec Inc., Laser Precision Corp., Northern Lights Cable Inc., Siecor Corp., Tektronix, Inc., and 3M.

Many individuals reviewed and contributed to the accuracy and quality of this book: Mr. Dan Beougher, Mr. Michael DiMauro, Mr. James E. Hayes, Mr. Nadji Salani, Mr. Larry D. Sellers, Mr. Dominick Tambone, and Mr. William Golias.

Other individuals contributed their knowledge and time in my pursuit of accuracy and relevance. To these, I give additional thanks: Mr. Alan Pallarito, Mr. John Kessler, Mr. Pieter deBruijn, Mr. Wayne Kachmer, Mr. Garfield D. Stoute, Mr. Robert Fleisch, Mr. Al Chaiken, Mr. Richard Scheer, Mr. Neil Weiss, and Mr. Dan Silver.

I appreciate the constructive comments made by the following reviewers, whose efforts resulted in improvement of the book: Mr. R. Allen Shotwell, Ivy Tech; Mr. Clay Laster, San Antonio College; Mr. Lonnie Lasher, Iowa Central Community College; Mr. Harbans B. Mathur, Kent State—Trumbull Campus; Professor Elias Awad, Wentworth Institute of Technology.

I thank Condux International, Fotec Inc., Greenlee Textron, Laser-Precision Corp., Tra-Con Inc., and Tektronix Inc. for supplying the photographs that saved me thousands of words.

Lastly, but most importantly, I thank my dear wife and two sons, for putting up with my terminal case of "fiber mania," which has afflicted me since my 1978 entry into the exciting and stimulating world of fiber optic communications.

Eric R. Pearson
Certified Professional Consultant
President
Pearson Technologies Inc.
Acworth, GA
FiberXpert@aol.com

Introduction

The installation of fiber optic communication systems is not simple. Plugging a power cord into an outlet is simple. But neither is this installation extremely difficult. When they were 11 and 13, my two sons achieved low loss and high reliability during connector installation. However, during installation, many of those new to the technology have problems, some of which are significant. I have heard hundreds of field "horror" stories from associates and from the thousands who have attended the training programs of Pearson Technologies Inc.

These problems result from a lack of understanding of the basics and a lack of essential information. This book provides both the basics and the information essential to be successful.

This book has five unique features:

- It addresses, in a single volume, the entire process of cable system installation, from installing the cable to testing for certification or troubleshooting of the cable system. The reader need not purchase a connector book for connector installation, a test book for testing, and a splicing book for splicing.
- It does not focus on a single technique for each operation. As such, this book is independent of the products to be installed and can be used by almost everyone who desires to learn proper procedures.
- It is based on significant field experience, significant experimentation, and significant testing. The procedures are not copies of those recommended by manufacturers; they have been developed and refined through experience in installations and in training with more than 18,400 connectors.
- It is extremely practical. I have limited the theory to what is required to install a fiber cable system successfully. Many of the seemingly minor details, such as the use of a digital electronic thermometer instead of a glass or metal thermometer or dry polishing of multimode connectors instead of wet polishing, are based on one simple objective: achievement of the desired results with no wasted effort.
- It incorporates troubleshooting as a part of each section. The most glaring element missing from manufacturer instructions is troubleshooting information. If a procedure does not yield the desired results, how does one interpret the results to learn what step to correct? From my field and training experiences, I've made and observed enough errors to recognize the causes. I've incorporated these experiences two ways: first, I've written the instructions as precisely as possible so that the reader can avoid incorrect procedures; and second, I've incorporated troubleshooting sections that describe common errors and the corrective actions.

This book can be used in three ways:

- As a cookbook for fiber optic installation
- As a text book for studying fiber optic cable system installation and testing
- As a guide for comparing and evaluating different products

The educational and training elements will enable the instructor and the student to realize the maximum benefit from this detailed and comprehensive text. These educational and training elements appear throughout the chapters. These elements include: Objectives, Guidelines, Procedures, Troubleshooting, and Review Questions.

This book is organized according to the sequence of the process of installation. Chapter 1, The Basics of Fiber Optic Systems, contains the information required to understand the basics of fiber systems, to understand how installation activities can impact system performance, and to follow the instructions in the rest of the book.

Chapter 2, Advantages and Types of Fiber Optic Systems, contains an overview of the advantages, topologies, and standards used in fiber optics.

Chapter 3, How to Install Cable and Prepare Ends, contains guidelines for cable installation and procedures for end preparation of cables. With minor modifications, these cable end preparation procedures can be used for all cable designs used worldwide. In addition, this chapter contains three exercises that will sensitize the reader to what the fiber will, and will not, withstand. With this sensitivity, the reader will be able to avoid fiber breakage.

Chapter 4, Connector Installation: Four Methods and Two Styles, contains procedures for installation and inspection of connectors. These procedures are for the four most commonly used procedures and for the two most commonly used connector styles. It also contains the detailed procedures for polishing singlemode connectors for low reflectance, which is required for high data rate and high bandwidth systems, such as those used in telephone and cable TV applications.

Chapter 5, How to Make Loss Measurements, contains the procedures for testing loss. These procedures include insertion loss, optical time domain reflectometry, and reflectance (or return loss). This chapter enables the reader to test installed cables, connectors, and splices to determine whether the end to end loss is acceptable and whether the cable system has been installed correctly.

Chapter 6, How to Install Splices Properly, contains the procedures for making splices. This chapter enables the reader to make both fusion and mechanical splices.

Chapter 7, How to Certify and Troubleshoot Fiber Systems, contains examples of test data and interpretations. These data can come from certification activities or troubleshooting activities. In short, Chapter 7 brings everything together.

CHAPTER

1

The Basics of Fiber Optic Systems

CHAPTER OBJECTIVES

From this chapter, you will be able to:

1. Understand the basics of fiber optic transmission technology.
2. Understand how installation procedures can affect system performance.
3. Follow the instructions in Chapters 3–6.

INTRODUCTION

In this chapter, we present an overview of fiber optic systems. This overview consists of five elements:

- Identification of system components and their functions
- Structures of different types of components
- Advantages of different types of components
- Types of performance that system components must have
- Typical values of performance

FIBER OPTIC TRANSMISSION SYSTEMS: COMPONENTS AND FUNCTIONS

Installation of fiber optic transmission systems involves installation of three groups of products:

- Major system components
- Passive optical components
- Hardware

The four major system components (Figure 1–1), which are required in all systems, are the *fiber*, *cable*, connections, and end electronics (*optoelectronics*).

Functions of the Four Major Components

Fiber. The function of the fiber is to guide the light from one end of the system to the other. The light signal is modulated to carry information.

Cable. The function of the cable is to protect the fiber. Such protection is required to allow the fiber to be installed and handled during its lifetime (estimated

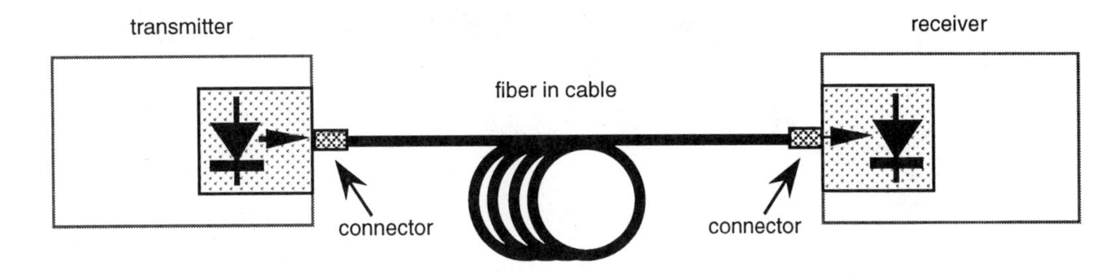

Figure 1–1 Four Major System Components

at more than 20 years) without breakage and without unnecessary loss of signal strength. All types of fibers require some cable materials in order to enable the fiber to achieve its function.

Connection Mechanisms. The functions of the connection mechanisms are alignment and end protection. Fibers, most of which are roughly the size of a human hair, must be aligned at light sources, at receivers, and at fiber-to-fiber connections. Precision alignment is required to minimize loss of signal strength.

Fiber ends must be protected from mechanical damage and from contamination by dirt. Damage and contamination can result in increased loss of signal strength, which can, in turn, result in a degradation of signal quality.

End Electronics. The function of the optoelectronic devices is conversion. The electrical signals must be converted to optical signals at the transmitter; the optical signals must be converted to electrical signals at the receiver. We will examine each of these four major components in more detail later in this chapter.

Functions of the Six Passive Optical Components

Some, but not all, systems require passive optical components. *Passive components* are those without electrical to optical or optical to electrical signal conversion. Such components are couplers and splitters, wavelength division multiplexors/demultiplexors, optical rotary joints, optical fiber amplifiers, and optical switches. The main installation concern with passive components is signal strength loss through such components. This signal strength loss is due to both connector loss and internal loss.

Couplers. Couplers are devices that combine different optical signals on different paths into a single optical signal on a single fiber. Each of these different optical signals has a different wavelength of light[1] (Figure 1–2). Each wavelength carries a different data stream. Use of couplers increases the capacity of the fiber. Couplers can be used to transmit full-duplex signals over a single fiber (Figure 1–3).

Splitters. Splitters are devices that split a single optical signal into multiple optical signals. Splitters can be used to increase system reliability and to reduce

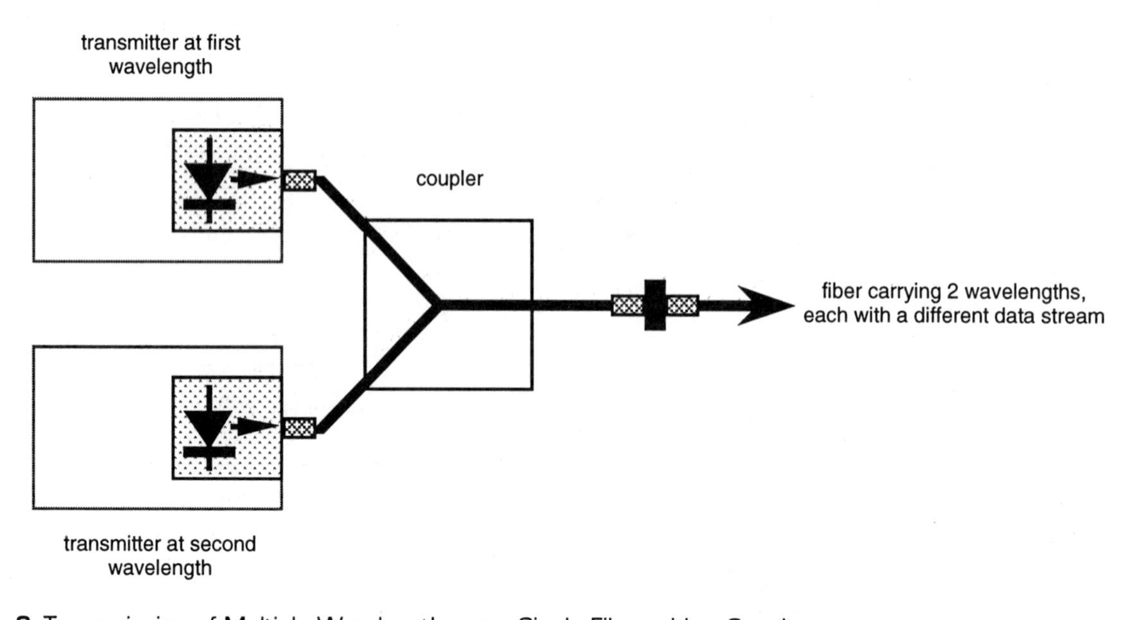

Figure 1–2 Transmission of Multiple Wavelengths on a Single Fiber with a Coupler

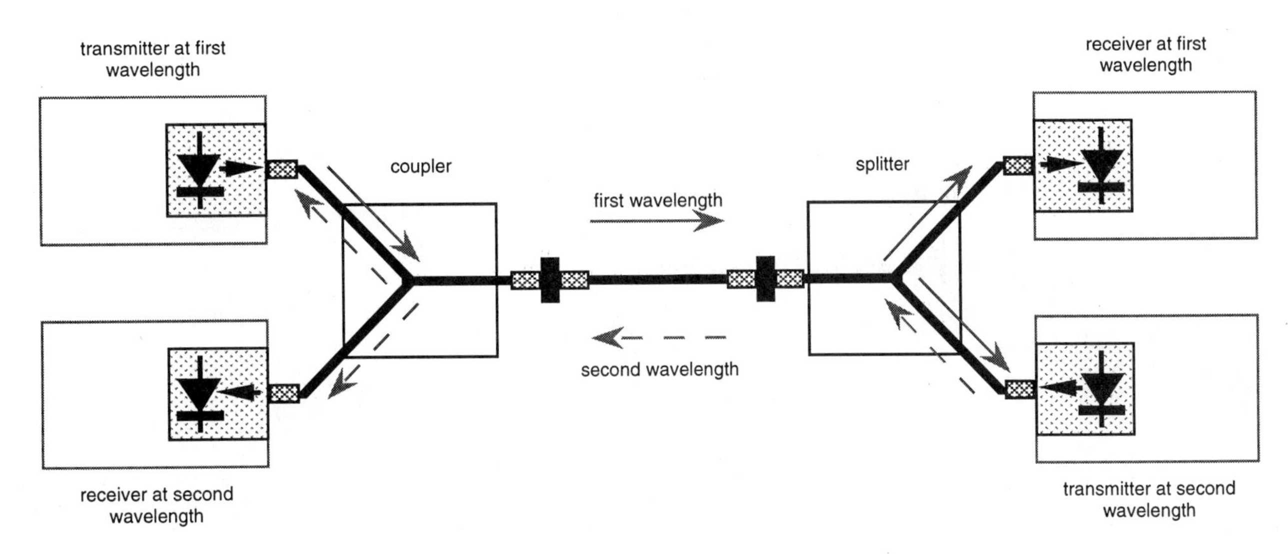

Figure 1–3 Bidirectional Transmission on a Single Fiber with a Coupler

system cost. The optical signal may consist of a single wavelength or of several wavelengths, each of which carries a unique signal. Increased system reliability results from the dividing, or sharing, of a single optical signal to redundant or multiple signal paths (Figure 1–4). Should one path be interrupted (a problem known as "back hoe fade"), a second path is immediately available. Applications of splitters to create redundant signal paths are mission-critical networks, such as those servicing NASA launches and those servicing the Federal Aviation Authority for control of airports.

Reduced system cost results from the delivery of a single information stream to multiple receiving sites (Figure 1–5). Cable TV systems use splitters to share the cost of relatively expensive laser transmitters with multiple, lower cost receivers.

Wavelength Division Multiplexors/Demultiplexors. Wavelength division multiplexors and demultiplexors are similar to couplers and splitters. The difference is that wavelength division demultiplexors direct each of the different colors to one of the outputs (Figure 1–6). In contrast, splitters split the light stream, delivering all colors to each of the outputs.

Optical Rotary Joints. Optical rotary joints serve the same function as do electrical rotary joints: optical rotary joints allow the transmission of an optical signal from a rotating structure to a stationary structure. Rotary joints are used on sensors that are towed behind ships and on rotating test equipment.

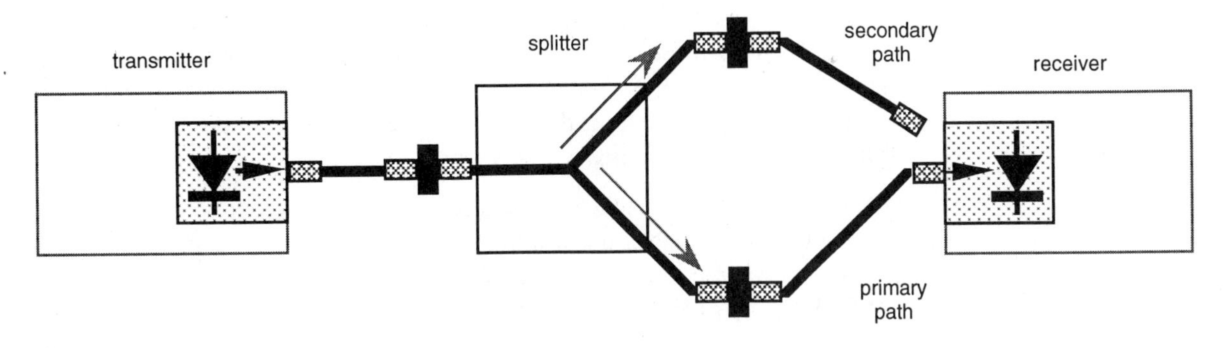

Figure 1–4 Redundant Optical Paths with a Splitter

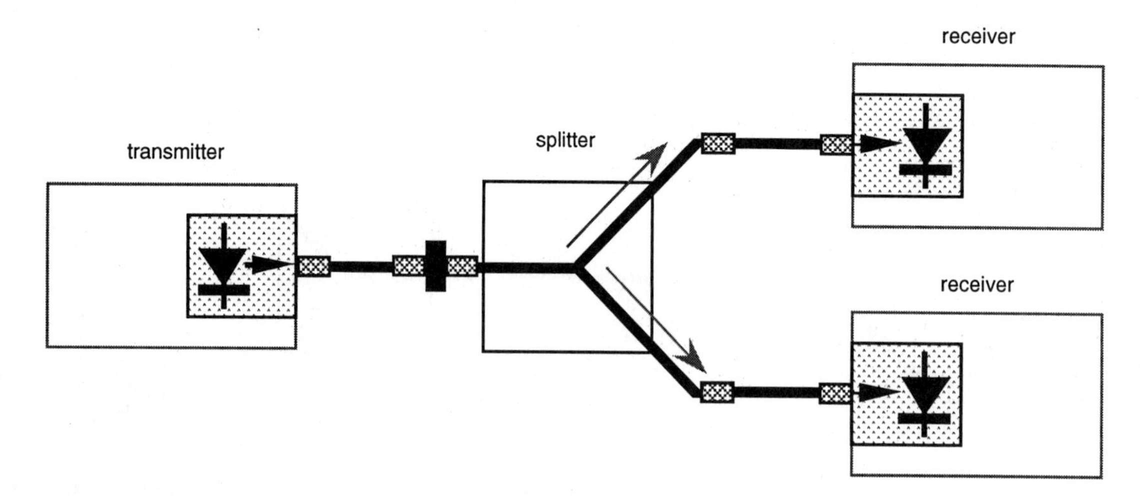

Figure 1–5 Multiple Receivers Driven from a Single Transmitter with a Splitter

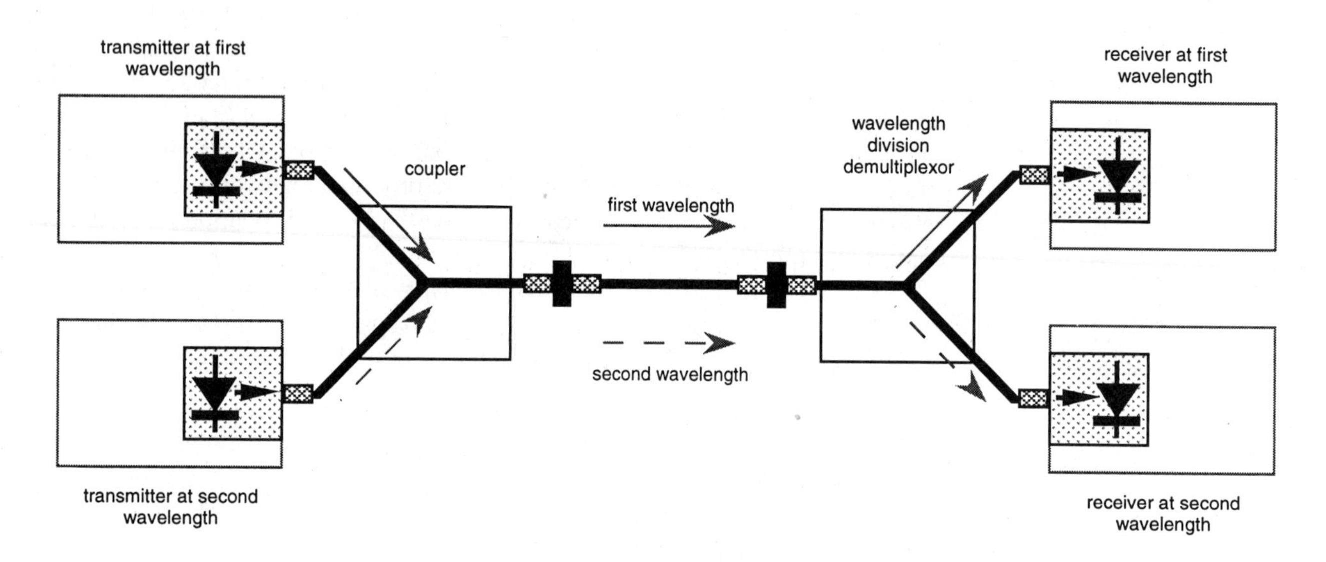

Figure 1–6 Function of a Wavelength Demultiplexor

Optical Amplifiers. Optical amplifiers allow for the amplification of an optical signal without requiring the optical signal to electrical signal to optical signal conversions. An optical amplifier contains a length of fiber that has a unique composition, usually erbium doped. The erbium doped fiber is excited with a laser having a wavelength (i.e., a second wavelength) different from that of the incoming signal. The incoming signal causes the light at the second wavelength to be converted to the wavelength of the incoming signal. Such amplifiers are used in telephone and CATV systems.

Optical Switches. Switches are used to switch optical signals from one path to another. Most switches use a moving fiber to deliver the signal to multiple outputs. An example of such a switch is an optical bypass switch, which can be used in FDDI (fiber distributed data interface) systems.

Hardware Components

Some systems require hardware. Types of hardware are:

- Inner duct
- Splice trays
- Splice enclosures
- Cable hangers
- Patch panels
- Cable end boxes
- Cable management systems

Such hardware generally increases system reliability.

Inner Duct. Use of inner duct has three functions: to allow segregation of fiber from copper cables; to create rodent resistance; and to ease the installation of additional cables.

Splice Trays and Splice Enclosures. Splice trays protect splices. Splice enclosures protect splice trays and splices from environmental and mechanical damage. Splice enclosures can have one shell or two shells. With two shells, the space between the shells can be filled with a reenter able water-blocking compound to prevent moisture damage to the splices.

Cable Hangers. Cable hangers grip cables without causing damage to the fibers. Such hangers need to be chosen for a specific cable, since the gripping force must be both high enough to grip the cable and low enough to avoid causing either excess loss of signal strength or fiber breakage.

Patch Panels. Patch panels serve the same functions in fiber as in copper systems, with one significant difference: in copper systems, the copper segments are installed on the back of the patch panel, with short jumpers on the front of the panel connecting the segments. In fiber systems, the input segments are often installed on the back of the patch panel, with the output segments on the front. The short jumper is eliminated to achieve three goals: reduced installation cost, reduced maintenance cost, and reduced optical power loss.

Cable Enclosures. Cable enclosures, or end boxes, protect the ends of the cable. The ends require protection because the fibers are exposed there. (This is true for loose tube and premise cable designs. Break out cable designs do not require enclosures, since the cable protects the fiber and the buffer tube.) Cable enclosures may include integral patch panels or splice trays.

Cable Management Systems. Cable management systems provide mechanisms for handling and storage of fibers and cables without violation of bend radius or tension limits.

THE BASICS OF OPTICAL FIBERS

The function of the optical fiber is to guide light from one location to another location. The fiber is the equivalent of a conductor in a copper wire system. However, the fiber guides rather than conducts the light. Because of this difference between a fiber and a conductor, fiber is also known as an *optical waveguide*. (Sometimes it is also called a "light pipe.")

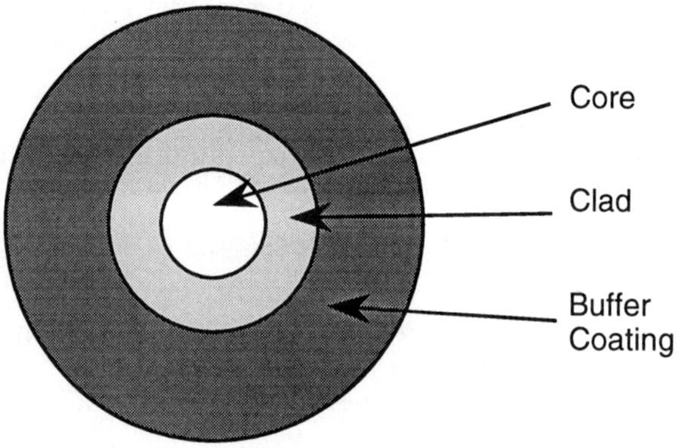

Core
Clad
Buffer Coating

Figure 1–7 The Three Regions in an Optical Fiber

The Three Regions of the Fiber Structure

The structure of an optical fiber has at least two, but usually three, regions:

- Core
- Clad (or cladding)
- Coating (or buffer coating) (Figure 1–7)

The *core* is the central region in which most of the light energy travels, similar to the central region of a water pipe. The *clad* is the region that surrounds the core. The clad causes the light to travel along the fiber by preventing the light from leaking out of or escaping from the core (Figure 1–8).

The clad serves two additional functions. First, the clad insulates and protects the core from the environment. Second, the clad increases the size of the fiber. By increasing the size, the clad increases the strength of the fiber and makes the fiber easier to handle and use. Fiber core and clad diameters depend on the materials of the fiber.

Some, but not all, fibers have a third region, called the *buffer coating* or *coating*. The function of the buffer coating is to allow the fiber to retain its intrinsic high strength. The buffer coating achieves its function by protecting the fiber from both chemical and mechanical damage.

The Three Fiber Performance Mechanisms

In order to install fiber optic systems successfully, you must understand how the optical fiber functions. The optical fiber has three key performance mechanisms:

- Total internal reflection
- Reduction in signal intensity
- Pulse spreading

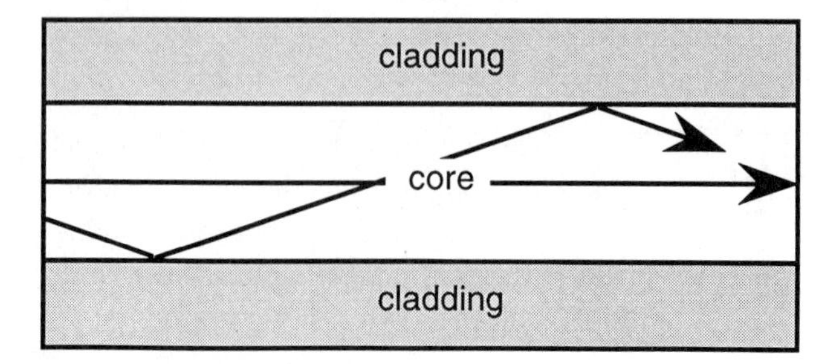

Figure 1–8 The Paths Taken by Light in an Optical Fiber Core

Total Internal Reflection. Light travels in straight lines. Yet, with an optical fiber, we can move light in paths that are not straight, just as we move water in water pipes that are not straight. Why, then, does light travel along a fiber? The answer is in everyone's experience.

Imagine that you are standing at the edge of a lake; as you look at the surface of the water at a distance from you, you see the sky reflected in the water. However, as you look closer and closer to your feet, you stop seeing the sky and instead see into the water. This change from reflection, from which you see the sky, to refraction, from which you see into the water, occurs at a certain angle to the surface of the water. We will call this angle the *critical angle*. (This angle is also called the "acceptance angle.") On one side of this critical angle, you see a reflection of the sky. On the other side of this critical angle, you see into the water. This critical angle occurs because air and water have a different optical property. This optical property is the speed at which light travels in these two materials. While we could use the term "speed of light," scientists have found it more convenient to use the term *index of refraction* (IR or η), which is a measure of the speed of light in a material.

$$\eta = (\text{speed of light in vacuum/speed of light in material})$$

In glass optical fibers, the index of refraction ranges from 1.46 to 1.51 (Table 1–1). The index of refraction is used to calibrate an optical time domain reflectometer (OTDR) so that the OTDR can provide accurate length measurements. (Note that the industry uses the core-clad diameter ratio to denote the size of the fiber [e.g. 50/125]).

Optical fibers work on the same principle as light reflecting or refracting at the air-water surface. At the cylindrical core-clad surface, the index of refraction changes. When a ray of light enters the end of a fiber, the light strikes the boundary between the core and the clad at an angle equal to or less than the critical angle

		Wavelength (in nm) of Fiber			
Supplier	**Product**	**850**	**1300**	**1310**	**1550**
Corning Inc.	SMF-28			1.4675	1.4681
	SMF-21			1.4680	1.4680
	SMF-DF			1.4718	1.4711
	Titan			1.4675	1.4681
AT&T				1.4660	1.4670
	2136E			1.4670	1.4670
	215E			1.4670	1.4670
Corning Inc.	50/125	1.4897	1.4856		
	62.5/125	1.5014	1.4966		
	85/125	1.4748	1.4692		
	100/140	1.4805	1.4748		
AT&T	50/125	1.4860	1.4810		
	62.5/125	1.4960	1.4910		
Plasma	50/125	1.4810	1.4773		
	62.5/125	1.4850	1.4910		
Spectran	50/125	1.4860	1.4810		
	62.5/125	1.4960	1.4910		

Table 1–1 Indices of Refraction for Commonly Used Fibers (Data courtesy of Northern Lights Cable, Inc., N. Bennington, VT)

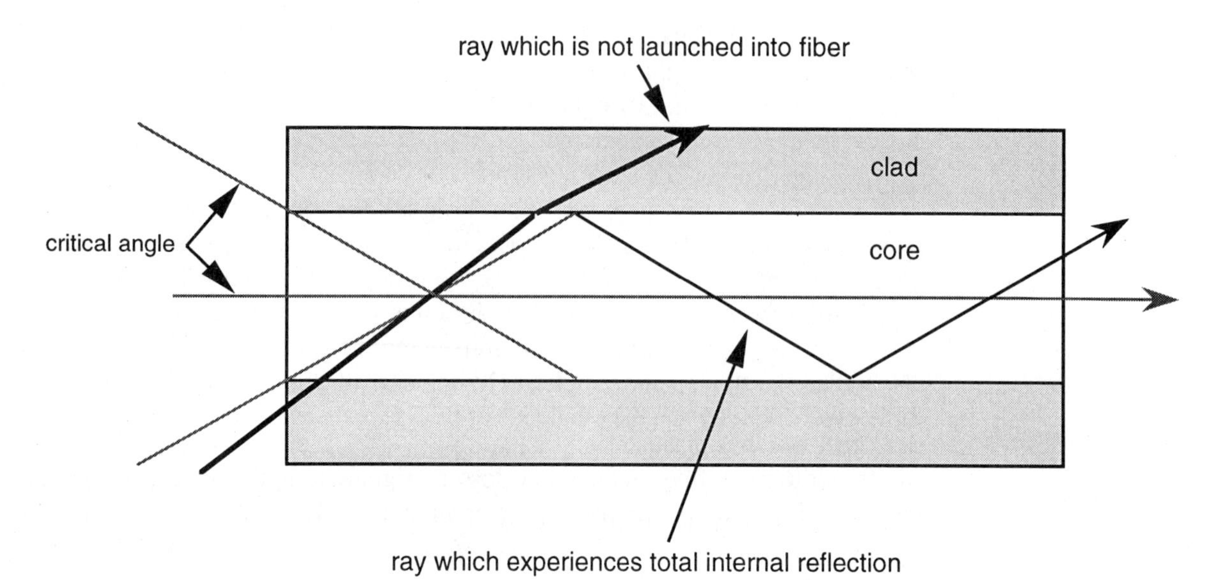

Figure 1–9 Different Paths Are Taken by Rays Within and Without the Critical Angle

or at an angle greater than the critical angle. If the angle is less than the critical angle, the light will be reflected, just as the image of the sky is reflected on the surface of the water. If the angle is greater than the critical angle, the light passes into and through the clad of fiber, just as the light from sky passes into the water of a lake (Figure 1–9).

Once light has entered a fiber within the critical angle, it is continuously reflected at the boundary of the core and clad, as long as the ray of light remains within the critical angle. We call this process *total internal reflection*. We use the term "internal reflection" because the reflection occurs inside the core. We use the term "total" because there is essentially no loss of signal strength at the reflection locations.

Just as we do not use the term "speed of light," we do not use the term "critical angle." Instead, we use the term *numerical aperture*, or *NA*, which is a measure of the critical angle. The NA is defined as:

$$\text{numerical aperture (NA)} = \text{sine (critical angle)}$$

Table 1–2 contains typical NAs for glass fibers.

Reduction in Signal Intensity. When light travels along a fiber, it becomes weaker. There are two mechanisms by which the signal becomes weaker:

- Loss of signal strength in the fiber (fiber attenuation)
- Connection loss[2]

Fiber Diameters, Core/Clad, in Microns	Most Common Numerical Aperture	Other Available Numerical Apertures
50/125	0.2 ± 0.02	0.23–0.25
62.5/125	0.29 ± 0.02	0.25–0.27
85/125	0.275 ± 0.015	
100/140	0.29 ± 0.02	0.25–0.27

Table 1–2 NA of All-Glass Fibers

Attenuation is the reduction in signal strength along a fiber and is measured in units of *dB* (decibels). When we refer to fibers and cables, we use the term *attenuation rate* in units of dB/km, to describe the loss of signal strength per unit length.

This reduction occurs at a rate low enough to allow transmission to long distances. Optical fibers exhibit a linear decrease in the logarithm of the power with an increase in transmission distance (Figure 1–10).

Note: This loss of light signal strength is called attenuation. Attenuation limits the distance of transmission, resulting in a distance beyond which we cannot transmit without regeneration or amplification. If we exceed this distance, there will be insufficient signal strength to allow the receiving electronics to convert the optical signal to an electrical signal accurately. This is one of the two performance mechanisms that limit the distance of transmission of fiber optic systems.

While fiber attenuation can result from a number of mechanisms, Rayleigh scattering is the most important. This scattering occurs when light in a fiber core strikes an atom. Some of this scattered light strikes the core-clad boundary at an angle greater than the critical angle. When this happens, this light exits the core and is lost (Figure 1–11). Table 1–3 contains typical values of the attenuation rates of various types of cabled fibers. Note that the fibers listed are glass, except for the final one, which is plastic.

In addition to fiber attenuation, loss of signal strength can occur whenever there is a connection, such as a splice or connector. Some of this connection loss depends upon the connection. We will present such losses later in this chapter.

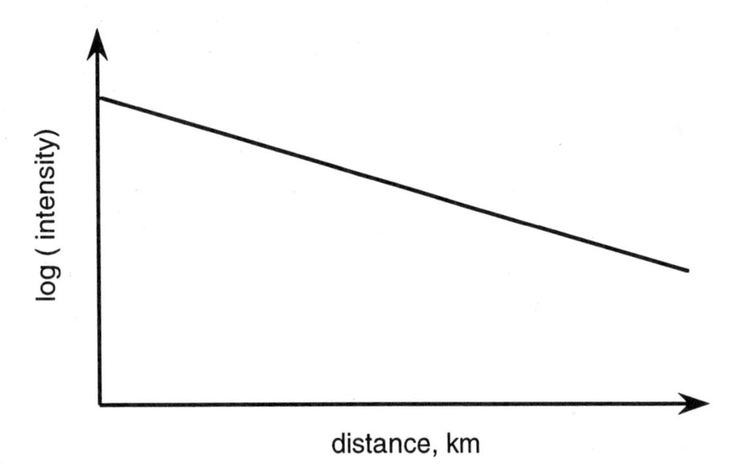

Figure 1–10 The Intensity of Light as It Travels Along a Fiber

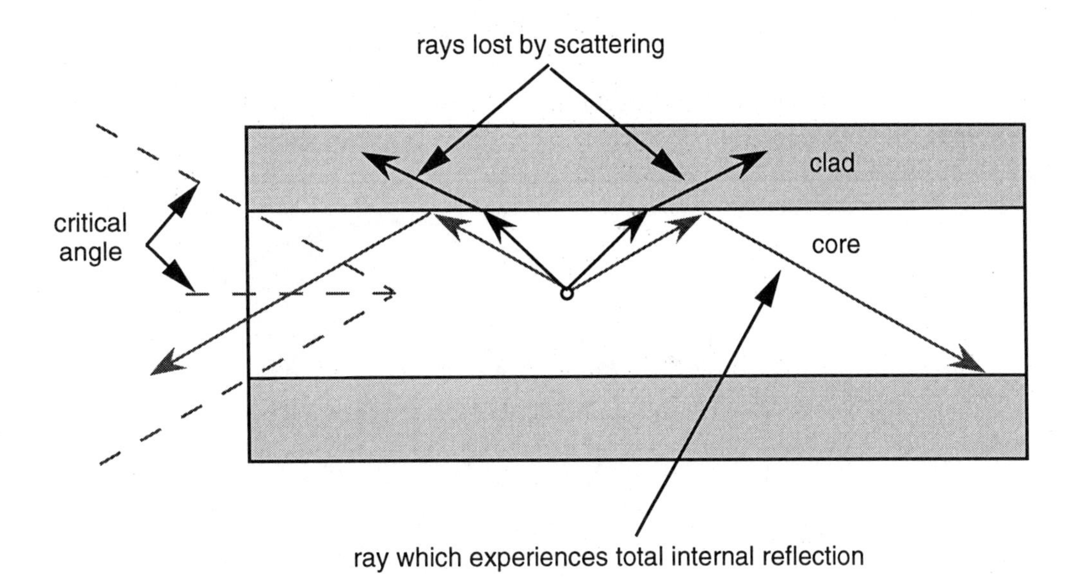

Figure 1–11 Light Lost Due to Scattering

Fiber Type	Core Diameter (µm)	Wavelength (nm)	Attenuation Rate (dB/km)
singlemode	9	1550	0.25
singlemode	9	1310	0.5
multimode	50	1300	1.0
multimode	62.5	1300	1.5
multimode	50	850	3.0
multimode	62.5	850	4.0
multimode	960	660	125–250

Table 1–3 Attenuation Rates of Fibers when Cabled

However, some of this connection loss depends upon two fiber characteristics, core offset and clad non-circularity.

The core offset is the distance between the center of the core and the center of the clad (Figure 1–12). The amount of this offset affects the performance of the fiber in connectors and in splices.

The clad non-circularity, or ovality, is the deviation of the core from perfect roundness (Figure 1–13). Both of these conditions can result in excessive loss of signal strength at connections. Table 1–4 contains typical values for offset and non-circularity from fiber data sheets.

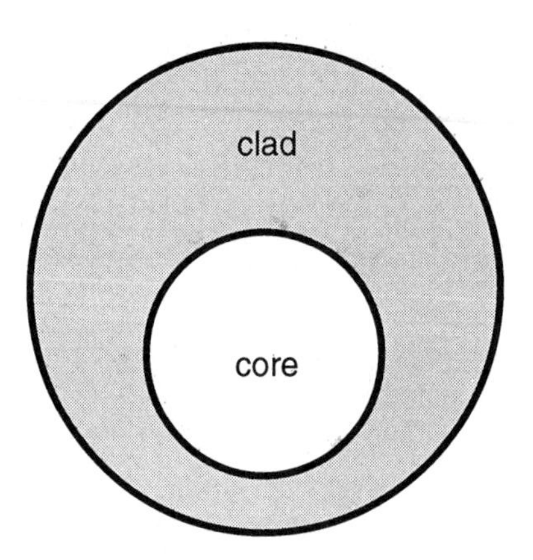

Figure 1–12 Core Offset

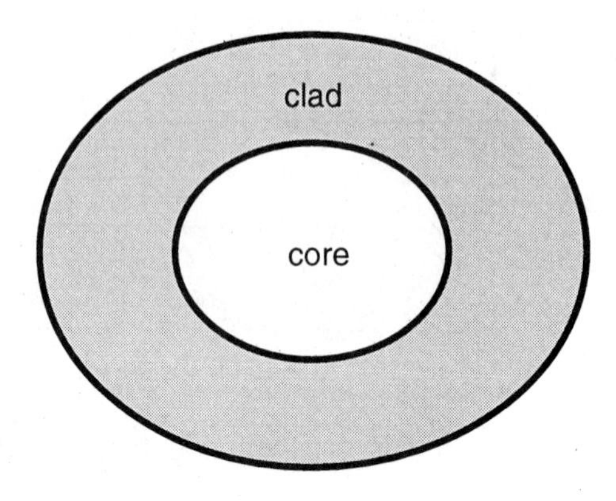

Figure 1–13 Clad Ovality

Fiber Type	Core Offset	Clad Non-Circularity
singlemode	@1310 nm <1 μm	≤2%
singlemode	@1550 nm<1 μm	≤2%
multimode	<3 μm	≤2%

Table 1–4 Fiber Concentricity and Non-Circularity Values (Data courtesy of Corning, Inc.)

Pulse Spreading. The shape of a pulse of light[3] in an optical fiber changes as it travels along a fiber. Specifically, a pulse of light gets broader in time. We call this broadening "pulse spreading" or "pulse dispersion." Let us examine how and why such pulse spreading or pulse dispersion occurs.

Consider a light source that has rays emerging at all angles (Figure 1–14). These rays can be emitting at all angles, up to and beyond the critical angle. Since those rays outside the critical angle will not experience total internal reflection in the fiber, we can ignore them.

We position this light source near the end of the fiber (Figure 1–14). We place a filter between the light source and the fiber. (Fiber systems do not use such a filter. We introduce it to simplify the explanation of pulse spreading.) This filter blocks all rays of light except those traveling parallel to the axis of the fiber (axial rays) and those traveling parallel to the critical angle (critical angle rays). When we pulse the light source, these axial rays and critical angle rays enter the fiber at the same time. The axial rays travel a path with a length equal to the length of the fiber. However, the critical angle rays reflect from clad to opposite clad as they travel along the fiber. This total internal reflection creates a longer path for the critical angle rays than for the axial rays. Because of this difference in path length, these two types of rays of light exit the output end of the fiber at different times, even though they entered the fiber at the same time.

For any length of fiber, the first ray of light to arrive at the end will be the axial ray. The last ray to arrive will be the critical angle ray (Figure 1–15). The axial ray will be represented by a point at the leading (left) side of each pulse; the critical angle ray, by the trailing (right) side of a pulse (Figure 1–15).

When we remove the filter, we allow rays of light to enter the fiber at all angles between the axis and the critical angle. These rays at all angles arrive at the

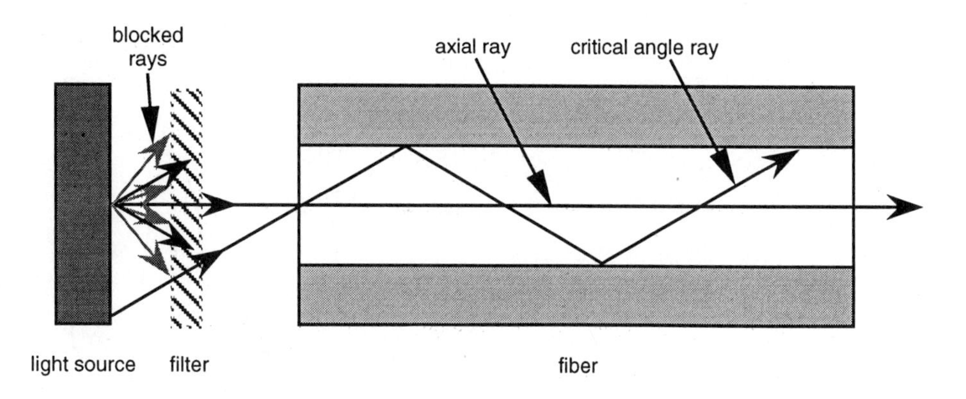

Figure 1–14 Axial Rays and Critical Angle Rays Travel Different Distances and Arrive at Different Times

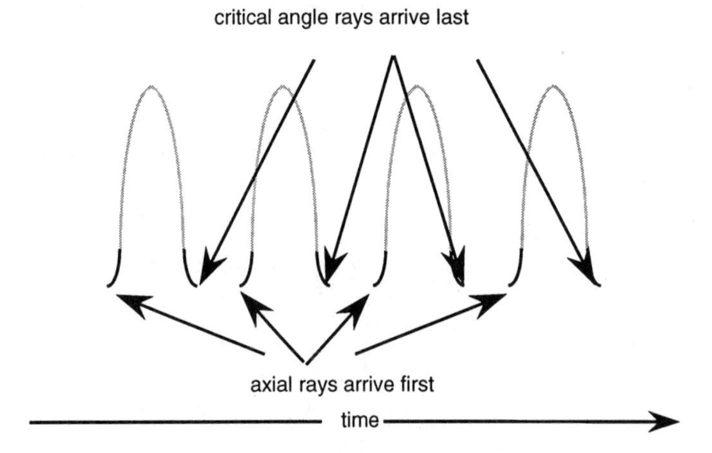

Figure 1–15 Arrival Times of Axial and Critical Angle Rays

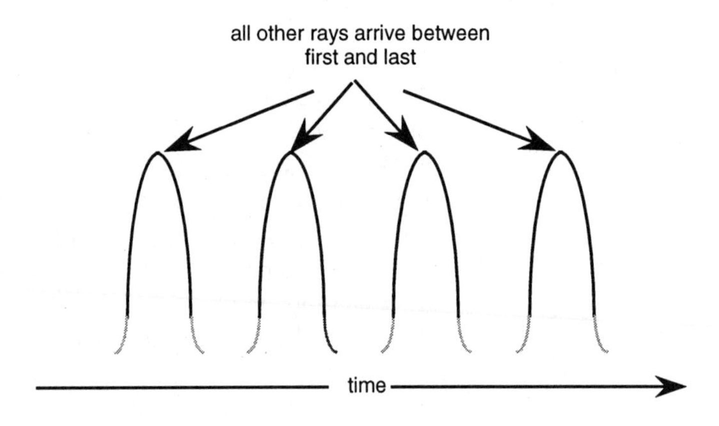

Figure 1–16 Arrival Times of All Rays

output end of the fiber at all times between the times of arrival of the axial and critical angle rays (Figure 1–16).

The difference between the time of arrival of the axial ray and the time of arrival of the critical angle ray is a measure of the pulse width. As the length of the fiber increases, the difference in the times of arrival of axial and critical angle rays also increases. Because this difference increases, the pulse width becomes larger (Figures 1–17 to 1–19).

Digital systems transmit multiple pulses, each of which experiences an increase in pulse width. As the transmission path becomes longer, the pulse width

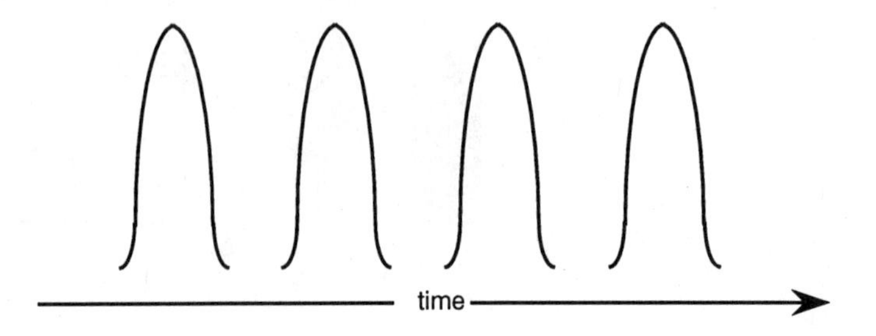

Figure 1–17 Pulse Width at Beginning of Fiber

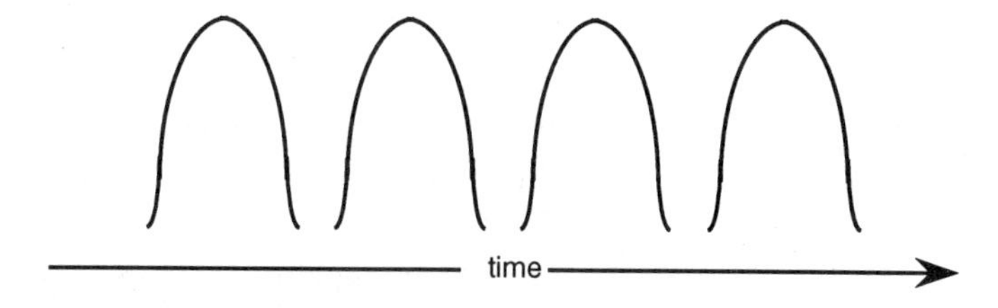

Figure 1–18 Pulse Width at Some Distance Beyond Beginning of Fiber

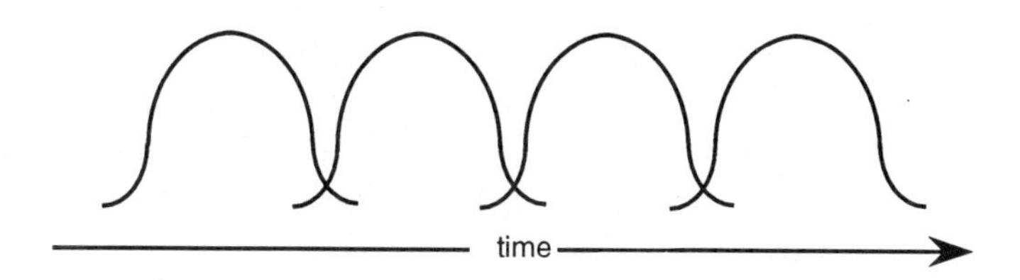

Figure 1–19 Pulse Width with Overlapping

becomes larger (Figure 1–18). As the pulse width becomes larger, the leading and trailing edges of successive pulses begin to overlap (Figure 1–19). When overlapping occurs, the receiving electronics add together the optical energy from the overlapping pulse regions, producing an output power waveform like that in Figure 1–20.

A digital system operates on the basic principle that a signal strength level below a certain level is a '0' and above that level is a '1.' The output power waveform in Figure 1–20 will be interpreted by the receiving electronics as a series of 1s. Note that the input signal in Figure 1–17 was a series of alternating 1s and 0s. Excessive pulse spreading results in loss of signal accuracy.

Note: Pulse spreading limits the distance to which we can transmit without regeneration. If we exceed this distance, successive pulses will spread sufficiently to overlap. This overlapping results in an output signal different from the input signal (Figure 1–17). This is the second performance mechanism that limits the distance of transmission for fiber optic systems.

Although pulse spreading does limit the distance of transmission, the rate of pulse spreading in optical fibers is so low that high bandwidths or bit rates can be transmitted over large distances.

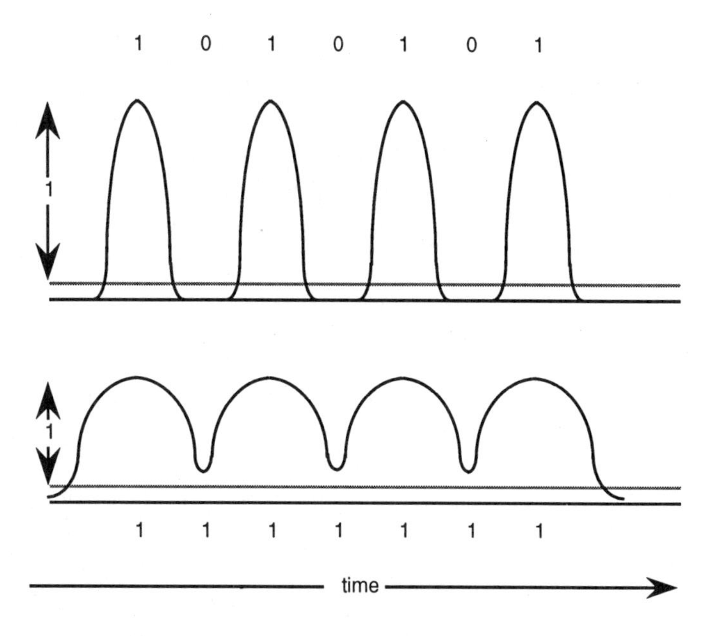

Figure 1–20 Output Pulse Intensity Waveform with Overlapping

We have just described one of the three causes of pulse spreading. This cause is called *modal dispersion* or *modal pulse spreading*. The term *mode* can be roughly defined as path. Modal pulse spreading is a result of different rays of light traveling different paths in a fiber. Since these different paths have different lengths, the time of arrival of different rays will be different.

The second and third causes of pulse spreading are *chromatic dispersion* and *material dispersion*. You have all observed the cause of chromatic dispersion in high school physics: white light passing through a glass prism separates into its component colors because different colors of light travel at different speeds in materials. Chromatic dispersion is the pulse spreading that occurs because all light sources, including those used in fiber optic communication systems, emit a range of colors of light. If the rays of light emitted by these sources travel parallel to one another, they will not experience modal dispersion. However, they will arrive at the end of the fiber at different times because they are traveling at different speeds. This difference in arrival times will result in pulse spreading.

You will be concerned with chromatic dispersion. You will address this concern by using a test light source that has a wavelength, or color, and spectral width, or range of colors, which are close to those of the light source of the optoelectronics to be attached to the cable system.

The third cause of pulse spreading is material dispersion. Material dispersion is pulse spreading resulting from different rays of light traveling in different regions of the fiber. Imagine two rays of light that are traveling parallel paths and have the exact same color of light. These two rays of light can, and do, travel in different regions of the core of the fiber. Since they do, and since optical fibers do not have an exactly uniform chemical composition throughout the entire core (because they are non-crystalline, or amorphous, materials), these two rays travel at different speeds and arrive at different times. Again, this difference in time of arrival results in pulse spreading.

There are two changes in the shape of a pulse as it travels along a fiber: first, the strength of the pulse gets lower; and second, the pulse gets wider. Both of these changes limit the distance to which a fiber optic system can transmit.

Reducing the Rate of Pulse Spreading. Clever engineers and scientists realized that pulse spreading would limit the usefulness of optical fibers as an information transmission medium. They reasoned that pulse spreading would cause them to space the pulses further apart at the input end in order to eliminate excessive overlap at the output end of the fiber. This increase in spacing would limit the amount of information, called *bandwidth* or *bit rate*, which could be transmitted. In addition, their analysis showed that the maximum bandwidth or bit rate would be much lower than they desired.

In order to increase the information-carrying capacity (bandwidth or bit rate) of fibers, these scientists made two changes to the core of the fiber. With the first change, scientists and engineers sought to reduce the amount of pulse spreading by compensating for the differences in path lengths. In the second change, scientists and engineers completely eliminated modal pulse spreading.

In order to compensate for differences in path lengths, scientists redesigned the core of the fiber. Instead of having a constant composition in the core, scientists designed a fiber with multiple compositions. The first composition, in the center of the core (Figure 1–21), was chosen to have the lowest speed of light. Around the first composition is a second composition, in which the speed of light is higher than in the first. Around the second composition is a third composition, in which the

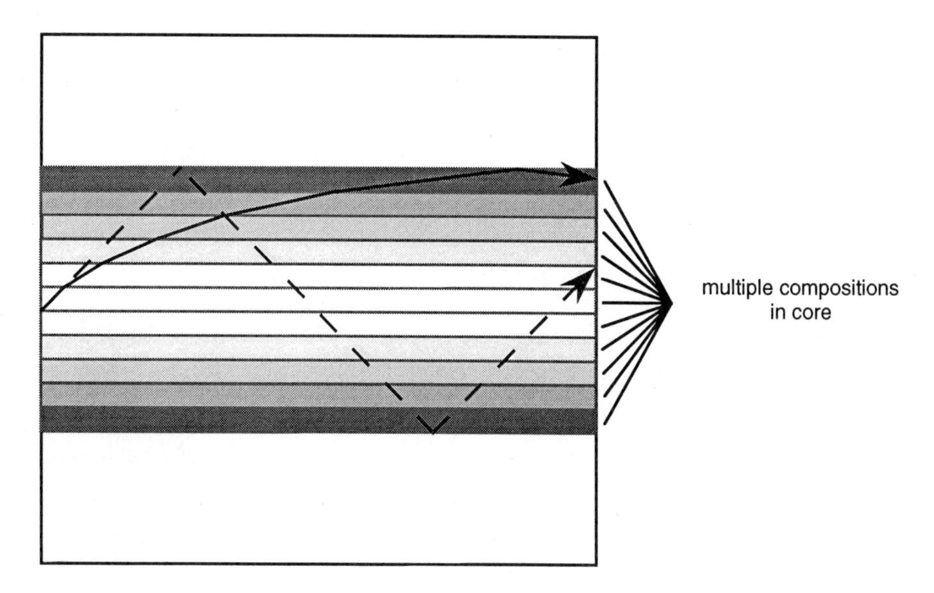

Figure 1–21 Path of Axial and Critical Angle Rays in Graded Index Core

speed of light is higher than in the second. This layering, or grading, of compositions is continued until the core has from 200 to 1,000 different compositions.

This grading of the core results in compensation: the farther away from the fiber axis a ray of light travels, the farther it goes. In addition, the farther away from the fiber axis a ray of light travels, the higher the average speed at which it travels. The longer distance of travel is offset, at least partially, by a higher average speed.

By grading the core in this manner, the scientists realized a second form of compensation: each time a ray of light crossed from a lower speed region to a higher speed region, the ray bends back towards the lower speed region (Figure 1–21). This bending creates a fiber path that appears to be curved (Figure 1–22), as compared to the straight path of light in fiber with a constant composition throughout the core (Figure 1–8). This second form of compensation results in the ray of light traveling a shorter distance than it would if the core had a single composition. These two compensations reduce the rate at which pulses spread in an optical fiber. This reduction in the pulse spreading rate results in an increase in the information handling capacity (i.e., the bandwidth or bit rate) of the fiber.

We must differentiate between cores with these two composition structures. The first type[4] is a *step index* fiber. A step index fiber has a core with a single composition (Figure 1–23). The second type is a *graded index* fiber (Figure 1–24). A graded index fiber has a core with a composition that gradually changes from the center of the core to the clad. Because these multiple compositions reduce the rate of pulse spreading, graded index fibers have a higher information-carrying capacity than step index fibers. “Index” refers to the index of refraction, which undergoes an

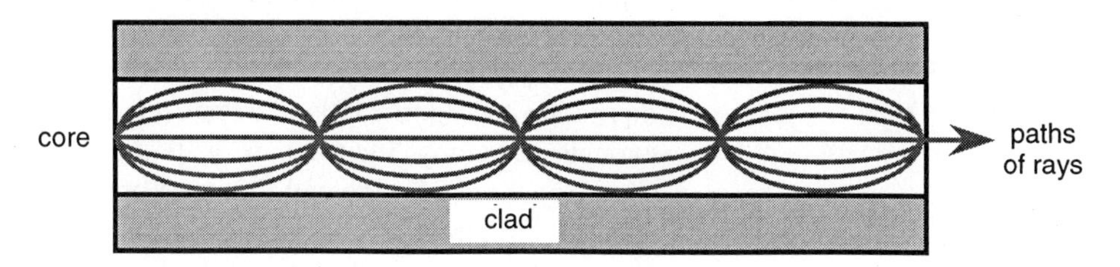

Figure 1–22 Curved Path of Ray in Graded Index Fiber

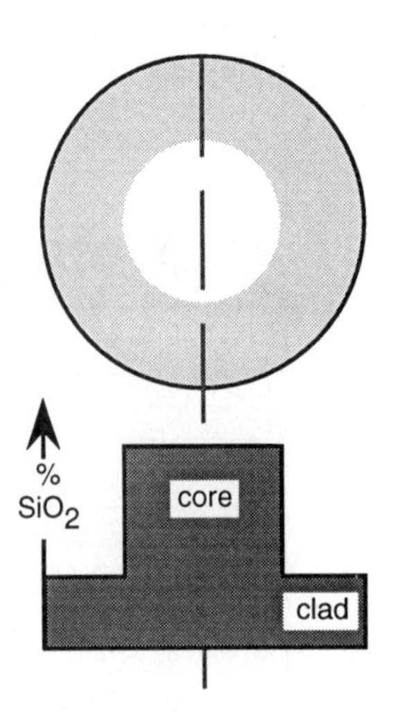

Figure 1–23 Composition Profile of Step Index Fiber

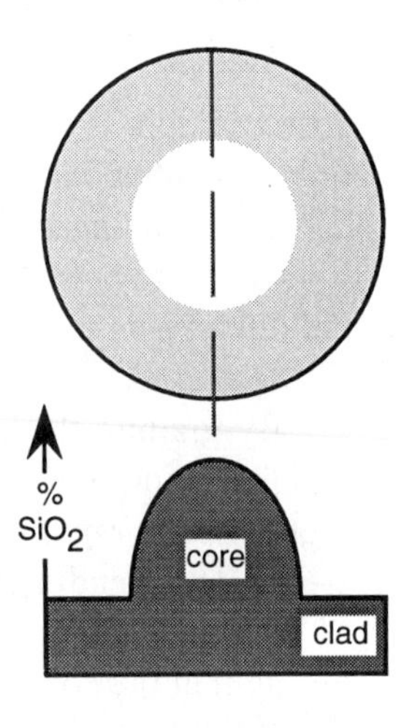

Figure 1–24 Composition Profile of Graded Index Fiber

abrupt change at the core-clad boundary of a step index fiber, but which changes gradually throughout the core of a graded index fiber. Both step index and graded index fibers (Figures 1–8 and 1–22) are called *multimode*, since light can travel multiple paths.

Note: Pulse spreading is not a concern during installation. Cable installation errors can result in a reduction in pulse spreading and an increase in information-carrying capacity. Because such errors result in a reduction in signal strength at the receiver, such errors are undesirable.

Although the change from step index fibers to graded index fibers resulted in a considerable reduction in the pulse spreading rate, all three dispersions (modal, chromatic, and material) were still active. Scientists wanted to increase the information-carrying capacity beyond the theoretical and practical limits of graded index fibers.

To learn how to achieve this additional increase in information-carrying capacity, scientists examined the quantum mechanics of light. From this examination, scientists learned how to eliminate modal pulse spreading, the largest cause of pulse spreading, completely. Scientists learned that they could eliminate modal dispersion by significantly reducing the core diameter. When the core is small enough, all rays of light behave as though they travel parallel to the axis of the fiber (Figure 1–25). We call this type of fiber a *singlemode* or *monomode* fiber, since all light travels in a single path. Singlemode fibers have significantly higher information-carrying capacities than do multimode fibers because the largest cause of pulse spreading, modal dispersion, does not occur.

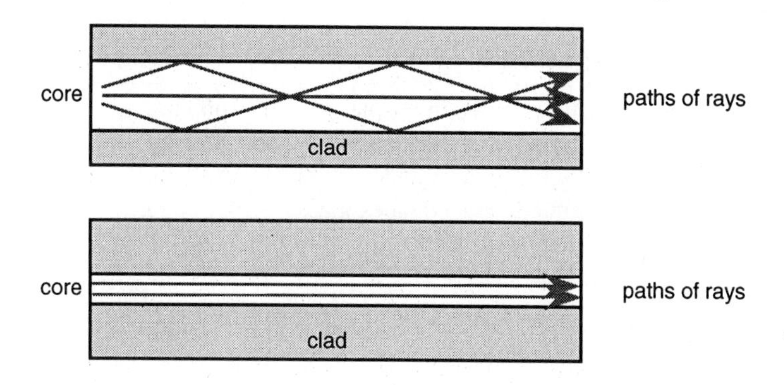

Figure 1–25 Comparison of Paths of Light in a Multimode Fiber (Top) and a Singlemode Fiber (Bottom)

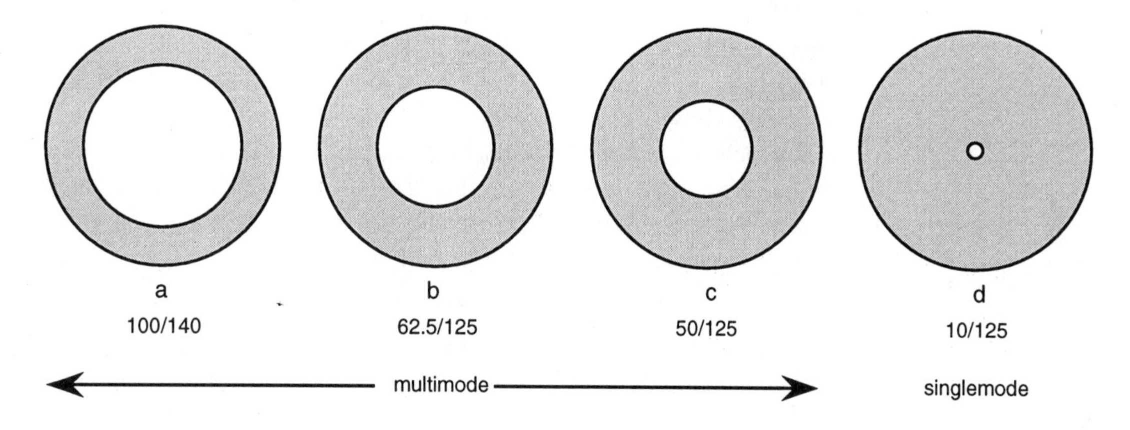

Figure 1–26 A Comparison of Core Diameters: Multimode and Singlemode Fibers, Drawn to Scale

This reduction in core size is significant: typical multimode core diameters are 50 μm, 62.5 μm, and 100 μm; typical singlemode core diameters range from 8 to 10 μm (Figure 1–26 a, b, c, and d respectively).

The core diameters of singlemode fibers are so small that some of the energy field travels in the clad. Because connections require alignment of the region in which the light travels, we need another term, the *mode field diameter*. The mode field diameter is the diameter within which most of the light energy travels. The mode field diameter is slightly larger than the core diameter, approximately 10 μm.

Pulse spreading rates are not our primary interest: information-carrying capacity is. The most commonly used information carrying capacity terms are "bandwidth" and "bit rate." It would make sense to use these two terms to characterize fibers. However, use of these terms is technically difficult because of two factors: link length and chromatic dispersion. In order to supply fiber with a desired bandwidth or bit rate, the supplier must know the length of the link(s). He or she must know this length(s) in order to calculate the pulse dispersion. In most situations, the supplier will not know such lengths.

In addition, the supplier usually does not know the nature of the light source to be used in the system. The nature of the light source determines the amount of chromatic dispersion that occurs in the link. Because of these two difficulties, the terms "bandwidth" and "bit rate" are not used as performance measures of optical fibers. Two other terms are used: *bandwidth-distance product*, in MHz-km, for multimode fibers, and *dispersion rate*, in picoseconds/nanometer/kilometer (ps/nm/km), for singlemode fibers.

Type of Fiber	Wavelength	Bandwidth-Distance Product
multimode step index plastic	660 nm	5 MHz-km
multimode step index glass	850 nm	20 MHz-km
multimode graded index glass	850 nm	600 MHz-km
multimode graded index glass	1300 nm	1000–2500 MHz-km
singlemode glass	1310 nm	76,800–300,000 Mbps-km

Table 1–5 Comparison of Typical Bandwidth-Distance Products of Fibers

The bandwidth-distance product (BWDP) is the product of the length of the fiber or cable and the maximum bandwidth that the fiber or cable can transmit under specified measurement conditions.[5] Roughly speaking, a fiber with a bandwidth-distance product of 20 MHz-km, can, under the precisely defined measurement conditions, transmit a signal of 20 MHz over a distance of 1 km, a signal of 10 MHz over a distance of 2 km, or a signal of 40 MHz over a distance of 0.5 km.

From Table 1–5, we can see that the information-carrying capacity of graded index multimode fibers is thirty to fifty times greater than that of step index multimode fibers. We can also see that the information-carrying capacity of singlemode fibers is approximately seventy-seven to 128 times that of graded index multimode fibers.

The Three Fiber Material Systems and Four Fiber Types

Most fibers have a glass core and a glass clad, but some fibers have a plastic core and/or a plastic clad. These various combinations result in three material systems and four types of fibers: all glass, plastic clad silica (PCS), hard clad silica (HCS), and plastic optical fiber (POF).

All glass and PCS fibers have a coating or buffer coating (Figure 1–7). On all-glass fibers, this buffer coating protects the fiber from becoming weakened by abrasions of the clad or by interaction of the clad with gases and liquids in its environment. In other words, the buffer coating allows the fiber to retain its intrinsic high strength.

On PCS fibers, the buffer coating prevents the clad, which is a weak silicone rubber, from being mechanically removed from the core. If such removal occurs, the light will escape from the core and the attenuation of the fiber will increase.

A buffer coating is usually absent from a POF fiber and may, or may not, be present on a HCS fiber.

All-Glass Fibers. According to Kessler Marketing Intelligence, Newport, R.I., most fibers (88%) are all-glass (glass core and glass clad), singlemode fibers. These fibers are used because of their high information-carrying capacity (high bit rate or bandwidth, Table 1–5) and low attenuation rates (for long distances of transmission, Table 1–3). Singlemode fibers are used by telephone companies, CATV companies, and corporations as part of backbone networks.

Singlemode fibers are less expensive than multimode fibers. However, the optoelectronics and connectors for singlemode systems are more expensive than those for multimode systems. Almost all singlemode fibers have core diameters of 8–10 μm and a clad diameter of 125 μm.

Most of the remaining 12 percent are all-glass multimode fibers, which are used by corporations for LANs, backbone networks, video transmissions (RGB and CCTV), process control applications, and security applications. These fibers are used because of their moderate bandwidth-distance products, moderate attenuation rates, reduced optoelectronic and connector prices (relative to those of singlemode systems), lower fiber prices (relative to those of HCS and PCS fibers), compliance with standards, and upgradeability.

Most multimode fibers (Table 1–3) are graded index fibers with a clad diameter of 125 μm, although there is some use of fibers with clad diameters of 140 μm and above. Core diameters with the 125 μm clad are: 50 μ, 62.5 μ, and 85 μ. The 62.5 μm fiber is the most frequently used because most standards (such as FDDI, 802.3, Fiber Channel, and EIA/TIA-568A) have been or are being developed around this fiber.

Although the 62.5 μm fiber is most frequently used, the 50 μm fiber has the advantages of comparatively reduced attenuation rate, increased bandwidth-distance product, and reduced cost. However, the 50 μm fiber has a smaller core size and numerical aperture (NA), both of which contribute to reduced light power launched or coupled into the core from typical LED light sources.

The 140 μm clad diameter fiber has a core diameter of 100 μm. This fiber is the most expensive of the multimode, graded index, all-glass fibers. This fiber experienced popular use in the early to mid-1980s, but has lost market share to the 62.5 μm fiber. It is sometimes used in LANs incorporating optical splitters.

Hard Clad Silica (HCS, by the Ensign-Bickford Co. and Spectran Corp., and TECS, by 3M). Hard clad silica consists of a step index silica (SiO_2) core surrounded by a proprietary hard plastic clad. This fiber combines the low attenuation rate of a glass core with the advantages of the plastic clad. This clad has five advantages relative to those of a glass clad: the fiber has very high strength, a very small bend radius, and a high resistance to surface damage, which would weaken an all-glass fiber. This clad allows a connector to be crimped directly to the clad. In addition, this clad allows the fiber to be cleaved as though it were an all-glass fiber. This direct crimping and cleaving allows for fast, and low-cost, connector installation. The increased costs of this fiber and its connectors and reduced optical performance (increased attenuation rate—6 dB/km at 850 nm—and reduced bandwidth-distance product—17 MHz-km at 850 nm) have resulted in limited usage of HCS and TECS.

However, HCS is used when the environment favors the cleaving operation, and in connector-intensive situations, in which reduced labor cost offsets higher fiber and connector costs.

Plastic Clad Silica (PCS). PCS consists of a soft plastic clad of silicone rubber, placed around a step index silica (SiO_2) core. This fiber combines the low attenuation rate of a glass core with a soft plastic clad to which a connector can be crimped. The soft plastic clad can allow the core to be eccentric in the clad and in the connector, resulting in undesirably high connector losses. This product requires a buffer coating to protect the soft clad. This product was popular in the early 1980s, but now is rarely used in communication applications.

Plastic Optical Fibers (POF). All plastic fibers consist of a step index, plastic core surrounded by a plastic clad. The materials used in POFs are usually polymethmethacrylates, although styrenes, fluorocarbons, and polycarbonates can be used. POF optoelectronics have a visible wavelength of approximately 660 nm.

POFs have technical and economic advantages. The two technical advantages are an extremely large core (typically 960 μ) and ruggedness. POFs do not suffer from the problem of breakage, as do all-glass, HCS, and PCS fibers.

The economic advantage of POFs is low cost. POF cables, fiber, connectors, connector installation, and LEDs all cost less than their glass fiber cousins. Low-cost, injection molded connectors can be crimped directly to the clad. Installation of such connectors takes less than two minutes, as opposed to 5 to 15 minutes for all-glass fiber connectors.

POF has five disadvantages. The first disadvantage is a high attenuation rate (Table 1–3), which limits the use of POF to short distance applications, typically less than 328 feet (though it is possible to use POF to 500 feet). This distance limitation suggests that POF is ideal for desktop to wiring closets, in which the typical distances are less than 328 feet. Low cost laser diodes may increase this limit to thousands of feet and bit rates to in excess of 100 Mbps.

The second disadvantage is low bandwidth-distance product (Table 1–5). The third, fourth, and fifth disadvantages are lack of knowledge on the part of the potential user, lack of support from those companies most likely to benefit from marketing of this product, and lack of standards, both for measurement and for product performance criteria. POF is likely to see increased usage in the next 10 years.

BASIC CABLE FACTS

The Ten Structural Elements and Functions

All structural elements of a fiber optic cable protect the fiber during installation and use. Many types of protection are required. The fiber, usually glass, cannot be subjected to excessive tension, bending, crushing, or to chemical or mechanical damage from the environment.

This protection comes from ten structural parts:

- Buffer tube
- Strength members
- Outer jacket
- Inner jacket
- Gel filling compound
- Water blocking compound
- Binding tape or yarn
- Armor
- Ribbon
- Fillers

Buffer Tube. The *buffer tube* is the first layer of plastic placed around a fiber by the cable manufacturer. There are two types of buffer tubes, *tight buffer tube* and *loose buffer tube*.

In the tight buffer tube, the inner diameter of the buffer tube is exactly the same as the outer diameter of the fiber (Figure 1–27). The tight buffer tube design has the unique advantage of allowing the ends of a broken fiber to remain

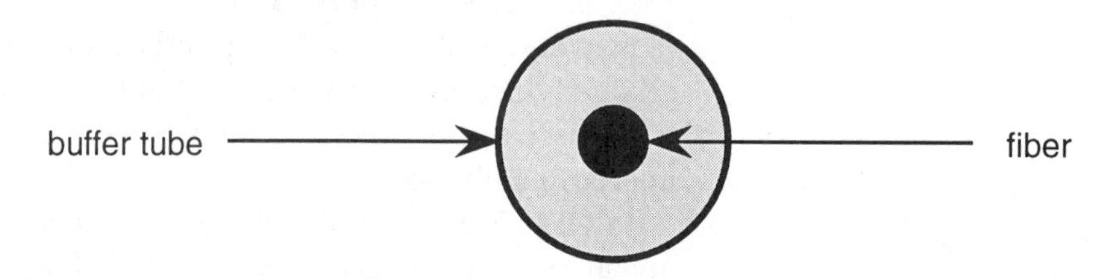

Figure 1–27 Cross Section of a Tight Buffer Tube

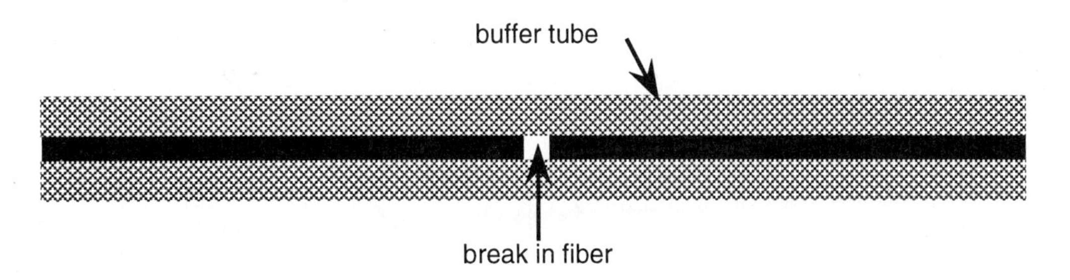

Figure 1–28 Broken Fiber in a Tight Buffer Tube

aligned and in contact (Figure 1–28). In this situation, a broken cable may not result in a lost signal. This advantage creates a highly reliable cable in severe environments. For this reason, tight tube designs are favored by the Army in field tactical applications, by the Navy for shipboard applications, and by the Air Force in aircraft applications.

A second advantage of tight tube designs is low labor cost for end preparation. The cost is low because there are no water blocking compounds to remove for splicing and connector installation. A third advantage is increased ruggedness: the buffer tube increases the overall diameter of the buffered fiber approximately 3.5 times.

In the loose buffer tube, the inner diameter of the plastic tube is much larger than the outer diameter of the fiber (Figure 1–29). In addition, the length of the fiber can be greater than the length of the tube. This difference between fiber length and buffer tube length results in the fiber having a helical path in the buffer tube. Loose buffer tubes can contain one or more fibers; tight buffer tubes contain only one fiber.

The loose tube design has two basic advantages: isolation from mechanical and environmental forces and ease of filling with water blocking materials. This isolation results from a mechanical dead zone (Figure 1–30). This mechanical dead zone isolates the fiber from mechanical and environmental forces.

The mechanical dead zone occurs because the fiber can both move from the center of the loose buffer tube to the inside edge when stress is applied to the cable, and straighten out from its normal helical path. Both of these features allow a fiber to be isolated from forces applied to a cable. Whenever a force is applied to a loose tube cable, there is a force level below which no force is applied to the fiber. If no force is applied to the fiber, it is impossible to break the fiber.

A second, generally accepted advantage of loose tube designs is good resistance to damage from moisture. Loose tube designs can be easily filled with water blocking materials.

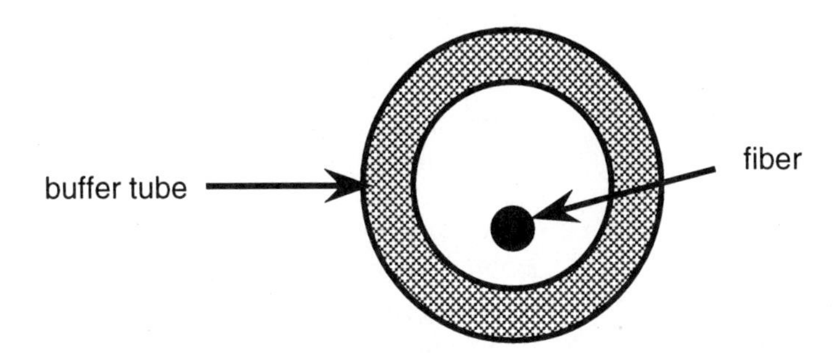

Figure 1–29 Cross Section of a Loose Buffer Tube

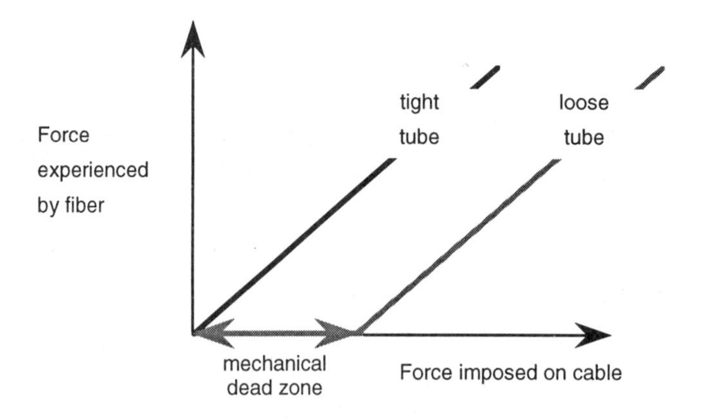

Figure 1–30 Mechanical Dead Zone in a Loose Buffer Tube

Strength Members. The second cable structural element is the *strength members.* Strength members are materials that restrict the elongation of the fiber when the cable is under load. Strength members are necessary in order to prevent fiber breakage, since fiber breaking loads are 2–10 pounds. The combination of a central strength member and buffer tubes is called a *cable core.*

Strength members can be metal wires, metal ropes, fiberglass-epoxy rods, flexible aramid yarns, such as Kevlar®, or flexible fiberglass rovings. Fiberglass epoxy rods and flexible aramid yarns are the most commonly used strength member materials. In the future, flexible fiberglass rovings will become more common because their use reduces the cost of the cable.

Outer Jacket. The third structural element is the outer jacket. The outer jacket is the outermost layer of the cable. This jacket protects the fiber and the cable from the environment and from the installation process.

Inner Jacket. The fourth structural element is the inner jacket, which is not included in all designs. The inner jacket separates subunits of fiber(s) from one another. The inner jacket can separate the cable core from other structural elements.

Gel Filling and Water Blocking Compounds. The fifth and sixth structural elements are the filling and blocking compounds. *Gel filling* and *water blocking* compounds provide protection against damage from water. The gel filling compounds fill the empty space in a loose buffer tube. This filling is necessary to prevent moisture from contacting fibers. Such contact can result in a reduction in the strength of the cable.

The water blocking compounds prevent water from traveling along the cable into the electronics. In addition, these compounds prevent water from entering a cable and freezing. The expansion of freezing water can result in an increase in the attenuation of the fiber.

Binding Tapes and Yarns. The seventh structural element is *binding tapes and yarns*, which hold the buffer tubes or sub cables together during jacket extrusion.

Armor, Ribbons, and Fillers. The eighth, ninth, and tenth elements are *armor, ribbons*, and *fillers*. Armor, usually corrugated stainless steel, provides resistance to damage from gnawing rodents. Ribbons are precision tapes on which multiple fibers are arranged. Fillers are used as a manufacturing convenience to allow for efficient use of cable processing equipment.

Structures, Advantages, and Disadvantages of the Seven Cable Designs

The two types of buffer tubes lead to seven designs: five loose tube designs, and two tight tube designs:

- Loose tube, single fiber per tube
- Loose tube, multiple fiber per tube
- Central buffer tube
- Ribbon
- Star or slotted core
- (Tight-tube) premise
- (Tight-tube) break out

The five most commonly used designs are: break-out cable; premise cable; loose tube, multiple fiber per tube; central buffer tube; and ribbon.

Loose Tube, Single Fiber Per Tube. This design consists of a fiber surrounded by a loose buffer tube (Figure 1–31). Multiple buffer tubes are stranded around a central filler or strength member to form a cable core. This cable core is surrounded by binder tape or yarn and by a jacket.

The advantages of this design are the two basic advantages of the loose tube design: isolation of the fiber from mechanical and environmental forces and good resistance to damage from moisture. An additional advantage is the protection of each fiber by a separate buffer tube. The disadvantages are relatively high cost, large size, high shipping cost, and relatively large bend radii.

When used, this cable design is used in outdoor applications with fiber counts less than twelve. However, this design is rarely used in the United States.

Loose Tube, Multiple Fiber Per Tube. The loose buffer tube, multiple fiber per tube (*MFPT*) design consists of a number of fibers surrounded by a loose buffer tube (Figure 1–32). The most common numbers of fibers are six and twelve. Multiple buffer tubes are stranded around a central filler or strength member to form a cable core. The MFPT cable core is surrounded by binder tape or yarn and by a jacket.

The advantages of the MFPT design are the two basic advantages of the loose tube design. Additional advantages are low cable cost, ease of water blocking, wide temperature range of operation, and convenient mid-span access. Convenient mid-span access is important when a system requires access to some, but not all, of the fibers at some location between the ends. The separate buffer tubes allow access to the required fibers without risk of damage to all fibers in the cable.

The disadvantages of the MFPT design are high labor cost for end preparation (exposing and preparing fibers for splices or connection) and high cost for end

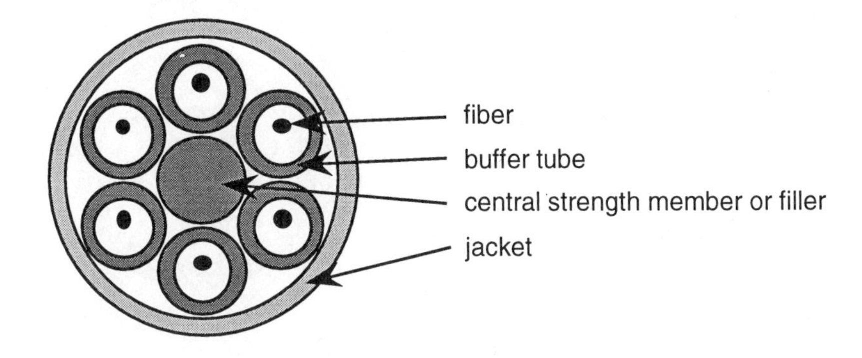

Figure 1–31 Cable Cross Section of Loose Tube, Single Fiber per Tube Design

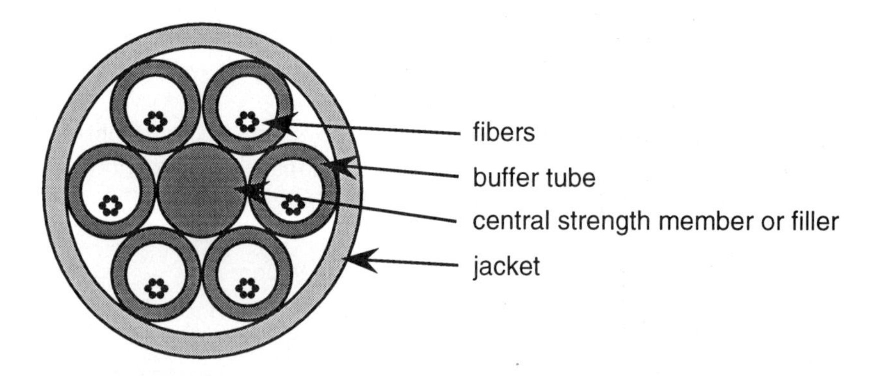

Figure 1–32 Cable Cross Section of Loose Tube, Multiple Fiber per Tube Design

hardware (wall or rack mounted boxes). The high labor cost for end preparation results from the difficulty of removing the heavy duty jacket and from the need to remove the water blocking grease and gel.

> ***Rule of Thumb:*** *End preparation time for this design is 0.11 hour/fiber/end, or 2.6 man-hours per end for a 24-fiber cable and 5.2 man-hours per cable segment (for all dielectric designs).*

The high cost for end hardware is a result of the fibers being exposed during the end preparation process. These exposed fibers must be strengthened or protected by hardware. The MFPT design is most frequently found in outdoor applications with fiber counts greater than twelve.

Central Buffer Tube (CBT). The *central buffer tube* design consists of a single, centrally located loose buffer tube (Figure 1–33). The buffer tube encloses all the fibers, usually in multiple bundles, each of which contains twelve fibers. The bundles are held together by color-coded binder threads.

This buffer tube is surrounded by strength members and a jacket. The advantages of the CBT design are the basic advantages of the loose tube design. Additional advantages are low cost, small diameter, small bend radii, and relatively low shipping cost.

The disadvantages are inconvenient mid-span access, high labor cost for end preparation, and high cost for end hardware. Some of these designs use high strength piano wire as strength members. Such designs must be bonded and grounded. This cable design is most frequently found in outdoor applications with fiber counts greater than twelve.

Ribbon. The *ribbon* cable design is similar to the CBT design, with a different arrangement of the fibers. Four to twenty-four fibers are precisely arranged on a substrate to form a ribbon (Figure 1–34). The ribbons are stacked and may be

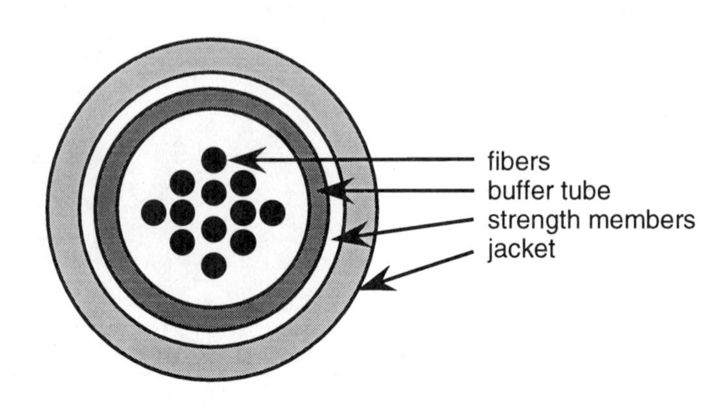

Figure 1–33 Cross Section of Central Buffer Tube Cable Design

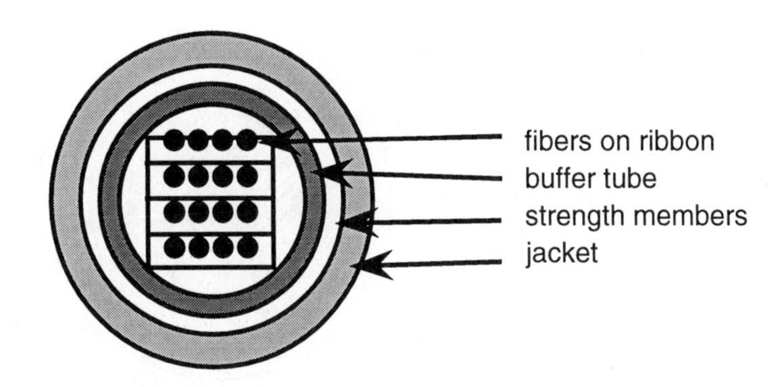

Figure 1–34 Cross Section of Ribbon Cable Design

twisted or buckled slightly. The stack of ribbons is enclosed in a centrally located buffer tube. The balance of the structure can be the same as that of the CBT design.

The advantages of the ribbon cable design are the advantages of the loose tube designs and the advantages of the central tube, multiple fiber per tube design. The major additional advantage is reduced splicing cost. This reduced cost results from the fiber alignment on ribbons, which allows for simultaneous splicing of up to twenty-four fibers.

The disadvantages of this design are the same as those for the central buffer tube design. In addition, there is a potential disadvantage of splice losses higher than those for the previous two designs (see Barbara Birrell and Sheila Cooper's "Practical Guidelines for Mass Mechanical Splicing," *Fiberoptic Product News*, April 1995, pp. 27–30). This increased splice loss is not significant in digital systems, but may be significant in high performance analog systems, like those of CATV companies. The final disadvantage is increased cable cost. This cable design is most frequently found in outdoor applications with fiber counts greater than twelve.

The Star or Slotted Core. The *star or slotted core* design consists of a core in which slots are formed (Figure 1–35). The slots take a helical path along the core. Fibers, or fibers in buffer tubes, are placed in the slots to form a cable core. The cable core is surrounded by a jacket.

The advantages of the slotted core, or star core, are the advantages of the loose tube designs, if the design is a loose tube. The disadvantages of this design are a limited number of suppliers and limited fiber counts. Typically, fiber counts are limited to twenty-four. This cable design is most frequently found in OPGW (optical power ground wire, used by power generating utilities) applications with fiber counts less than twenty-four. With the exception of the OPGW cable, this design is rarely used in the United States.

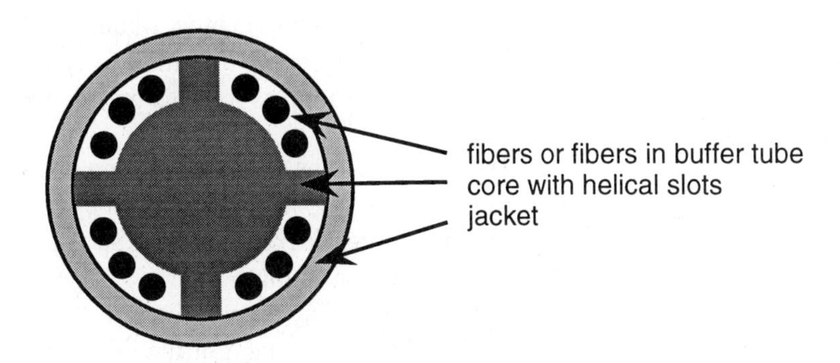

Figure 1–35 Cross Section of Star Core, or Slotted Core, Cable Design

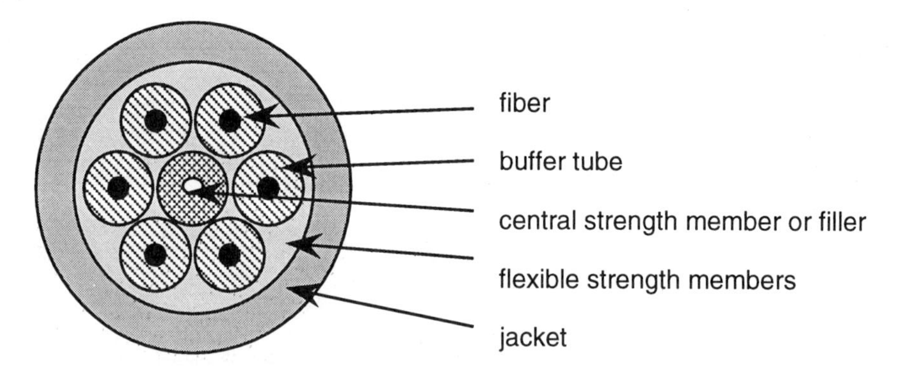

Figure 1–36 Cross Section of Premise Cable Design

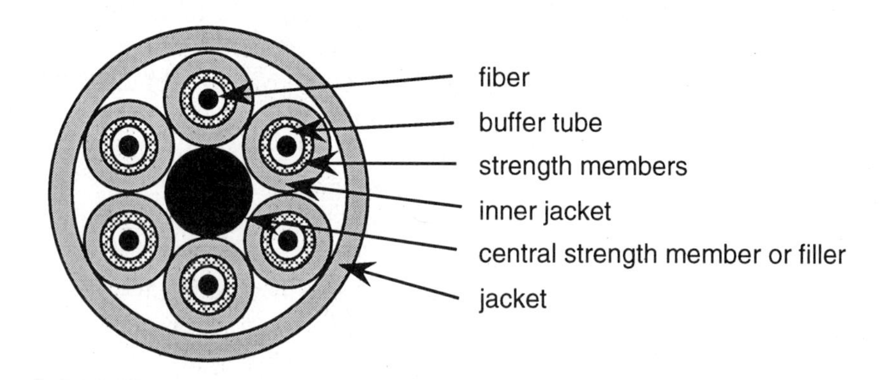

Figure 1–37 Cross Section of Break Out Cable Design

Premise. The *premise* design consists of multiple, tight buffer tubes stranded around a central strength member or filler (Figure 1–36). The buffer tubes are surrounded by flexible strength members, usually Kevlar®. The strength members are surrounded by a jacket.

The main advantage of the premise cable design is the unique advantage of tight tube designs, which is alignment of broken fiber ends. Additional advantages relative to the break out design are reduced end preparation labor cost, reduced cable cost, reduced diameter, reduced minimum bend radii, and reduced shipping cost.

The disadvantages of the premise design are that it is less rugged than the break out design and that it has high cost for end hardware. When the jacket is removed from this design, the tight buffer tubes are exposed. Because these buffer tubes can be damaged, they must be protected with the same type of hardware used to protect the ends of the loose tube cable designs.

This cable design is most frequently found in medium length, indoor applications with fiber counts less than forty-eight. This design can be used in outdoor applications as long as it meets the environmental requirements of the application.

Break Out. The *break out* design consists of fiber surrounded by a tight buffer tube (Figure 1–37). The buffer tube is surrounded by flexible strength members and by an inner jacket to form a sub cable. The sub cables are stranded around a central filler or strength member. These sub cables can be surrounded by binder tape or yarn and by an outer jacket.

The main advantage of the break out design is the unique advantage of tight tube designs. Additional advantages are rugged design, relatively low cost for labor of end preparation, and no cost for end hardware. The strength member and inner jacket of each fiber provide enough protection so that end hardware is not required.

The disadvantages of the break out design are high cost, large diameter, large bend radii, and high shipping cost. This cable design is most frequently found in short length, indoor applications with fiber counts less than forty-eight. The break out design can be used in outdoor applications as long as it meets the environmental requirements (water resistance and temperature range) of the application.

All of these designs are available with modifications. Modifications include an armor and an additional jacket for rodent resistance; an additional layer of strength members, for increased installation load or for increased long-term use load; and additional jackets for resistance to abrasion or to chemical attack.

The Five Most Important Installation Performance Characteristics

The five most important installation performance characteristics of the cable are:

- Maximum recommended installation load
- Minimum recommended long-term bend radius
- Minimum recommended short-term bend radius
- Installation and storage temperature ranges
- Moisture resistance

The Maximum Recommended Installation Load. This is the maximum load, in pounds force, that can be applied during installation. Exceeding this load can result in breakage of the fibers or a permanent increase in the attenuation. Typical ranges are 150–600 pounds.

> ***Rule of Thumb:*** *Most, but not all, fiber optic cables have a minimum recommended, long-term, bend radius equal to ten times the diameter of the cable.*

> ***Rule of Thumb:*** *Most, but not all, fiber optic cables have a minimum recommended short-term bend radius equal to twenty times the diameter of the cable.*

The Minimum Recommended Long-Term Bend Radius. This is the minimum radius to which the cable can be bent for its lifetime without causing breakage of the fibers or a localized increase in attenuation. This performance characteristic must be observed during the installation process. This limitation requires proper installation of cable in all locations, including cable trays, cable boxes, splice enclosures, and patch panels.

The Minimum Recommended Short-Term Bend Radius. This is the minimum radius to which the cable can be bent while under the maximum recommended installation load. Violation of this bend radius can result in breakage of the fibers, a localized increase in attenuation, or localized weakening of the fibers. Localized weakening can result in a reduction of the reliability of the cable.

This limitation requires selection of pulleys, slings, or sheaves with the proper diameter for installation. Pulley diameters may not fit in manholes.

The Temperature Ranges for Installation and Storage. These ranges must be observed so that the cable materials will not experience any degradation. Excessively low temperatures will result in cracking of plastics; excessively high temperatures will result in degradation of plastic properties. Cracking and degradation can result in exposure of the fiber to the environment.

Moisture Resistance. Moisture resistance is important for two reasons. First, moisture that can travel along a cable can be piped to the electronics. Moisture and electronics usually do not mix. Second, moisture that freezes inside a cable expands and causes an increase in the attenuation of the cable. Such an increase can shut down the transmission system.

BASIC CONNECTION FACTS

The Three Functions of the Connection System

Fiber connections have three functions:

- Precise alignment
- Fiber retention
- End protection

Precise Alignment. Most fiber cores are smaller than the human hair. This extremely small size requires precise alignment to achieve efficient transfer of light at light sources, at light detectors, and at fiber-to-fiber connections.

Fiber Retention. Fiber retention prevents fibers from moving toward or away from one another. Such movement can result in excessively high signal intensity loss or damage to protruding fiber ends.

End Protection. End protection ensures reliability through preventing unnecessary damage. Such damage can also result in excessively high loss of signal strength.

Eight Causes of Signal Strength Loss in Connections

A connection can cause loss of light intensity in eight ways (Figure 1–38):

1. Radial offset or radial misalignment
2. Lateral separation or lateral misalignment
3. Angular mismatch or angular misalignment
4. NA mismatch
5. Core diameter mismatch
6. Lack of cleanliness
7. Lack of smooth surface
8. Surface, or Fresnel, reflection

Radial Offset or Radial Misalignment (Figure 1–38 a). If the axes of the cores of two connections are offset in the radial direction, light traveling from the core of the transmitting fiber will enter the clad of the receiving fiber. This light will be lost, since light experiences total internal reflection only when it travels in the core.

Lateral Separation or Lateral Misalignment (Figure 1–38 b). If the spacing between the ends of the two cores is excessive, light traveling at or near the critical angle will strike the clad of the receiving fiber and will be lost.

Angular Mismatch or Angular Misalignment (Figure 1–38 c). If the axes of two fibers are not coincident, light traveling from the transmitting fiber to the receiving fiber can be lost by two mechanisms. First, light traveling at the critical angle from the transmitting fiber will strike the clad of the receiving fiber. Second, light traveling at the critical angle (or the NA) can enter the core at an angle greater than the critical angle of the receiving fiber. This light would be lost because light that strikes a core at an angle greater than the critical angle does not experience total internal reflection at the boundary of the core and clad.

NA Mismatch (Figure 1–38 d). If the NA of the transmitting fiber is larger than that of the receiving fiber, light will be lost, because the receiving fiber cannot accept such light.

Core Diameter Mismatch (Figure 1–38 e). If the core diameter of the transmitting fiber is larger than that of the receiving fiber, some light will be lost, because it will strike the clad of the receiving fiber.

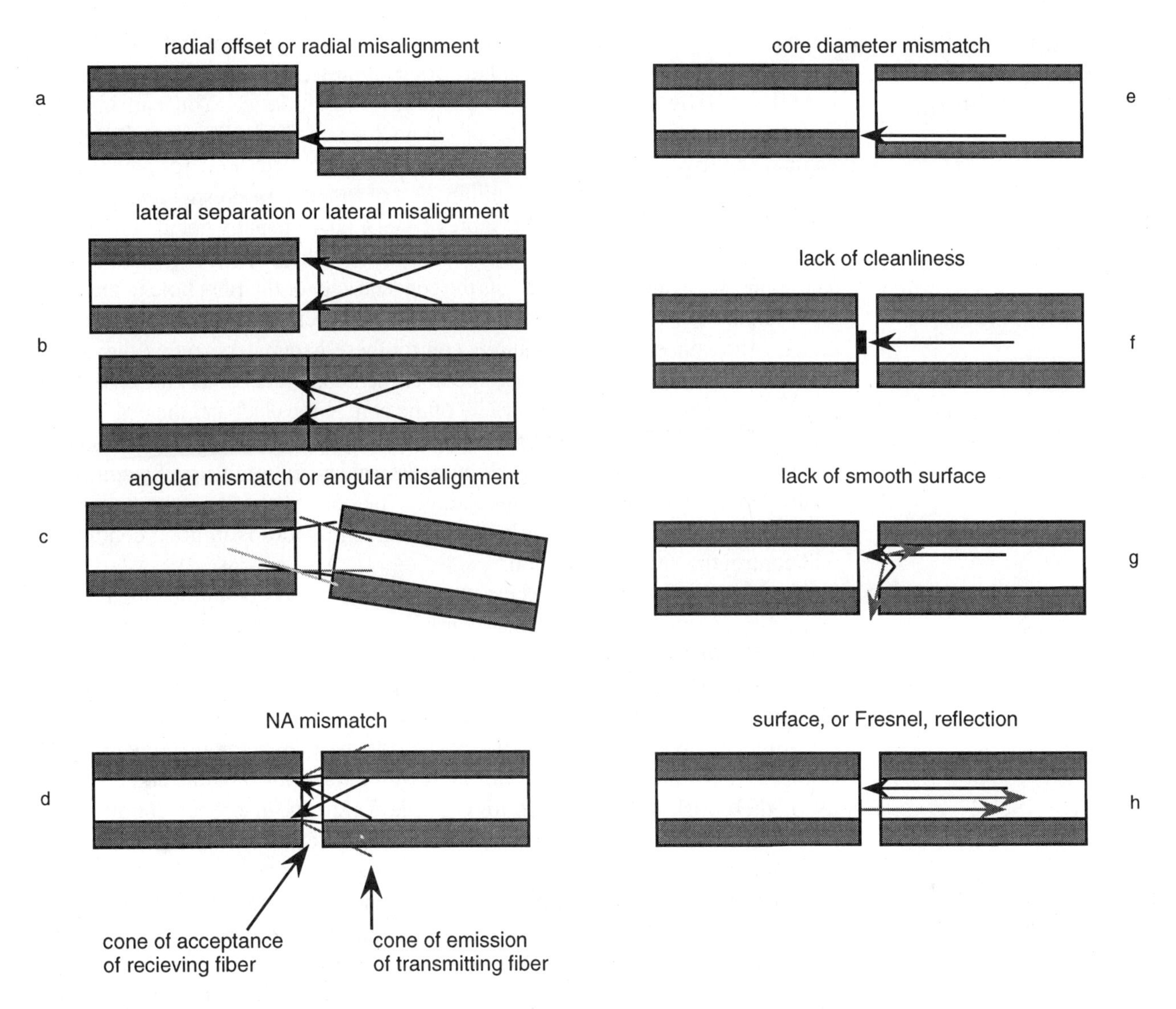

Figure 1–38 The Eight Causes of Loss in Connection Mechanisms

Lack of Cleanliness (Figure 1–38 f). Any dirt or particulate impurity on the surface of either the transmitting or receiving core will block some of the light. This problem is severe in singlemode fibers because their cores are much smaller (8–10 µm) than those of multimode fibers (50–100 µm).

Lack of Smooth Surface (Figure 1-38 g). If the surface of either the transmitting or receiving core is not smooth, the light will be refracted or reflected. In either case, the light can be redirected so that it strikes the clad of the receiving fiber or so that it enters the core at an angle greater than the critical angle. Such light will be lost. This problem is severe in singlemode fibers because of their small core diameters.

Surface, or Fresnel, Reflection (Figure 1–38 h). Whenever light travels through a surface, it might be partially reflected. If the surface has different indices of refraction on either side, as do air and fiber cores, light will experience a partial reflection. This light, which is reflected back into the fiber, represents a reduction in strength of the signal, and, therefore, a loss.

Four Causes of Loss That You Can Control or Influence

Of these eight causes of loss, three are under your direct control and you can influence a fourth. You can control the lack of cleanliness, lack of smooth surface, and lateral separation through proper installation techniques. You can influence Fresnel reflection through choice of connector and through proper installation techniques. If you use a contact connector (usually called "PC" for physical contact), you will experience a near-elimination of loss due to Fresnel reflection.

The other five causes of loss (Figures 1–38, a–e) are usually not under your control. Radial offset is determined by the connector, splice, and fiber manufacturers. The connector manufacturer controls the concentricity of the fiber hole in the ferrule and the diameter of the fiber hole. These two parameters will influence the radial offset. Most connector manufacturers control these parameters so that the connectors meet their stated performance. The fiber manufacturers control the clad diameter, the core diameter, the concentricity of the core in the clad, and the ovality of the core well enough that connectors meet their stated performance specifications.

In the case of non-contact connectors, lateral separation is usually completely controlled by the connector manufacturer. The connector manufacturer controls the dimensions of the polishing tool. As long as this tool is in good condition, it will control the lateral separation.

In the case of contact connectors, lateral separation is not a problem, as long as you follow proper installation procedures. In this type of connector, the ferrule length is not critical. The low loss is a result of contact of the ferrules. However, if excessive polishing time and/or polishing pressure is used with ceramic ferrules, the fiber surface can be under cut, so that an air gap, and Fresnel reflection, will result.

Angular mismatch is under the control of the connector manufacturer. The connector manufacturer controls the angle of the fiber hole and the perpendicularity of the polishing fixture. As long as the fixture is in proper condition, the installer will not experience high loss due to this mechanism.

NA is controlled by the fiber manufacturer. As long as the fibers being connected have been specified with the same NA, this will not be a problem. We have observed that the range of NA values produced by fiber manufacturers does not result in significant problems due to proper NA selection by cable manufacturers (Table 1–6).

Core diameter mismatch is under the control of the fiber and cable manufacturers. Fortunately, fiber manufacturers control the core diameter tightly enough that core diameter mismatch is rarely a problem.

Splices

Types and Advantages. Connections can be splices or connectors. *Splices* are permanent or semi-permanent connections between fibers. Splices are used when rerouting of optical paths is not required or expected. Splices are used in two situations: mid-span splices, which connect two lengths of cable; and *pigtails*, at the ends of a main cable.

Average	0.2009
Standard Deviation	0.0026
Range	0.197-0.206
Number of fibers	60

Table 1–6 Average Loss and Range of NA of 50 μm Fibers (Data courtesy of Pearson Technologies Inc.)

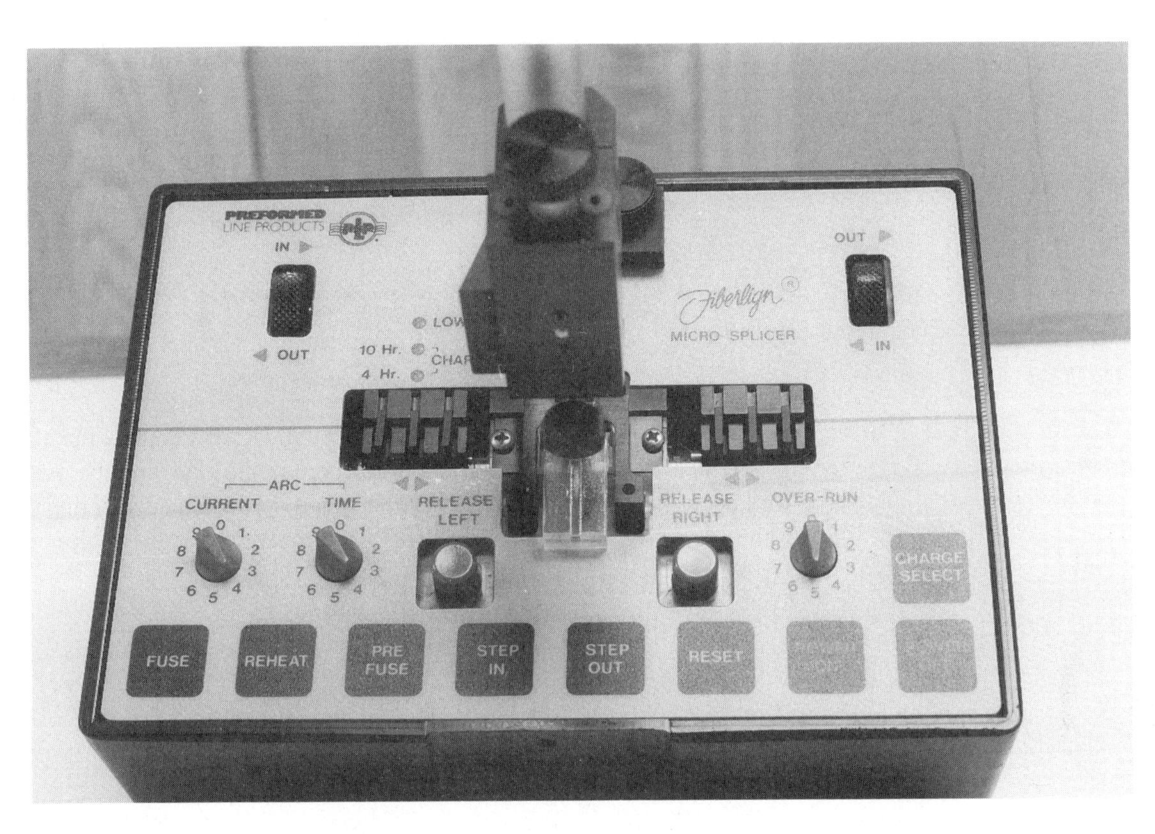

Figure 1–39 A Fusion Splicer (Courtesy of Pearson Technologies Inc.)

Pigtails are lengths of cable with a factory-installed connector on one end. Pigtails are spliced onto the end of a main cable in order to reduce system cost or improve reflectance performance. Reduced cost can result from use of factory-installed connectors, which are one-half to one-third the cost of field installed connectors.

Splices are of two types: fusion and mechanical. *Fusion splices* are made by aligning and melting together (i.e., fusing) two fiber ends to form a single fiber of increased length. The alignment and fusing are performed by a fusion splicer (Figure 1–39). Fusion splices are permanent, have a low consumable cost/splice, and, in general, have lower losses than do *mechanical splices*. However, the capital investment in equipment required to make fusion splices is significantly higher ($6,000–$36,000) than that for mechanical splices. In addition, fusion splices cannot be made in an atmosphere that contains explosive gasses. Typical fusion splice losses are less than 0.2 dB, with frequent values less than 0.1 dB.

Mechanical splices are made by inserting two ends of fibers into a piece of hardware that aligns the ends. This hardware is a consumable part, which results in a higher cost of consumables than for fusion splices. Mechanical splices have typical costs of $8–$24. Typical mechanical splice losses are in the range of 0.3–0.5 dB.

Mechanical splices (Figure 1–40) can be permanent or semi-permanent. Some permanent mechanical splices (from Norland Products Inc. and AT&T) require the use of an adhesive to permanently fix the fibers in the splice. Other permanent mechanical splices (from AMP and 3M) require the crimping of the splice to the fiber. Semi-permanent mechanical splices (from GTE, Advanced Components Applications Inc., and AMP) do not require the use of an adhesive, but use a mechanism to grip the fibers mechanically.

All mechanical splices use an index-matching gel or oil to reduce loss by reducing Fresnel reflections. This gel is installed in the splice by the manufacturer.

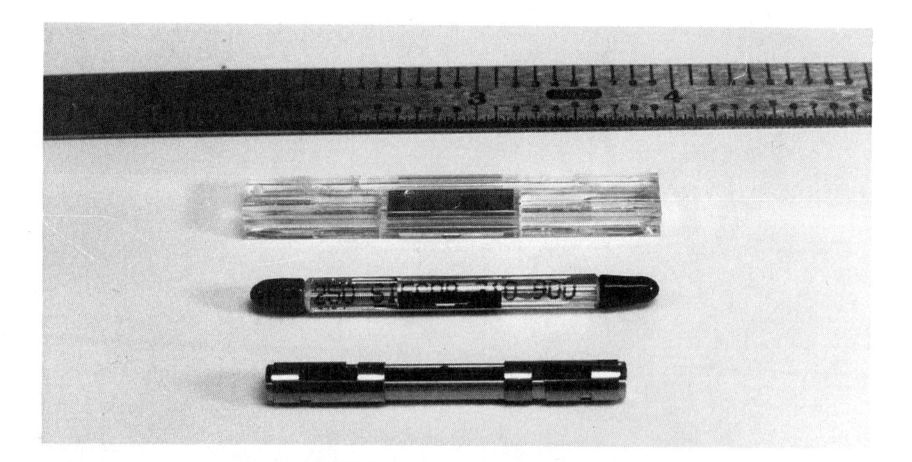

Figure 1–40 Types of Mechanical Splices (Courtesy of Pearson Technologies Inc.)

Splices have four advantages over connectors:

- Low reflectance
- Lower loss
- Better immunity to connection damage
- Lower maintenance cost

Hardware Required for Splice Protection. The installation of all splices requires exposing the fibers. These exposed fibers need to be protected from the environment. This protection is provided by splice covers (for fusion splices), splice trays, and splice enclosures.

Connectors

Advantages. In most cases, the choice between splices and connectors is clear cut. If there is a need to reroute the signal, connectors are the only choice, since the splice tray, splice case, and enclosure are not designed to allow for convenient or reliable rerouting, whereas connectors provide for rerouting of optical signals. Connectors have two additional advantages: low to moderate loss and ruggedness of the connection to the cable.

The Three-Part Connector System. Connectors are part of a three part system. The connector system includes the connector or plug, a receptacle, and an adapter (Figure 1–41). The connector aligns the fiber, holds the fiber and the cable, and protects the end of the fiber. The receptacle aligns the fiber to the source or detector and accepts the connector. The adapter, also called a feed through, a barrel, or a bulkhead, allows two connectors to be aligned to one another.

Connector Structural Elements and Functions. Regardless of the specific product or style, connectors have five basic structural elements:

- Ferrule
- Retaining nut
- Backshell
- Boot or heat shrink tubing
- Cap

Some connectors also have a crimp ring or crimp sleeve (Figure 1–42), and the SC style has an outer housing (Figure 1–43).

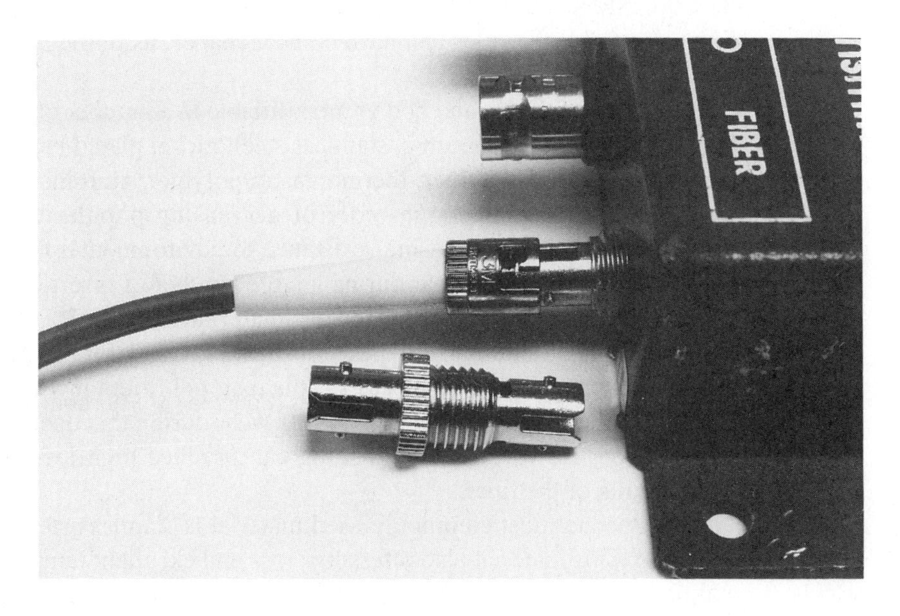

Figure 1–41 The Three Parts of the ST-Compatible Connector System (Courtesy of Pearson Technologies Inc.)

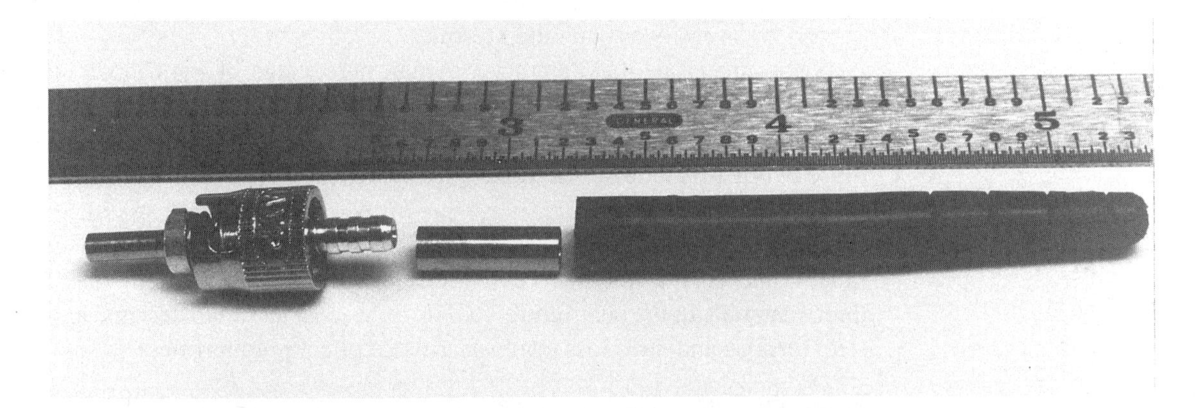

Figure 1–42 The Structure of the ST-Compatible Connector (Ferrule, Retaining Nut, Backshell, Crimp Sleeve, Boot) (Courtesy of Pearson Technologies Inc.)

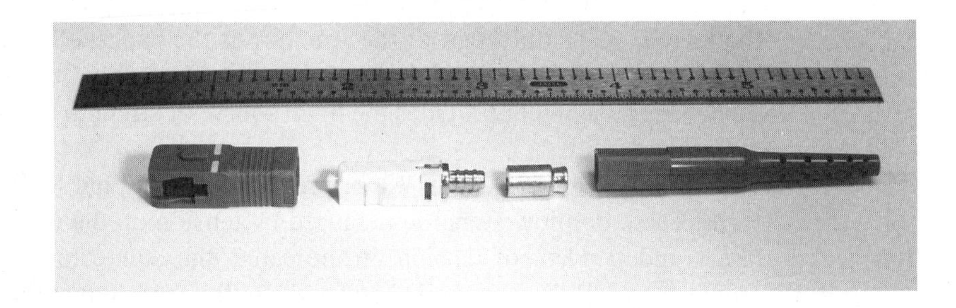

Figure 1–43 The Structure of the SC Connector (Housing, Ferrule, Retaining Nut, Backshell, Crimp Sleeve, Boot) (Courtesy of Pearson Technologies Inc.)

Ferrule. The *ferrule*, or tip, is the most important part of the connector (Figure 1–42), because it precisely aligns the fiber. Ferrules can have straight tips as in the ST®, 905 SMA, FDDI, and FC styles. Ferrules can have stepped tips, as in the 906 SMA connector. Ferrules can have conical shapes, as in the biconic style and the AMP Optimate®.

Connector ferrules and bodies are available in a number of different materials. Ferrule materials are: ceramic, stainless steel, nickel plated brass, thermoset polymer, liquid crystal polymer, thermoplastic polymer, aluminum, and zinc. These materials are listed in rough order of decreasing popularity, decreasing cost, decreasing softness, decreasing resistance to environmental factors, and decreasing resistance to ferrule wear during a large number of insertion cycles.

The most commonly used ferrule material is ceramic. Ceramic materials, either zirconia or alumina, are used almost exclusively for singlemode connectors and in many multimode connectors. This material offers low loss. Polishing time is low because the ferrule material resists wear during this operation. Epoxy adhesion is good. However, over polishing can produce high loss connectors due to undercutting of the fiber.

The second most commonly used material is stainless steel, especially in the ST® style. This material also offers low loss and excellent temperature stability. In addition, undercutting of the fiber does not appear to be a problem.

Metal ferrule contact connectors can be repolished. Shallow scratches and damage of the fiber core can be polished out, lengthening the lifetime of these products. Ceramic ferrules, however, cannot be repaired by repolishing as frequently or as easily as can metal ferrules.

Other materials make up a very small percentage of multimode connectors. Plastic ferrules have the advantages of low cost, but also have higher losses than do metal and ceramic ferrules. In addition, the plastic ferrules wear faster than those of other materials. Plastic ferrules can be used in indoor applications requiring few insertion cycles.

While ceramic and stainless steel are now the two most commonly used materials, liquid crystal polymer (LCP) ferrules are likely to become one of the top three materials in the near future. LCP ferrules cost less than ceramic and stainless steel ferrules and have loss levels near those of ceramic ferrules.

Retaining Nut. The second part of the connector structure is the retaining nut, which provides the mechanism by which the connector is secured to a receptacle or an adapter (and thus, to another connector). Retaining nuts are either threaded (SMA, FC, D4) or bayonet (ST®, mini-BNC). The SC style has a push-on, push-off latching mechanism instead of a retaining nut.

Backshell. The third part of the structure is the backshell, which is the portion of the connector in back of the retaining nut. The cable is attached to the backshell. This attachment provides the main source of strength between the cable and the connector.

All connector backshells either prevent excess signal loss caused by tension on the cable or allow signal loss caused by tension on the cable. If the connector loss is independent of tension on the cable, the connector is considered "pull-proof." The SC, FC, and the D4 styles are pull-proof, in that their backshell structure eliminates separation of ferrules.

If the backshell design creates pull-proof performance, it can also create "wiggle proof" performance. Wiggle proof performance is achieved when the connector is immune to excess loss caused by a lateral pressure applied to the backshell. In a non–pull-proof design, lateral pressure on the backshell results in an increased signal loss.

Boot or Heat Shrink Tubing. All connectors have a boot or heat shrink tubing, which fits over the backshell. Both of these components serve a cosmetic function by concealing the backshell, cable, and strength members. The boot also limits the bend radius of the cable as the cable exits the backshell. By limiting the bend radius, the boot increases the cable's reliability through reducing breakage.

Cap. The final basic part is the dust cap. This cap prevents dust contamination and abrasions of the core, both of which can cause significant increases in loss.

In addition to the five basic parts, some connectors have a crimp ring or sleeve. The crimp ring fits around the backshell and traps the cable strength members. Connectors that do not have a crimp ring use other mechanisms for attaching the cable to the connector.

The Five Basic Features. Connectors can be grouped by five basic features:

- Method of optical coupling
- Keying
- Contact of fiber cores
- Style
- Technique of installation

Optical Coupling. *Optical coupling* refers to the manner in which light is efficiently transferred from the fiber in one connector to the fiber in a second connector. There are two methods of optical coupling: *butt coupling* and *expanded beam* or *lensed coupling*. In butt coupling, the connectors are mechanically positioned close enough for the light to pass from one fiber to another. Most connectors used today are of this type.

Expanded beam, or lensed coupling, requires a lens system in the connector to increase the size of the beam of light at one connector and reduce it at the other. Such connectors are not frequently used, because the cost and performance advantages envisioned by the manufacturers did not materialize. In addition, such connectors use index matching oil, which attracts dust; dust creates maintenance problems.

Keying. The second basic feature is *keying*, or lack thereof. Keying enables connectors to go through multiple cycles of connection and disconnection (an insertion cycle) with minimal variation in the signal strength loss across the connection. Unkeyed connectors can, and do, rotate relative to one another, resulting in a larger range of losses during multiple insertion cycles than that experienced by keyed connectors (Table 1–7). Most connectors in use are keyed.

Connector Type	Average Range (dB/pair)
Unkeyed	1.00
Keyed	0.31

Table 1–7 Comparison of Ranges of Loss for Keyed and Unkeyed Connectors (Data courtesy of Pearson Technologies Inc.)

Connector Type	Average Loss (dB/pair)
Non-contact	0.97
Contact	0.40

Table 1–8 Comparison of Loss for Contact and Non-Contact Connectors (Data courtesy of Pearson Technologies Inc.)

Contact of Fibers. The third basic feature is contact of fibers, or lack thereof. Contact connectors have lower signal strength losses (Table 1–8) and lower reflectance values than non-contact connectors. Low reflectance is an important performance consideration for singlemode connectors. Because of this improved performance, contact connectors are more frequently used than non-contact connectors.

While contact connectors have improved performance, non-contact connectors can be more reliable than contact connectors in environments of high vibration, shock, or dirt. Such environmental factors can cause shattered or scratched fiber ends. In addition, contact connectors must be mated more carefully than non-contact connectors to avoid damage to the fiber ends. Most of the connectors designed and developed in the early 1980s were non-contact connectors. On the other hand, most of the connectors developed and marketed since 1985 are contact connectors.

Style. The fourth basic feature is the *style* of a connector. The style, or type, of a connector is the aggregate of design features that make a connector product unique. Connectors of the same style from different manufacturers are usually compatible. Connectors of different styles are not compatible, but can be connected with special adapters (available from AT&T and Storm Products). The style is an important factor since optoelectronics are not available for all styles of connectors. The ten non-proprietary styles are:

- ST®-compatible
- SMA 906
- SMA 905
- biconic
- mini-BNC
- FDDI
- ESCON
- SC
- FC, FC/PC
- D4

The *ST-compatible* design (ST is a trademark of AT&T) (Figure 1–42) is a keyed, contact connector with low average loss (0.3-0.5 dB/pair), and low cost. The ST-compatible is relatively easy to install. However, the ST-compatible is not pull-proof or wiggle proof. If the ferrule is stainless steel or liquid crystal polymer, the connector can be repolished to remove damage from the core of the fiber.

The ST-compatible style is presently the most commonly used connector style in both data communication and high performance (telephony and CATV) applications. This design is expected to continue to be commonly used into the late 1990s.

The *SMA 906* and *905* styles (Figure 1–44) are non-keyed, non-contact, with moderate average loss (1 dB/pair) and moderate cost. The SMA styles are somewhat more expensive to install than the ST and ST-compatible styles. The SMAs

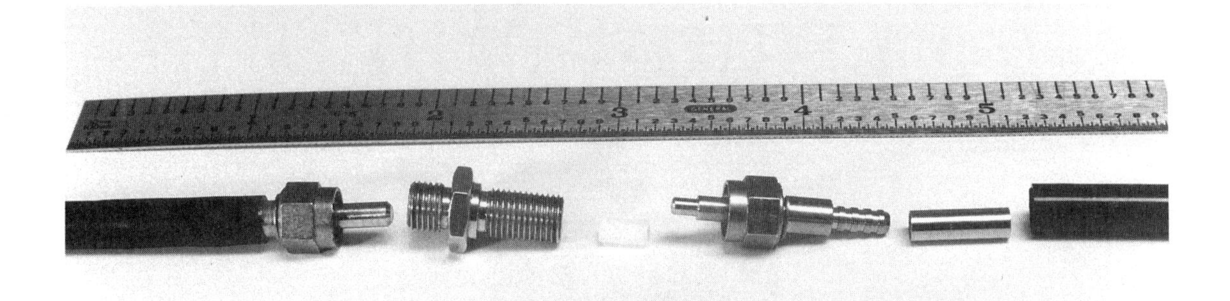

Figure 1–44 The 905 SMA Connector, Barrel with Alignment Sleeve, and 906 SMA Connector (Courtesy of Pearson Technologies Inc.)

Figure 1–45 The Biconic Barrel and Connector (Courtesy of Pearson Technologies Inc.)

are pull-proof but not wiggle proof. These connectors cannot be repolished to remove damage from the core of the fiber, since the non-contact feature requires a fixed air gap. Repolishing increases the air gap and the loss. A unique feature of the SMA 906 connector is a precision alignment sleeve. This sleeve, which fits into the barrel, is removable, replaceable, and forgettable. Without this sleeve, the SMA 906 connector style exhibits very high loss.

Since they are unkeyed, the SMAs have noticeable loss variability, also called rotational sensitivity. The SMA style has experienced a significant decrease in usage since the 1986: many consider the SMAs a dinosaur. However, the non-contact feature of these styles favor their use in high shock and high vibration environments.

The *biconic* style (Figure 1–45) is non-keyed, non-contact, with moderate loss and relatively high cost. Pearson Technologies has found the biconic to be much more difficult to install than the ST or SMA styles.The biconic style was originally developed as a multimode connector for telephony applications. It has found some use in non-telephony multimode applications, but is expected to be used less in the future. It suffers from rotational sensitivity.

The *mini-BNC* style (Figure 1–46) is non-keyed, non-contact with moderate loss and moderate cost. It is relatively easy to install. However, the mini-BNC is not pull-proof or wiggle proof. This style is used on IBM Token Ring optoelectronics.

The *FDDI* style (Figure 1–47) is a two-fiber, fixed shroud, keyed, contact connector with moderate loss and moderate cost. It is pull-proof and wiggle proof. The ferrule of this connector has the same dimensions as the ST-compatible connectors. While the FDDI style (the official name is MIC or media interface connector) meets the specifications of the FDDI standards, this connector is not always used in FDDI systems. The ST-compatible style is often used instead.

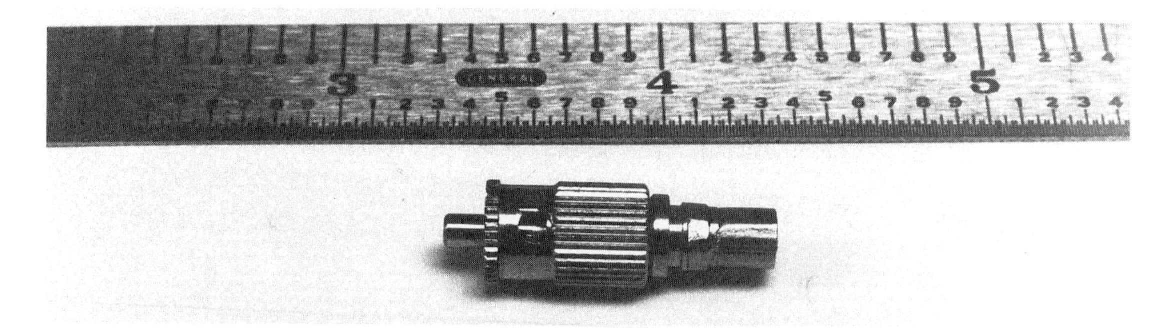

Figure 1–46 The Mini-BNC Connector (Courtesy of Pearson Technologies Inc.)

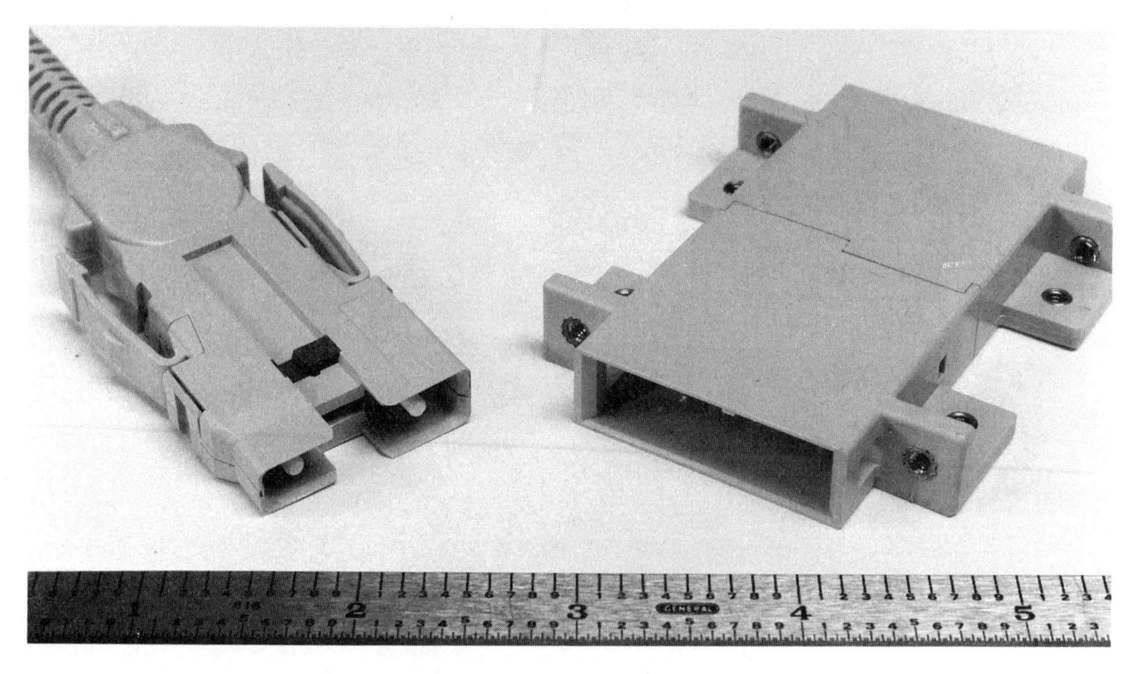

Figure 1–47 The FDDI Connector and Barrel (Courtesy of Pearson Technologies Inc.)

The *ESCON*™ connector style (Figure 1–48) is an IBM standard connector used in IBM's ESCON fiber system, which replaces the tag and bus communication system used in IBM computers. It is similar in style and dimensions to the FDDI MIC. This connector has a retractable shroud and is pull-proof and wiggle proof.

The *SC* style (Figure 1–43) is a keyed, contact connector, with low average loss (0.3 dB/pair) and a cost higher than that of the ST-compatible style. The SC style is slightly more difficult to install than the ST-compatible style, but is pull-proof and wiggle proof. The SC has a push-on, push-off insertion mechanism. This mechanism allows for reduced connector spacing and high density patch panels, which are preferred in telephony and other high fiber density applications. The SC is the style of choice in the EIA/TIA-568A building wiring standard and the Fiber Channel standard. Finally, multiple SCs can be clipped together to create duplex connectors. The forecasters at Kessler Marketing Intelligence expect the SC style to become the second most commonly used style by the year 2000.

The *FC*, *FC/PC* (Figure 1–49), and *D4* styles (Figure 1–50) are keyed, contact designs with low average losses (0.3 dB/pair) and relatively high costs. They are slightly more difficult to install than the ST-compatible designs and are pull-proof and wiggle proof. The FC/PC is the successor to the FC. The D4 has a unique feature: a key that can be tuned to the position of lowest loss and locked into that position.

Figure 1–48 The ESCON Connector and Barrel (Courtesy of Pearson Technologies Inc.)

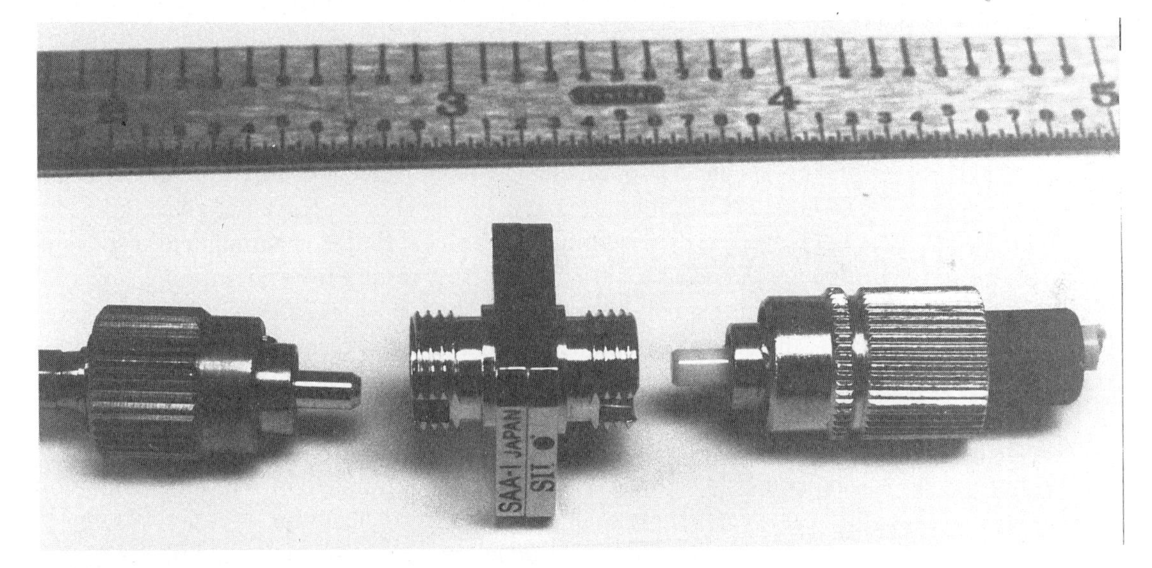

Figure 1–49 The FC/PC Connector and Barrel (Courtesy of Pearson Technologies Inc.)

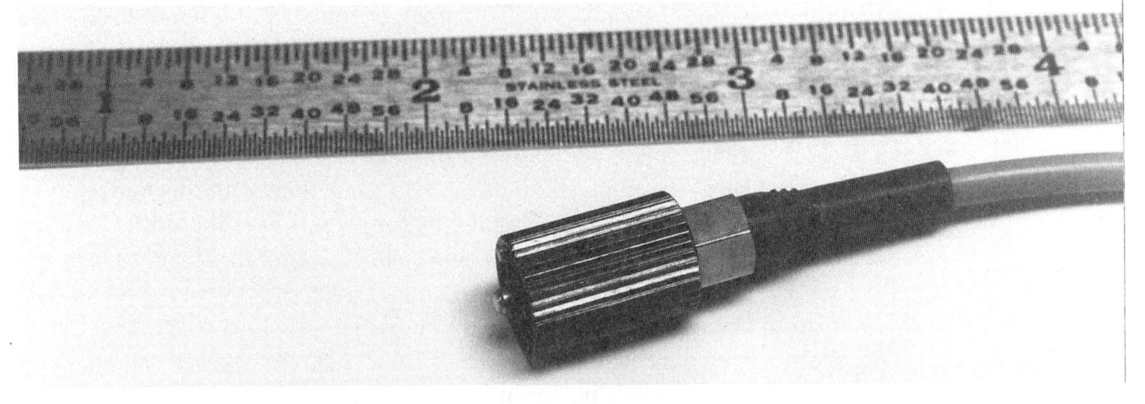

Figure 1–50 The D4 Connector (Courtesy of Pearson Technologies Inc.)

Installation Techniques. Installation of connectors, the final basic feature of a connector, requires four steps:

- Preparing the cable end
- Attaching the connector to the fiber
- Attaching the connector to the cable
- Finishing the end of the fiber

Since the preparation of the cable ends depends on the cable design and not the connector, we will present such steps in Chapter 3. The remaining three steps result in eight commonly used installation techniques (Figure 1–51): three techniques for attaching the fiber to the connector, three techniques for attaching the cable to the connector, and two techniques for finishing the end of the fiber.

There are three techniques for attaching the connector to the fiber:

- Inject an epoxy or adhesive into the backshell and ferrule of the connector.
- Use a connector that has been preloaded with a hot melt adhesive, which requires preheating.
- Mechanically grip the fiber with part of the connector structure.

The most commonly used technique for attaching the connector to the fiber is epoxy. A two-part, premeasured epoxy or adhesive is injected into the ferrule (and, in some cases, the backshell). This epoxy is then allowed to cure, either at room temperature or in a temperature-controlled oven. The advantages of this technique are a strong and reliable bond between the ferrule and the fiber, elimination of concern for pistoning (the movement of fiber into or out of the ferrule), and the fact that it is comparatively forgiving of installation errors.

However, this technique is the most time-consuming and inconvenient of the three techniques. This inconvenience results from the mixing, injecting, curing, and cleaning up of the epoxy or adhesive.

The second technique for attaching the connector to the fiber is the use of a hot melt adhesive (or polymer), which is preloaded in the connector. This connector, pioneered by 3M, is preheated prior to insertion of the fiber and cable. When the hot melt adhesive cools, the adhesive develops a strong bond between the fiber and connector. This technique has the advantage of reduced installation cost and time—since there is no epoxy, syringe, or clean-up.

The third technique for attaching the connector to the fiber uses one of three sub-techniques to grip the fiber mechanically. The first sub-technique involves a ferrule preloaded with a proprietary polymer. This polymer is deformed by a plunger inside the backshell. This plunger forces the polymer to flow around the fiber. The plunger is locked in place, causing the polymer to function as an epoxy or an adhesive. Surface friction between the polymer and the fiber holds the fiber in place.

A second sub-technique uses crimping. The backshell is crimped directly to the buffer tube of the fiber. If the cable is incompatible with this connector, this technique may allow the fiber to withdraw (or piston) into the ferrule. If the cable and connector are incompatible, pistoning can occur after the fiber has been polished.

The third sub-technique uses a chuck-like structure to grip the fiber buffer coating or buffer tube. When this chuck-like structure is screwed into the backshell of the connector, the fingers of the chuck are compressed against the fiber. This sub-technique is not commonly used in the United States.

method of fiber to connector bond	epoxy or adhesive			preload & preheat	mechanical gripping of fiber: polymer	mechanical gripping of fiber: crimp		mechanical gripping of fiber: chuck
method of cable to connector bond	crimp		epoxy or adhesive	epoxy or adhesive	crimp	crimp		crimp
method of end finishing	polish	cleave	polish	polish	polish	polish	cleave	cleave
STYLES	← **MANUFACTURERS** →							
ST® and compatibles	AT&T, AMP, Methode Electronics, Augat, Seiko, MA/Com, GTE, Siecor, Amphenol, 3M, ATC, Thomas & Betts		Molex	MA/Com, 3M	AMP	ATC, 3M	ATC, Siecor, EBOC	Leetec
SMA 906	AMP, GTE, MA/Com, ATC, Augat, Seiko, Amphenol					ATC		
SMA 905	AMP, ATC, Amphenol, MA/Com	Thomas & Betts	Molex			ATC	EBOC	Leetec
SC	Hirose, Augat, ATC, AMP, Molex, 3M, Methode Electronics			3M		ATC, 3M		
FC/PC	AMP, MA/Com, Hirose, Seiko, Siecor, Amphenol, Augat, ATC, 3M		Molex	3M				
BICONIC	AT&T, MA/Com, 3M		Methode Electronics					
D4	AMP, GTE, Seiko, ATC, MA/Com							
mini BNC	AMP, Sumitomo, Seiko, Hirose							
FDDI	AMP, Molex, Augat, Hirose, Methode Electronics							
ESCON	AT&T, AMP, Methode Electronics							

EBOC=Spectran Specialty Optics Co.

ATC=Automatic Tool & Connector

Figure 1–51 Eight Connector Installation Techniques

There are three techniques for attaching the connector to the cable:

- Mechanically grip the cable strength members (crimping).
- Use epoxy or two-part adhesive, or adhesive, or hot melt adhesive on the cable strength members.
- Trap the cable strength members.

The most commonly used technique is crimping. In this technique, the cable strength members are trapped between the crimp sleeve and the backshell. This technique is simple, fast, and reliable. It does require a crimp tool with the appropriate crimp nest. Increased connector to cable strength can be achieved by adding epoxy or adhesive to the strength members.

The second technique is the use of adhesive or epoxy. In this technique, the cable strength members are inserted into the backshell and "glued" to the connector by the "glue" in the backshell. This technique is fast and reliable. However, it can be awkward, since the fiber and cable can slide out of the connector during handling prior to curing.

The third technique involves trapping the strength members between the two parts of the backshell. These parts usually have a threaded juncture, in which the strength members are trapped. These two parts of the backshell have a specific clearance, into which the strength members must fit. If the amount is insufficient, the 'joint' will be loose, and the cable may pull out of the connector. Clients of Pearson Technologies report that they completely avoid this potential problem by adding epoxy to the strength members. With the use of epoxy, this technique becomes similar to the second technique. If the amount of strength member is excessive, some strength member material will need to be removed, without leaving an insufficient amount.

There are two techniques of finishing the end of the fiber: polishing and cleaving. The first, and most commonly used, is polishing. Polishing requires abrading the fiber and tip of the ferrule with fine polishing papers, films, or pastes until the surface is flush with the ferrule and smooth. This polishing can be done by hand, on glass, hard plastic, or hard rubber plates, or by automatic machines, which use either films or diamond slurries or pastes.

There are two types of polishing: flat and radius polishing. Flat polishing can be performed on multimode fibers but is not usually performed on singlemode fibers. Radius polishing can be performed on contact type multimode connectors (i.e., ST®), and is performed on contact type singlemode connectors to achieve low reflectance.

Cleaving is performed on multimode connectors. It is a fast and simple procedure. However, the signal strength loss of the connector will be determined by the quality and condition of the cleaving tool. This tool can be expensive, needs periodic maintenance, and must be handled with care.

Cleaving is a preferred procedure in environments in which polishing cannot be successfully performed or is impractical. Cleaving is also preferable in environments in which the speed of the technique is needed, either for its low cost, high speed (restoration), or convenience (dirty environments).

The Four Most Important Connector Performances. Four optical performances are important during the installation of fiber optic connectors:

- Maximum loss/pair, dB/pair
- Average loss/pair, dB/pair
- Repeatability of loss, maximum change in dB/pair
- Reflectance (also known as return loss and back reflection)

The *maximum loss/pair* (dB/pair) is the maximum loss you will experience when you install the connectors properly. This value will be value against which you accept or reject installed connectors.

Most connectors exhibit a typical loss/pair close to the *average loss/pair* stated by the connector manufacturer. This average loss/pair will be the value you expect to see most of the time. If you see losses consistently in excess of this value, you will suspect either errors in the installation procedure or in the test procedure.

Repeatability is the maximum change in loss/pair that can occur between successive measurements. Repeatability (roughly one-half the range) is important in maintenance and troubleshooting activities and in situations in which the optical power budget required by the system is close to the optical power budget available from the electronics.

Reflectance, which is also called *return loss* and *back reflection*, is the ratio of light intensity reflected from the connector back towards the light source to the intensity incident to the connector. It is a Fresnel reflection that occurs at any interface (such as that of a glass fiber core and air) at which the index of refraction changes (Figure 1–52). Reflectance causes signal errors and laser malfunctions. Reflectance needs to be measured for connectors in systems with singlemode fibers and laser diodes operating at high bit rates (400 Mbps). Reflectance is not a concern with most multimode systems.

In the ideal situation, fiber ends could be polished so that they were perfectly perpendicular to the fiber axis and perfectly smooth (Figure 1–52 a). This ideal

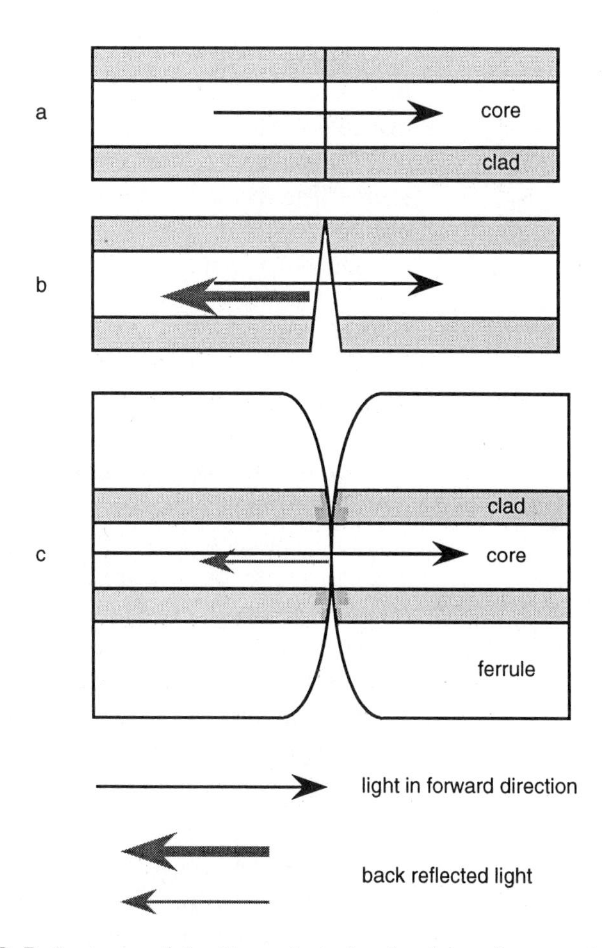

Figure 1–52 Reflectance: (a) with perfect cleaving (no reflectance), (b) with angle on fiber ends (high reflectance), (c) with radiussed ferrules (low reflectance)

situation eliminates the air gaps that could result in high reflectance. However, achieving perfect perpendicularity is nearly impossible: most connector pairs will have a slight angle on one or both of the ferrules (Figure 1–52 b). To minimize reflectance, physical contact of ferrules with a radius on the tip (i.e., a radiussed tip) is most often used (Figure 1–52 c). Use of radiussed tips requires a high degree of surface smoothness to achieve low reflectance. This high degree of surface smoothness is achieved through multiple polishing steps with successively finer abrasives. Typical values of low reflectance connectors are <-30 dB, although values as low as -45 dB are advertised.

BASIC TRANSMITTER AND RECEIVER FACTS

Although transmission and receiving optoelectronics must be installed, such installation is straightforward. However, the testing and troubleshooting of installed cables, connectors, and splices requires an understanding of the characteristics of these optoelectronic devices.

Four Important Characteristics of Transmitters

Transmitting optoelectronics have four important characteristics:

- Wavelength, in nm
- Spectral width, in nm
- NA
- Spot size, in microns

The *wavelength* of the light source is a measure of the color of the emitted light. This wavelength must be close to the wavelength at which the fiber performance is specified.

The *spectral width* is a measure of the range of colors emitted by the light source. The spectral width influences the amount of chromatic dispersion that will occur in the fiber.

The NA is a measure of the angle in which most of the light is emitted from the light source. A large NA can overfill the NA of the fiber. A small NA will under fill the NA of the fiber. The NA of the light source can influence loss measurements: a low-NA light source can result in reduced losses in both fiber and connections; a high-NA light source can result in increased losses at connections.

The *spot size* is the size of the area from which the light is emitted. A large spot size can fill the entire core of the fiber. A small spot size will partially fill the core. Like the NA, the spot size can influence the loss measurements of fibers and connections.

Transmitter optoelectronics are based on LED or laser light sources. LEDs are relatively low cost, low power, wide spectral width, low bandwidth or bit rate light sources. As such, LEDs are used for relatively short distance (<7 km or 22,960 feet), relatively low bandwidth (<200 MHz) or bit rate (<200 Mbps) applications.

Laser diodes are expensive, high power, narrow spectral width, high bandwidth or bit rate light sources. Laser diodes are used for high bandwidth and/or long distance applications, such as those in telephone and cable TV applications.

The Optical Power Budget

The transmitter-receiver electronic pair have one important characteristic: the *optical power budget* (OPB, in dB). The optical power budget is the maximum loss in optical signal strength that the transmitter receiver pair can withstand while still functioning at the specified level of accuracy. The end to end power loss of all

Wavelength (nm)	Type of Source	Fiber Core Diameter (µm)	Typical OPB (dB)	Spectral Width (nm)	Spot Size (µm)
850	LED	50	12		
	LED	62.5	16		
	LED	100	21		
	LED	any		30–50	150–250
1300	LED	any	16	60–190	50
1300	laser (singlemode)	9	25–45	0.5–5.0	2x10

Table 1–9 Typical Optical Performance Characteristics of Transmitter-Receiver Pairs

optical components in a fiber link (called the "link power budget" or "link loss") must be less than or equal to the optical power budget. Table 1–9 contains typical values for these optoelectronic characteristics.

Testing of Installed Systems

The testing of installed systems must simulate a transmitter-receiver pair. Ideally, this simulation requires using a testing light source with the same wavelength, spectral width, NA, and spot size as those in the transmitter-receiver pair. Ideally, the testing cables must have same core diameter, core offset, clad ovality, and NA as the cable under test. The power meter must be calibrated at the wavelength of the transmitter. Any significant differences between the characteristics of the transmitter-receiver pair and those of the test equipment can give test results that will not accurately indicate the performance of the transmitter-receiver pair.

REVIEW QUESTIONS

1. What are the three regions of the fiber structure?
2. What is/are the function(s) of each of the regions of the fiber structure?
3. What are the units of measure of the inner two regions of the fiber structure?
4. What are typical values of the inner two regions of the fiber structure?
5. What are the three performance mechanisms of fibers?
6. What is the significance of each of these three performance mechanisms?
7. What are the three units of measure of two of these performance mechanisms?
8. What are typical values of these two performance mechanisms?
9. What is the index of refraction?
10. What are the units of measure of the index of refraction?
11. When is the index of refraction used?
12. What is the NA of a fiber?
13. What are the units of measure of the NA?
14. What are typical values of the NA?
15. Why is the NA important in testing?
16. Why is the core diameter important in testing?
17. Why is the clad never removed?
18. What does excessive lack of concentricity cause?
19. What does excessive non-circularity cause?
20. What are the three causes of pulse spreading?
21. Describe the three causes of pulse spreading.
22. What type of fiber has three causes of pulse spreading?

23. What type of fiber has two causes of pulse spreading?
24. Why is the pulse spreading rate in an optical fiber dependent on the nature of the light source?
25. Describe the path of a ray of light in step index fibers.
26. Describe the path of a ray of light in graded index fibers.
27. Describe the path of a ray of light in singlemode fibers.
28. What is the practical significance of the difference between step index and graded index fibers?
29. What is the practical significance of the difference between multimode and singlemode fibers?
30. What are the two basic cable designs?
31. What are the differences in structure of these two basic designs?
32. What are the advantages of each of these two basic designs?
33. What are the seven cable designs?
34. Describe the structure of the seven cable designs.
35. Which designs are commonly used indoors?
36. Which designs are commonly used outdoors?
37. What is/are the advantages of each of the five most commonly designs?
38. What can happen if the long-term or short-term bend radii are violated?
39. What can happen if the maximum recommended installation load is exceeded?
40. Your cable has a diameter of 0.565″. What are the minimum recommended long-term and short-term bend radii?
41. What are the two types of connections?
42. Where are these two types of connections used?
43. What are the two types of splices?
44. What are the advantages and disadvantages of the two types of splices?
45. What three connector system parts must be compatible?
46. What are the two connector styles most commonly used in the United States?
47. What features are the same for these two connector styles?
48. What features are different for these two connector styles?
49. What are the four performance characteristics of connectors?
50. What are the units of measure and typical values for connectors?
51. What are the four important transmitter performance characteristics?
52. Why are each of these characteristics important?
53. What are typical values of these characteristics?
54. What is the most important transmitter-receiver pair performance characteristic?
55. Why is this characteristic important?
56. What is a typical value of this characteristic for a LED light source used with a 62.5 μm fiber at 850 nm?

CHAPTER

2

Advantages and Types of Fiber Optic Systems

CHAPTER OBJECTIVES

From this chapter, you will be able to:

1. Identify the reasons for use of fiber optic systems in your environment.
2. Recognize the possible topologies, systems, networks, and standardized products.

INTRODUCTION

In this chapter, we will examine three subjects: the advantages of using fiber optic communication systems; the topologies of fiber optic systems; and the standards that exist or are being developed.

WHY USE FIBER OPTICS?

Twelve Reasons for Use

There are twelve reasons for using fiber optic transmission:

- Long distance of transmission without repeaters or regenerators
- High capacity
- Reduced system cost
- Reduced maintenance cost
- Upgrading ease
- Lowest life cycle costs
- Small size
- Light weight
- Dielectric nature
- Immunity to EMI/RFI
- Intrinsic security of transmission
- Synergistic interaction of the properties of fiber with the cost factors of application

Long Distance of Transmission. Because of their extremely low attenuation rates (Table 1–3), optical fibers can transmit to long distances without repeaters or regenerators. Typical, off-the-shelf optoelectronics can transmit 40–80 km (24–48 mi.). Some telephone systems are transmitting to at least 137 mi. This regenerator spacing greatly exceeds the 6 mi. typical distance for copper telephone systems.

Use of fiber extends the distance of data communication systems to at least 10,000 ft. With this extended distance, networks can be designed for large geographical areas without the need for signal regenerators or repeaters. Copper cables in data communication applications are often limited to 100 m (328 ft.).

This long-distance capability may result in a change in the network topology used in corporate networks. This capability will allow building networks to be designed with a single, central concentrator location in a building instead of multiple, or intermediate, concentrators in wiring closets. This "collapsed backbone" concept results in reduced network costs, reduced maintenance costs, reduced conduit requirements, and reduced floor space requirements.

High Capacity. Because of their extremely low pulse spreading rates (Table 1–5), fiber optic systems have high capacity. At this time, telephone systems can carry approximately 30,000 telephone conversations per pair of fibers. A typical, 1-inch diameter, 144-fiber cable can carry approximately 2,100,000 simultaneous conversations. In the future, this same fiber pair will be able to carry 300,000 telephone conversations per fiber pair or about 21 million simultaneous conversations per telephone cable. Twenty-one million simultaneous conversations is the approximate capacity required for everyone in Los Angeles to talk with everyone in New York—at the same time over the same 1″ diameter cable!

Reduced System and Maintenance Costs. The elimination of repeaters, regenerators, and intermediate concentrators results in a reduced system cost. In some applications, such as telephone and CATV, fiber systems have an initial installed cost less than those of copper-based systems.

Because the number of active devices is lower in a fiber system than in a copper system, the number of points of potential failure are fewer. This reduction in failure points results in reduced maintenance costs. For example, elimination of coax amplifiers in CATV systems eliminates the need for weekly adjustment of the gain of each of these amplifiers. This adjustment is required to compensate for temperature changes in the environment. This reduction in maintenance cost is also realized in LANs, when a collapsed backbone topology is implemented.

Upgrading Ease Results in Lowest Life Cycle Cost. Because of the high capacity of fiber, we can increase the capacity of a fiber optic transmission system simply by changing the electronics. For example, telephone companies have increased capacities from 45 Mbps to 2.7 Gbps on the same cable plant. Data communications users can increase from 4–16 Mbps to 100 Mbps, and beyond, in the same manner. Cable TV companies can increase their capacity from 40 to 160 channels.

This capability makes optical fibers a unique transmission medium. With this capability, you can leverage your investment in a cable plant for a longer time with fiber optic options than you can with most competing technologies. Because of this leverage, the life cycle cost of fiber optic systems is usually lower than that of competing technologies.

This lower life cycle cost will be achieved even if the initial installed cost is higher for fiber than for copper. For example, a building back bone fiber system may have a cost premium of 10–20 percent, but it will last for two to three times as long as the copper system.

If your networks will need capacity increases from a range of 4–16 Mbps to greater than 100 Mbps, fiber is your best choice. Such increases in capacity will occur if you are in a growing bandwidth environment.

A growing bandwidth environment is one that includes any of the following elements: use of digitized images, medical environments, CAD/CAM applications, multi-media applications, and growth in the number of users on your network.

The high-speed transmission technologies of FDDI, ATM, Fiber Channel, SONET, ESCON, and HiPPI all use fiber. If your bandwidth needs or future plans include implementation of these protocols, using fiber now will be more cost-effective than using copper now and using fiber in the future.

Small Size. Optical fibers and their cables are the smallest cables available. In crowded conduits, as many as 100 3-inch diameter twisted pair cables can be replaced with a single 0.75-inch diameter cable. When the choice is installation of new ducts, with the expense of digging trenches or ripping up streets in cities, the small size of optical fibers often means lower installed cost than for the equivalent capacity using competing technologies.

Light Weight. In many high-rise office buildings with extensive floor loads, the light weight of optical fibers has been the driving force for their use. In addition, the light weight is the advantage that drives the Army, Navy, and Air Force to use fiber for command, communications, and control.

Dielectric Nature. Dielectric nature results in four advantages. First, there are no ground loops to cause problems in process control applications. Second, dielectric fiber optic cables do not attract lightning and, therefore, have lower maintenance costs. Third, dielectric fiber optic cables have few restrictions as to the locations in which they can be installed in buildings. Fourth, dielectric fibers do not radiate signals, resulting in secure communication systems.

Immunity to EMI/RFI. Because the information in fiber optic systems is being transmitted with light, the signal is immune to electrical noise in the environment of the cable. Thus, there will be no noise picked up, even with long transmission distances and electrically noisy environments.

This immunity results in improvement in effective capacity of existing, low-speed (or legacy) networks. For example, token rings and Ethernet networks will experience improvements in utilization and in effective throughput when copper cables are replaced with fiber cables.

Intrinsic Security of Transmission. Fiber optic transmission is intrinsically secure. There are two aspects of this security. First, since there is no electrical signal, there is no signal that can be radiated from a fiber optic cable. Therefore, a detector placed in the vicinity of a fiber optic cable will pick up no signal. Second, although fiber optic cables can be tapped without breaking the fiber, such tapping will result in a reduction in the signal strength at the output end of the fiber. Such output intensity can be monitored with simple circuitry. This circuitry will activate an alarm or shut down the system in the event of a drop in output.

Synergistic Interaction of the Properties of Fiber with the Cost Factors of Application. The aforementioned eleven reasons for use are based on the properties of optical fibers. These properties can interact with the cost factors of the application to result in an installed system cost lower than that of copper cables. One example of this synergism is in an Army ammunition depot, which was due for an upgrade in the intrusion detection system. Initial plans were for a buried 4000-volt power cable and a separately buried copper, RS-422 cable. The dielectric nature of fiber cables enabled the planners to install the RS-422 fiber cable in the same trench with the power cable. By eliminating the second trench, the planners were able to save an estimated $80,000.

A similar example is in products from a manufacturer of modular furniture for open offices. These products are less expensive when they are to be used in a fibered office than when used in a wired office. Again, the reason is the ability to place power cables and fiber optic signal cables in the same conduit in the modular furniture. Such joint placement cannot be done with copper signal cables.

Three Potential Disadvantages

Fiber optics sometimes, but not always, has a higher cost when compared to many copper cable solutions; these are usually not apples to apples comparisons.

Because most fiber optic cables include glass fiber, fiber optic cable is less forgiving of abuse than is copper cable. This difference in resistance to abuse is due to the difference in bend radius and "stretchability" of glass and metal materials. Fiber optic connectors are also less forgiving of abuse than are copper connectors. Again, this difference in resistance to abuse is due to the differences between glass and metal materials. However, by following proper installation and handling procedures, you can install and use fiber optic cables and connectors without problems.

HOW FIBER OPTICS IS USED

Seven System Topologies

Topology is the description of the connections in a network. Each network has a physical topology and a logical topology. The physical topology is the description of the physical connections in the network. The logical topology is the description of the logical connections in the network. The logical connection describes the manner in which data moves in a network. For example, in a logical ring network, data is transmitted from one station to a second station. After processing the data, the second station transmits data to the third station, and so forth.

In a logical bus network, data are transmitted by one station to all other stations on the network. A logical bus network is analogous to a long corridor with many doors. Each of the other stations on the network receives the transmission and reads the address of the transmission. If the transmission is addressed to the receiving station, the station processes the data. If not, the station ignores the data.

A physical ring topology requires that any two nodes on the network be connected by a cable that runs directly between the two nodes. A physical star topology requires that any two nodes on the network be connected by a cable that runs to a centrally located wiring closet, at which all nodes can be cross connected. There has been a strong trend towards physical star networks.

There are seven combinations of physical and logical topologies:

- Point-to-point
- Logical ring, physical ring
- Logical ring, physical star
- Active star
- Passive star
- Passive bus
- Active bus

The rules for choosing the physical topology for fiber systems are not the same as those for copper systems.

- Most fiber optic system designs consist of a series of point-to-point links.
- The point-to-point design is driven by a limitation in the optical power budget.
- Deviation from a point-to-point design drives up the system cost.

To understand the reason for this preferred physical topology, we will design two systems for the same application, a fiber-to-the-desktop system. These two systems are: a point-to-point system and a point-to-multipoint system. If we design this system as a series of point-to-point links (Figure 2–1), the optical power budget requirement is low (Table 2–1).

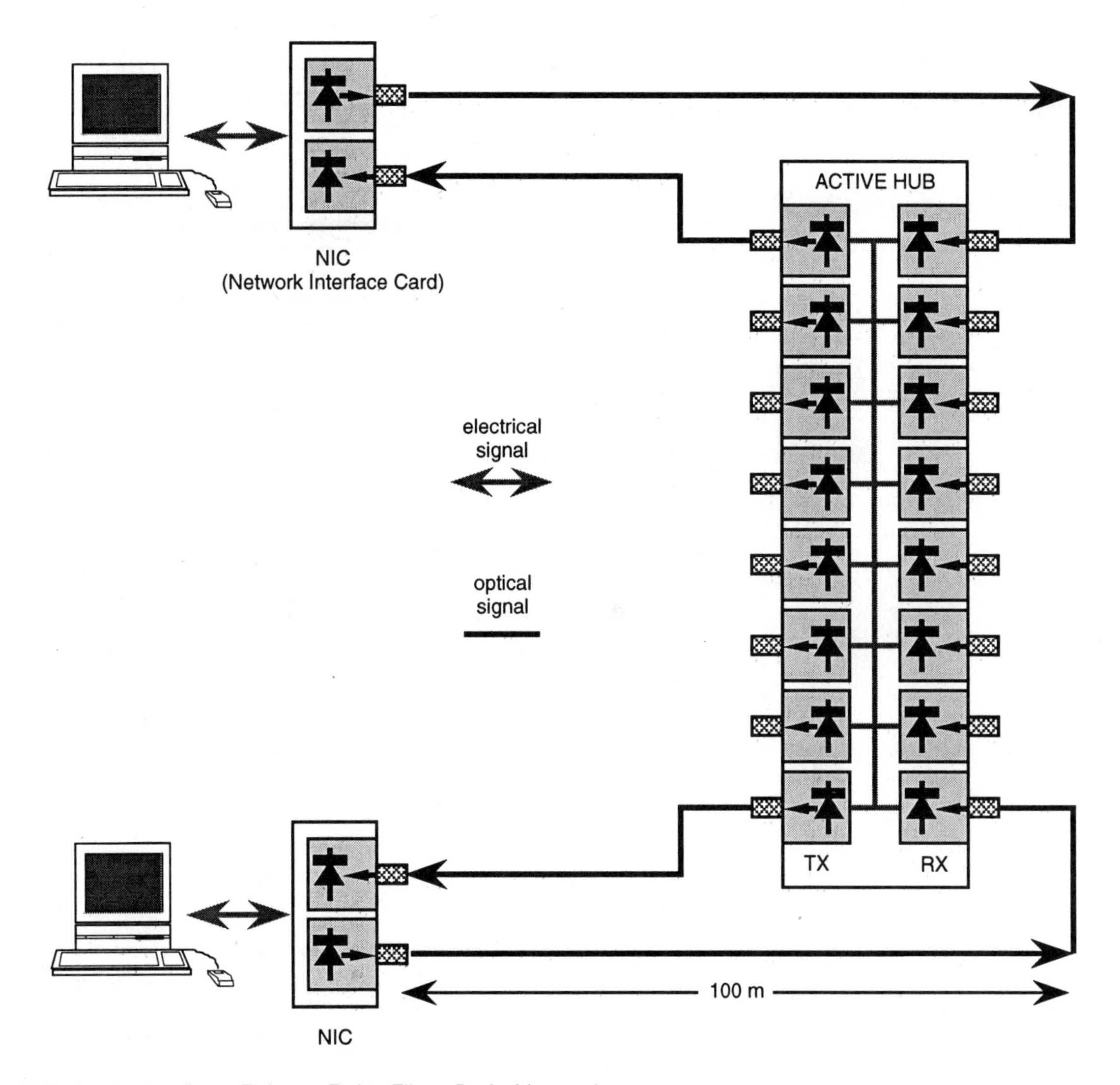

Figure 2-1 An Active Star, Point-to-Point Fiber Optic Network

attenuation in cable = 0.100 km × 4 dB/km =	0.4 dB
loss in connectors =	0.0 dB
aging or safety margin =	3.0 dB
optical power budget requirement =	3.4 dB

Table 2-1 Optical Power Budget Calculation for a Point-to-Point Link

A few comments on our calculation in Table 2–1: It is based on a multimode fiber with an 850 nm light source. There is no connector loss because the connectors are connected directly to the active devices. The aging margin is commonly assumed to be 3 dB. If an aging margin is not required, a safety margin (typically 3 to 9 dB) may be.

Comparison of this requirement to the values in Table 1–9 reveals that most optoelectronics will function properly with significantly higher optical power budget requirements. This point-to-point system can be implemented easily.

If we design this same system as a point-to-multipoint network (Figure 2–2), the optical power budget requirement is high (Table 2–2).

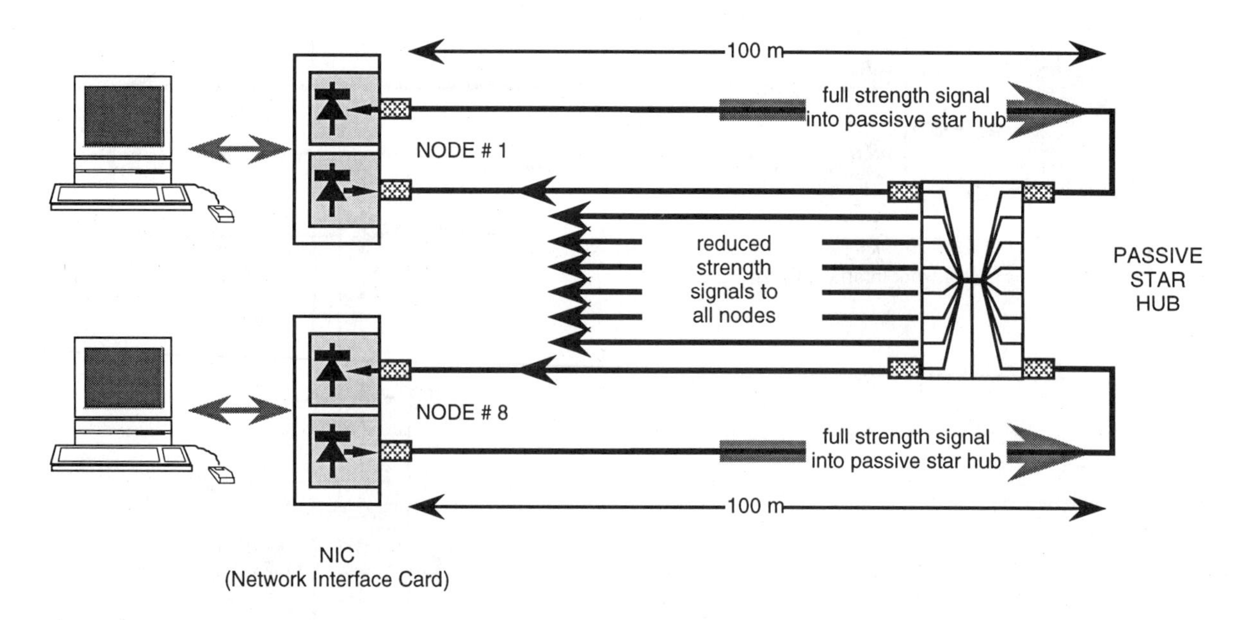

Figure 2–2 A Passive Star, Point-to-Multipoint Fiber Optic Network

attenuation in cable = 0.200 km × 4 dB/km =	0.8 dB
loss in connectors = 2 pair × 1 dB/pair	2.0 dB
intrinsic loss of splitter=	9.0 dB
extrinsic loss of splitter =	3.0 dB
aging or safety margin =	3.0 dB
optical power budget requirement =	17.8 dB

Table 2–2 Optical Power Budget Calculation for a Point-to-Multipoint Link

A few comments on our calculation in Table 2–2: It is based on a multimode fiber with an 850 nm light source. The intrinsic loss is the reduction of the signal strength between the input to and the output from the passive splitter. Consider a pulse of light consisting of eight photons entering the splitter. One of these eight photons will exit the output port. Thus, the strength of the output signal will be ⅛ of the strength of the input signal. By the equation we use for calculating optical power loss, this loss is 9.03 dB. The extrinsic loss is due to the loss of light energy in the splitter. A 3dB extrinsic loss is typical for a 1 × 8 splitter.

Comparison of this requirement to the values in Table 1–9 reveals that most 850 nm optoelectronics will not function properly with this high optical power budget requirement.

We could implement this topology with the 100/140 μm fiber (Table 1–9). However, doing so would increase the system cost, since this fiber is one of the most expensive multimode fibers available. We could implement this system with a high output power light source or with a high sensitivity receiver. However, again, use of these increased performance products will result in increased system cost.

Systems and Networks

Fiber optic systems and networks are being used for essentially all applications in which copper and microwave are used. With the exception of the passive star products and some CATV systems, all are point-to-point optical topologies. These applications include low speed data transmission applications, such as RS-232,

RS-422, and IEEE-488; closed circuit TV, in both AM and FM versions; cable TV, in AM and some FM versions; high-speed networks, such as Fiber Distributed Data Interface (FDDI); very high-speed data transmission, such as HiPPI, Fiber Channel, and asynchronous transfer mode (ATM); time division multiplexing of multiple signals; wavelength division multiplexing; channel extenders, such as HIPPI, ESCON, and IBM 32xx protocols; medium-speed, legacy data networks, such as Ethernet and Token Ring; and telephone networks, both proprietary and SONET. For a complete listing of optoelectronics suppliers, see *Worldwide Fiberoptic Suppliers Directory* (Kessler Marketing Intelligence, Newport, R.I.) or Fiberoptic Produce News Technology Reference (Gordon Publications, Morris Plains, N.J.).

STANDARD PRODUCTS

Some of these systems and networks are based on national or international standards. Eight product standards exist or are in development at this time:

- Ethernet
- Token Ring
- FDDI
- FDDI II
- HiPPI
- ESCON
- Fiber Channel
- SONET

Ethernet

There are three Ethernet fiber optic standards: 10Base-FL (Link), 10Base-FB (Backbone), 10Base-FP (Passive) (Figure 2–3). All use CSMA/CD protocols of Ethernet. 10Base-FL and 10Base-FP use asynchronous clock, while FB uses a synchronous, centralized clock. This standard specifies 62.5 μm core fibers, ST-compatible connectors, and the requirements in Table 2–3.

The FL standard is used for 2-km active links to connect repeaters and DTE and is a follow-on to the previous FOIRL standard. FB is similar to FL, but allows cascading of a larger number of repeaters, more than 15 in some cases, because of its centralized clock.

The lowest cost approach is the FP, which is a passive hub approach. The 10Base-F standard includes a systems-design guide and a comparison of coaxial, FP, and fiber approaches.

Wavelength, range (nm)	**800–900**		
Spectral width, maximum (nm)	**75**		
	10Base-FP	**10Base-FL**	**10Base-FB**
launch power (dBm)	-15 to -11	-20 to -12	-20 to -12
received power (dBm)	-41 to -27	-32.5 to -12	-32.5 to -12
maximum length (m)	500	2000	2000

Table 2–3 The Optical Performance Requirements of 10Base-F (From Donald J. Sterling, Jr., Technician's Guide to Fiber Optics, 2nd ed. [Albany, NY: Delmar Publishers, 1993], 223.)

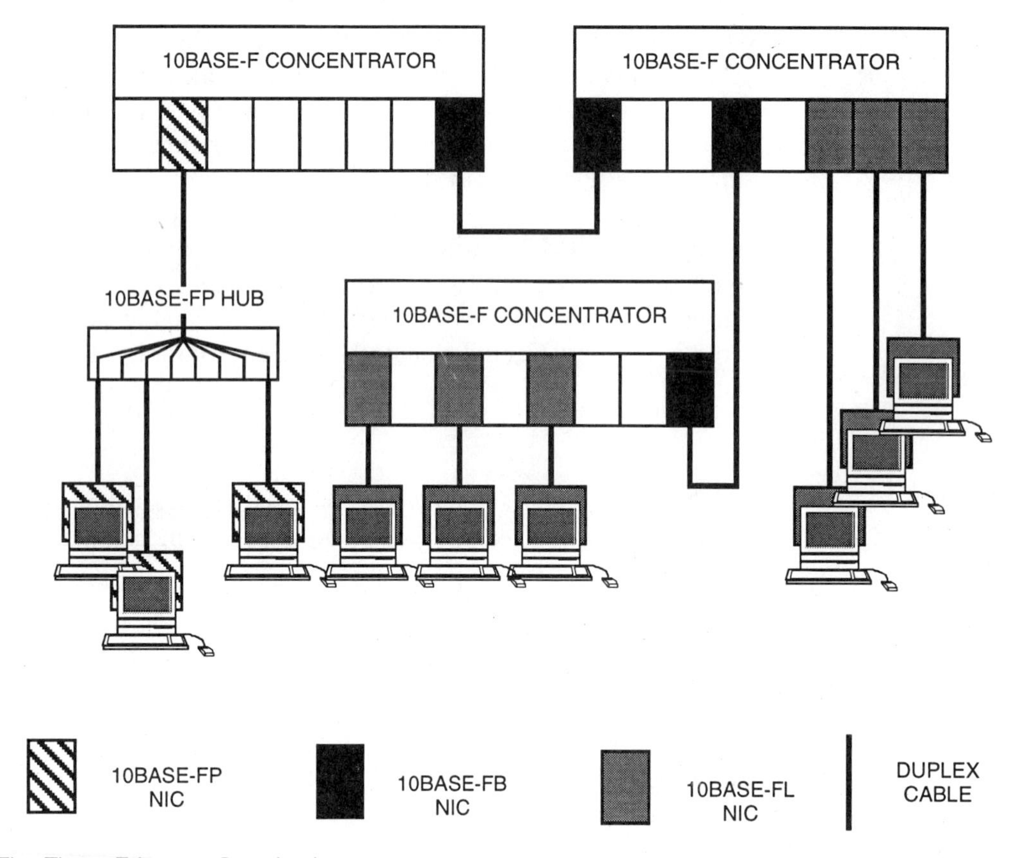

Figure 2–3 The Three Ethernet Standards

Token Ring

The IEEE 802 committee is developing a standard for fiber optic implementation. Although no formal standard exists, such products have been available since 1988.

The Fiber Distributed Data Interface (FDDI)

General Description. FDDI is an approved international digital transmission standard for a "high speed," token passing LAN. The standard is 75 percent complete (Figure 2–4 and Table 2–4). FDDI allows for an electrical data transfer at a rate of 100 Mbps (Table 2–5). The electrical signal is encoded with a 4-bit to 5-bit scheme, so that the optical data rate is 125 Mbps.

FDDI networks can operate at an effective electrical data transfer rate of 100 Mbps, but only if all stations are processing FDDI protocol. If some of the stations are processing foreign protocols, such as Ethernet or Token Ring, the translation time will reduce the effective throughput to lower rates (Table 2–5).

Section	ANSI Number	ISO Number
PMD	X3.166 (1990)	IS-9314-3 (1990)
PHY	X3.148	IS-9314-1 (1989)
MAC	X3.139 (1987)	IS-9314-5 (1990)
SMT	X3T9.5 Rev. 7 (1992)	IS-9314-6 WD

Table 2–4 The Four FDDI Specifications

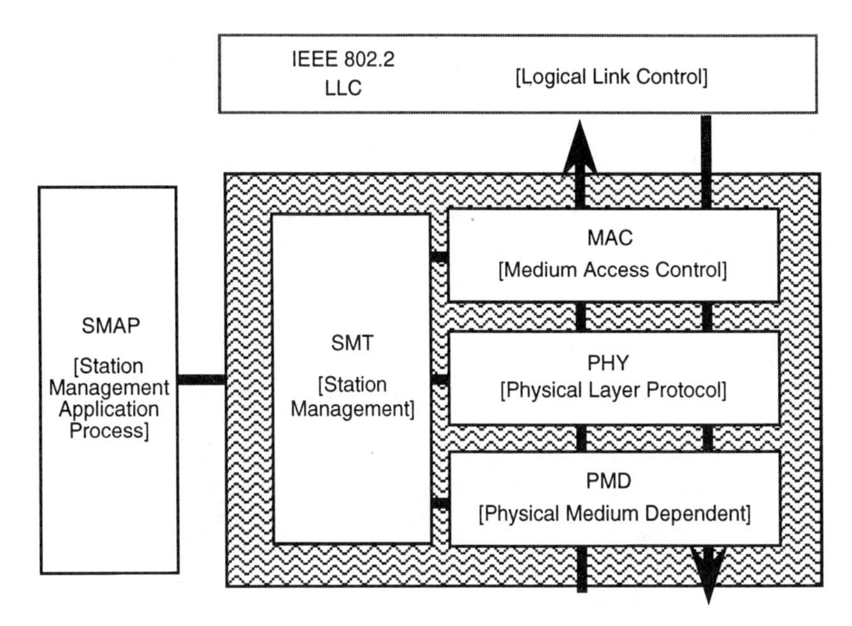

Figure 2–4 The Structure of the FDDI Standard

Type of Speed	Rate (Mbps)	Comment
electrical data transfer rate	100	
optical transfer rate	125	
effective throughput	100	no translation
effective throughput	60–85	translation required; i.e., 10Base-F

Table 2–5 The Four Speeds of FDDI

The topology is a dual counter rotating ring (Figure 2–5) with a maximum transmission distance of 2,000 m between adjacent stations. This ring can have a maximum circumference of 100 km. Each station acts as a signal regenerator; that is, each station receives an optical signal from the previous station on the ring, converts that optical signal to an electrical signal, processes the electrical signal, and creates a new optical signal for transmission to the next station on the ring. Thus, an FDDI token passing network is a series of point-to-point links that form a ring.

This topology creates two data paths, each of which can support a data transfer rate of 100 Mbps. It is possible to configure station management software to support different data streams on the two data paths. In this manner, the network will support a transfer rate of 200 Mbps. However, this configuration does not comply with the FDDI specifications.

The topology creates one of the four design elements that provide reliability in an FDDI network. In the event of a failure of any link between adjacent stations, the ring will self-reroute, providing the first element of reliability (Figure 2–6).

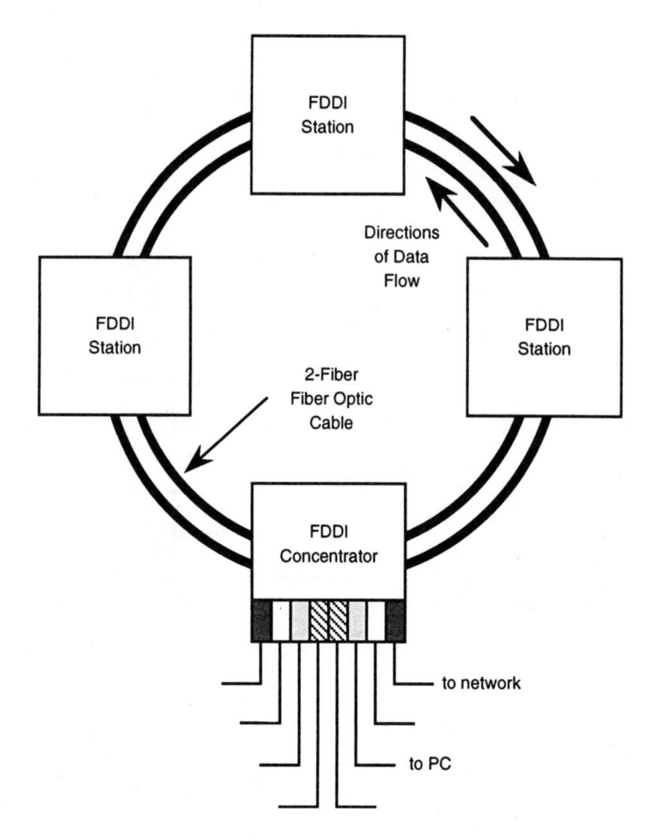

Figure 2-5 The Dual Counter Rotating FDDI Ring

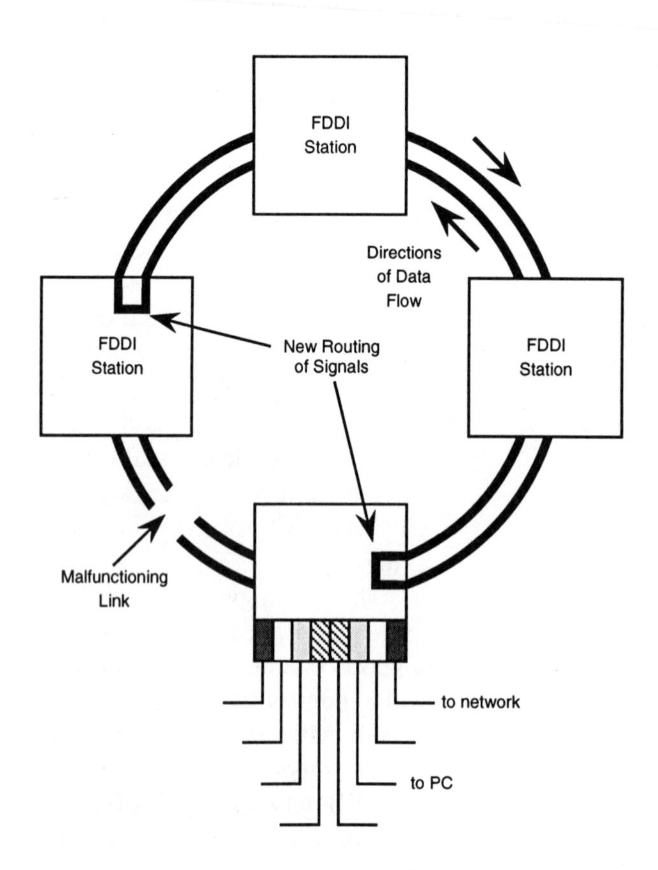

Figure 2-6 Automatic Loop Back in an FDDI Ring in the Event of a Cable, Connector, or Station Malfunction

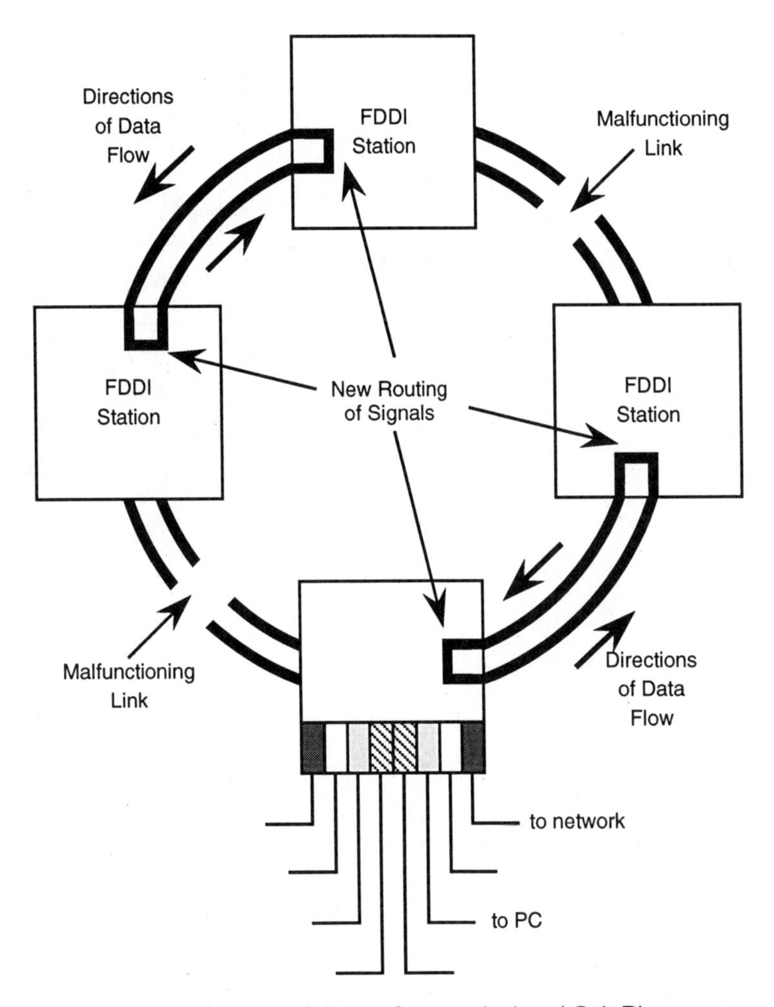

Figure 2–7 A Ring with Multiple Failures Creates Isolated Sub-Rings

However, multiple simultaneous failures can create separate sub-rings which cannot communicate with one another (Figure 2–7). If an FDDI network controls a complex process, multiple failures will cause process control to be lost.

The optical bypass switch option in FDDI addresses this problem and provides the second element of network reliability. The moving fiber optical bypass switch is activated during station failure. When the switch is activated, the station is dropped from the ring and the optical signal passes through the station (Figures 2–8 and 2–9) without the usual optical to electrical to optical conversion. When a bypass switch is activated, the optical signal can travel up to 4 km between active stations.

Number of Stations vs. Number of Nodes. The maximum number of stations on a single FDDI ring is 500. By the definition of an FDDI station, a station may, or may not, be a node. If the FDDI station is a concentrator with sixteen ports, each of these ports can connect to a lower speed, "legacy" network. With a configuration of concentrators and of 100 nodes, or users, per legacy network, a single FDDI ring can support more than 46,600 users or nodes.

Uses of an FDDI Network. FDDI is best suited for three uses:

- Backbone use (for interconnection of lower speed LANs)
- Interconnection of high-speed terminals
- Back end, or host-to-host connections

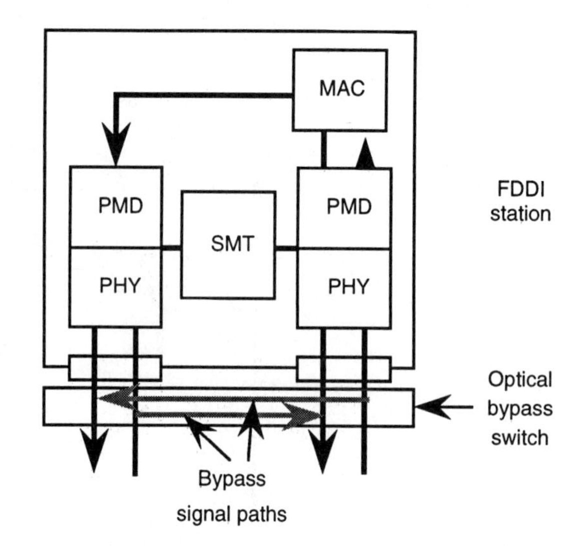

Figure 2-8 A Dual-Attachment Station with an Optical Bypass Switch

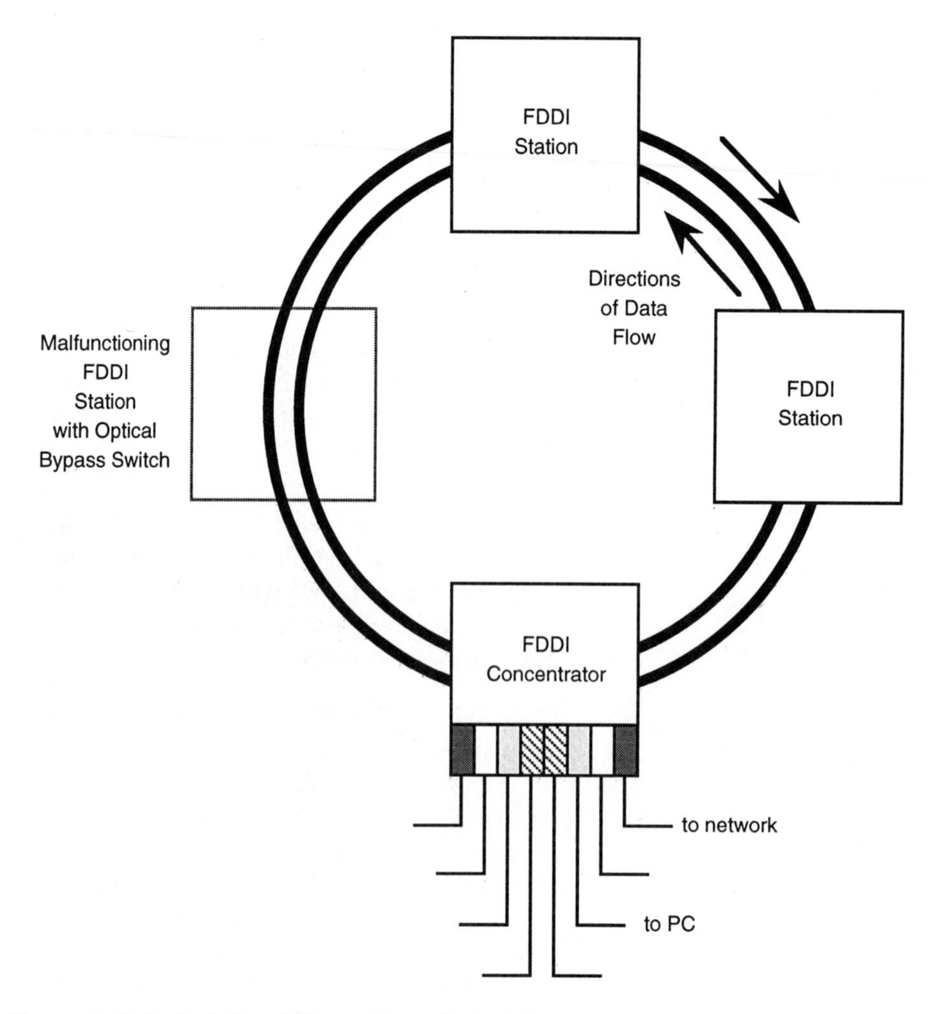

Figure 2-9 Optical Signal Flow with an Optical Bypass Switch

Type of Specification	Value	Comment
fiber core/clad diameters	62.5/125	µm
nominal fiber NA	0.275	test per RS-455-177
minimum bandwidth distance product	500	MHz-km @1300 nm
attenuation rate@1,300 nm	none	typically 1-1.5 dB/km
maximum loss between stations	11	dB @1300 nm
connector type	MIC	can use ST®

Table 2–6 Specifications for Migrating to FDDI

These three uses justify the investment in the relatively high cost interface optoelectronics. Because of this high cost, FDDI is not well-suited as a local area network.

Specifications for the FDDI Cable Plant. The FDDI cable consists of fiber and connectors described in Table 2–6.

Five Station Types and Four Physical Topologies. FDDI defines five station types: dual attachment station, dual attachment concentrator, single attachment station, single attachment concentrator (Figure 2–10), and the dual-homed, single attached station. The dual attachment station, or Class A attachment, is connected directly to the ring. The single attachment, or Class B attachment, must be attached to a Class A station in order to gain access to the ring. These five station types provide a high degree of design flexibility by allowing configuration of the ring in any type of physical network: ring (Figure 2–5), bus (Figure 2–11), star (Figure 2–12) and/or tree (Figure 2–13).

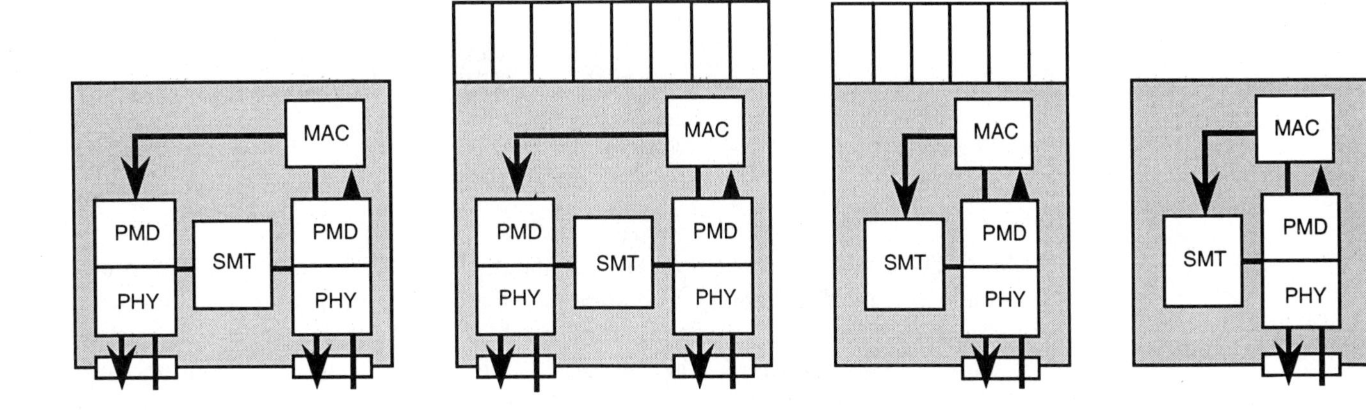

Figure 2–10 Four of the Five FDDI Station Types

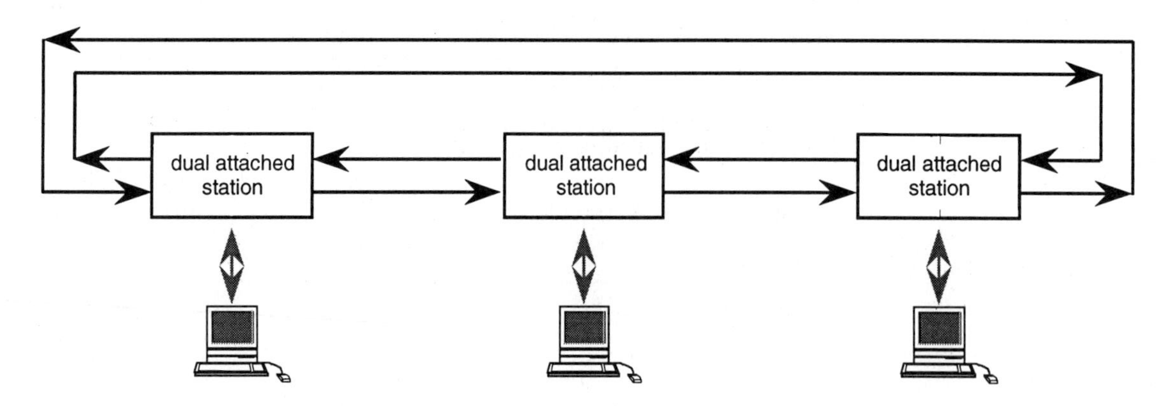

Figure 2–11 An FDDI Bus Network

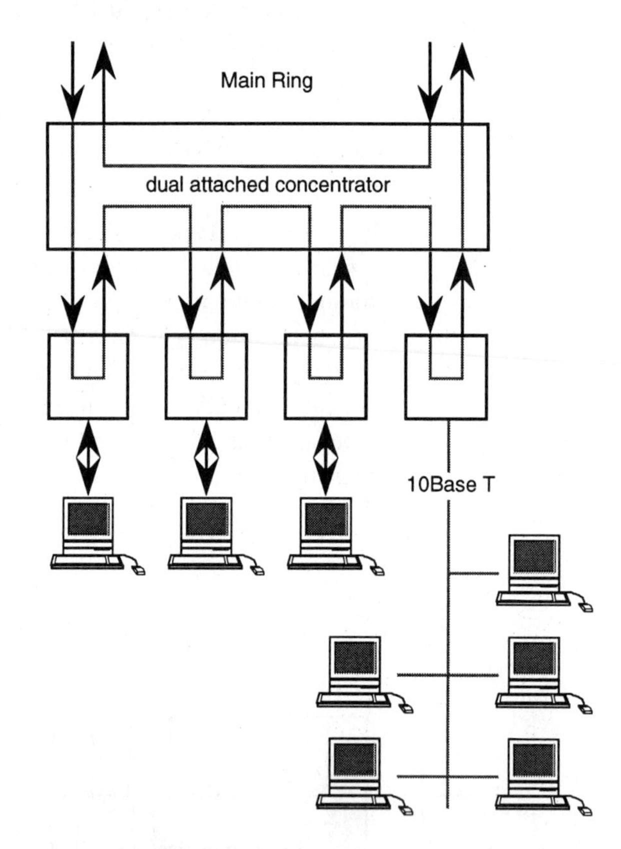

Figure 2–12 An FDDI Star Network

The dual-homed, single attached station (Figure 2–14) provides a third element of network reliability. This station type looks like a dual attached station, but functions like a single attached station. As does a dual attached station, this station type has two transmitter-receiver pairs. As in a single attached station, only one transmitter-receiver pair is active at any time. In the event of failure of the primary path of connection to the network, the secondary transmitter-receiver pair becomes active. The primary and secondary transmitter-receiver pairs are connected to the network at different locations.

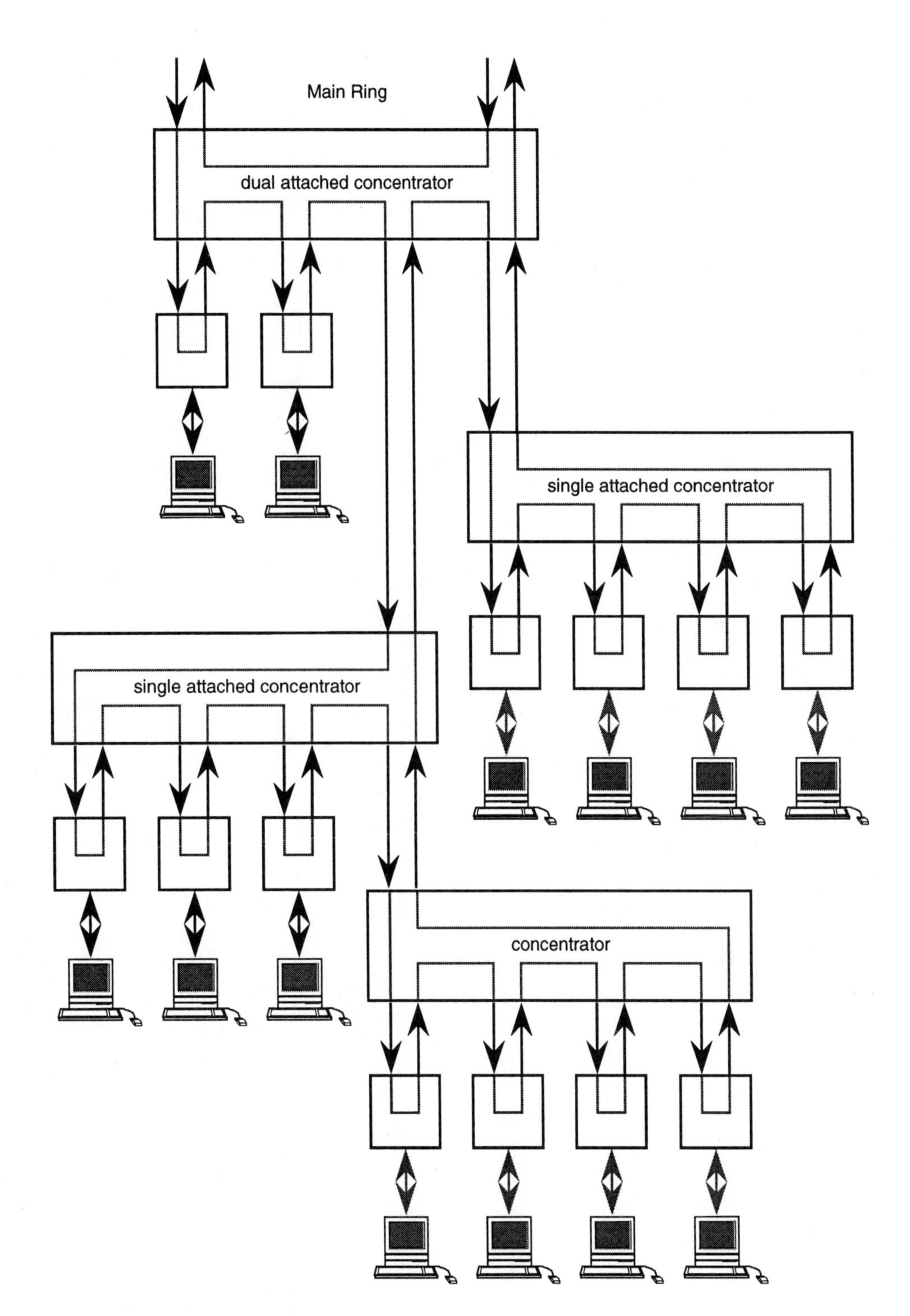

Figure 2–13 An FDDI Tree Network

The use of concentrators in an FDDI network provides the fourth element of network reliability. Since concentrators contain the intelligence to bypass malfunctioning stations, such stations will not bring down the network (Figure 2–15). Because of this increased reliability, there has been a trend towards increasing use of concentrators in FDDI networks.

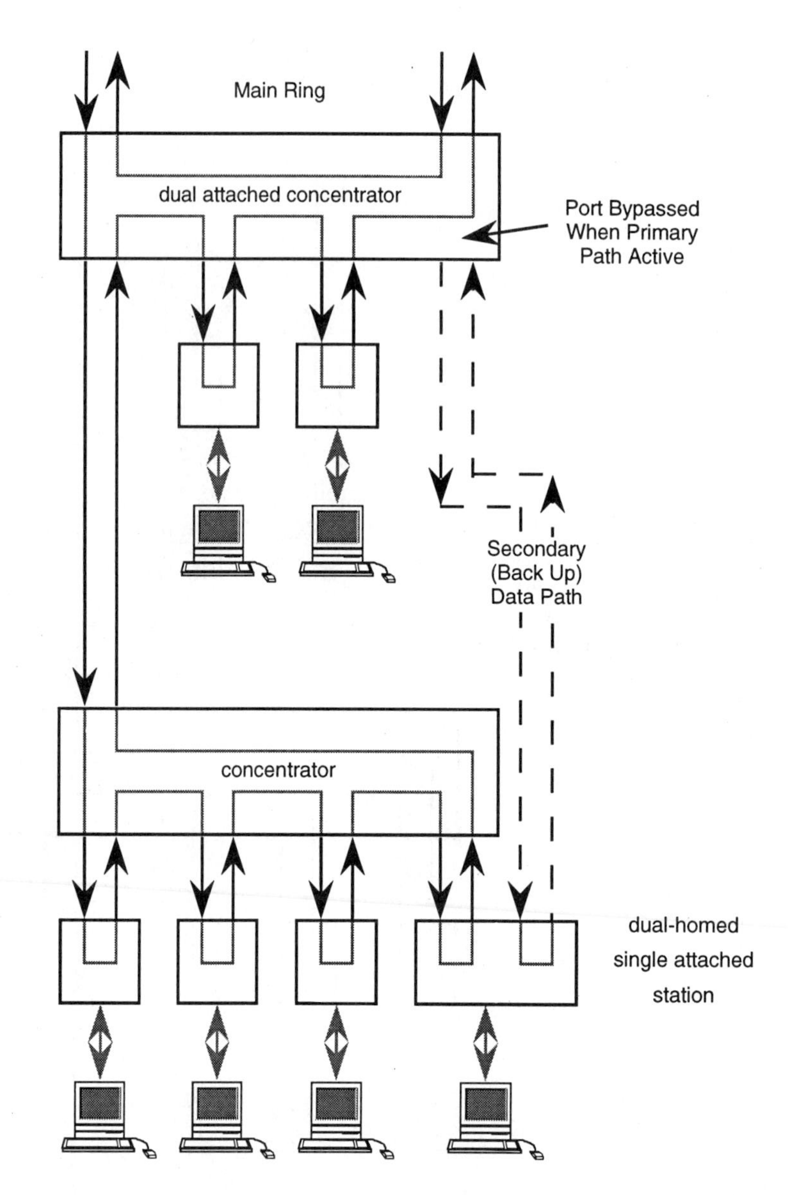

Figure 2-14 The Fifth Station Type, a Dual-Homed, Single Attachment Station

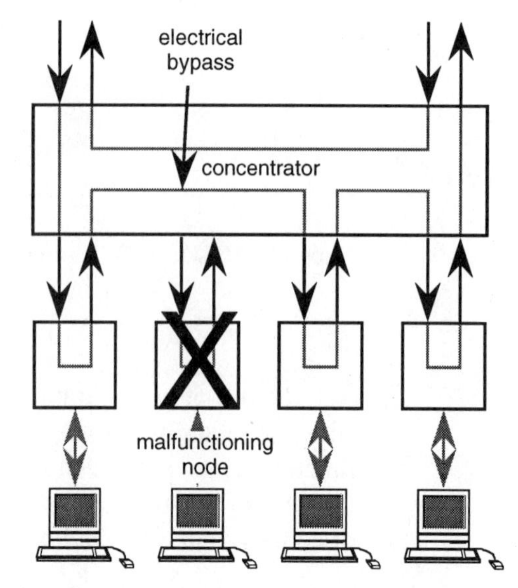

Figure 2-15 Electrical Bypass in an FDDI Concentrator

FDDI-II

FDDI-II is a mixed, multiple-service standard to allow the transmission of voice and video over an FDDI network. It is being handled under the auspices of IEEE 802.6. The standards are being developed by the X3T9.5 Standards Committee. It was expected that FDDI-II will add circuit switching capability to the basic FDDI network. FDDI-II would be "downward" compatible with FDDI, but FDDI would not be upward compatible with FDDI-II.

Industry has, apparently, lost interest in FDDI-II, in favor of ATM (asynchronous transfer mode). ATM is expected to provide many of the services originally envisioned to be provided by FDDI-II.

HiPPI

HiPPI, or high speed, parallel processor interface, was a standard developed for coaxial cable. HiPPI is well suited for high-speed transmission between massively parallel processors, mainframes, and high-speed storage devices, and between high-speed workstation clusters. This standard specifies 800–1,600 Mbps rates to distances of 25m.

To overcome this severe distance limitation, forty manufacturers issued an *ad-hoc* standard for a HiPPI fiber optic channel extender. Extenders manufactured to meet this *ad hoc* standard are interoperable and compatible. Such products extend the transmission distance to 40 km.

At first, the fiber HiPPI was a point-to-point communication link. As such, it was not intended for use in a network. However, interfaces between HiPPI and FDDI, HiPPI and ATM, and HiPPI and SONET exist. (Gary S. Christensen and James P. Hughes, "Complementary Technologies: HiPPI in the ATM Environment," *Fiberoptic Product News* [June 1995]: 39–41.) In addition, HiPPI cross point switches have become available, allowing HiPPI to be used in extended networks.

ESCON

Enterprise Systems Connection (ESCON®) is a fiber optic back end channel for IBM mainframes. This channel is used for communication with peripheral control units. These units control high-speed peripheral devices, such as tape and disk drives.

ESCON specifies ESCON connectors, 200 Mbps, 8 dB optical power budget, 1325 nm light sources, 150 nm spectral width, 3 km transmission distances with 62.5 μm fiber (or 2 km transmission distances with 50 μm fiber), and a transmission distance of up to 9 km with two directors. Use of singlemode fiber and laser diodes allows for transmission distances of up to 60 km (Donald J. Sterling, Jr., *Technician's Guide to Fiber Optics*, 2nd ed. [Albany, NY: Delmar Publishers, 1993], 234). A director is an active, switched star electronic subsystem that directs requests from the mainframe to the appropriate peripheral.

Fiber Channel

Fiber Channel is a developing standard designed to provide scalable (265.6, 531.25, 1062.5 Mbps), protocol independent, high rate data transmission over long distances (175 m–10 km, Table 2–7).

Fiber Channel supports both channel (point-to-point) and network connections (Charles Bazaar, "Fiber Channel: The Future of High Speed Connectivity," *Fiberoptic Product News* [April 1995]: 36–37 and [June 1995]: 7–8). SC connectors are specified.

Fiber, μm	Maximum Distance, km	Maximum Data Rate, Mbps	Wavelength, nm
9	10	1062.5	1310
50	2	265.6	780
62.5	1.5	265.6	1300
50	1	531.25	780
50	0.5	1062.5	780
62.5	0.350	531.25	780
62.5	0.175	1062.5	780

Table 2–7 Fiber Channel Capabilities

SONET

Synchronous Optical Network (SONET) is a developing international standard for bridging private and public networks. SONET requires use of singlemode fiber to achieve transmission rates of 51.84 Mbps (OC-1) to 9.952 Gbps (OC-192). The maximum rate will be 13.22 Gbps.

SONET has six advantages:

- It allows asynchronous LANs to access the synchronous public network.
- It is reverse compatible with DS-3 and will interface with T1 and T3.
- Customer premise equipment will be standardized, promising lower cost equipment with higher data rates and easier expandability.
- Use of synchronous transmission allows for ease of drop and add capabilities, which reduce network cost.
- Compliance with SONET specifications results in interoperability and compatibility of equipment from different vendors.
- A SONET network can be managed from end to end, even if the network spans thousands of miles. Such management is not possible or convenient with equipment from different vendors.

In Europe, SONET is known as Synchronous Digital Hierarchy (SDH). SONET and SDH are compatible but cannot be exactly mapped into each other.

SONET phase 1, introduced in 1984, was approved in 1988. SONET phase 2 is in development. Phase 2 includes standardization of network messages and commands, definition of the operation functions needed, protocol and message sets of operations channels to be used, analysis of the jitter and wander implications of pointer adjustments, and pointers added to the optical interface to allow frequency adjustment to occur between network elements running at slightly different clock rates while preserving payload structure.

SONET specifications appear in ANSI T1.1.5-1988, in ANSI T1.1.6-1988, and in Bellcore specifications TA-TSY-000755 (main specification), TA-TSY-000253 (transmission system), and TR-TSY-000496 (add/drop multiplexer).

Other Standards

Test Standards. Because light is a subtle medium, all performance numbers will be highly sensitive to the method of testing. Because of this subtlety, performance statements of all fiber optic products must be tied to the method used to test the performance of the product. Different test methods result in different performance statements; comparing the performance of similar products tested by different techniques is like comparing apples to oranges. Fortunately, the

Electronic Industries Association has created an extensive list of approved test methods (Appendix 3).

Building Wiring Standards. The TR 41 committee has issued EIA/TIA-568A, a building wiring standard that includes fiber optic cable. This document specifies 62.5/125 fiber and SC connectors for both singlemode and multimode fibers. This specification allows for use of FC for singlemode and ST for multimode connectors.

National Electrical Code (NEC®). The NEC addresses the requirements for cables used inside buildings. These requirements limit the rate at which flames spread along a burning cable. In addition, the plenum rating limits the amount of smoke generated by a burning cable.

The tests specified by the NEC are UL-1666, for riser-rated cables and UL-910, the Steiner Tunnel test, for plenum-rated cables. The NEC creates two general types of cables, each of which has three levels of performance (Table 2–8). The two types are the all-dielectric, or non-conductive, and the conductive. The all-dielectric or "N" type contains no conductive elements. Conductive elements can be conductors, metallic armor, or metallic strength members. The conductive or "C" type contains conductive elements.

A strong trend has developed in the preferred use of N series cables. The reason for this trend is the limited number of restrictions in the NEC for placement of this series. C series cables have a larger number of placement restrictions.

Military Standards. There are numerous standards for products used in military applications (Appendix 3).

Use	Rating	Designation Non-Conductive	Designation Conductive
horizontal	general	OFN	OFC
risers between floors	riser	OFNR	OFCR
air-handling plenum	plenum	OFNP	OFCP

Table 2–8 National Electric Code Fiber Optic Cable Categories

REVIEW QUESTIONS

1. What are the advantages of fiber optic transmission?
2. What is a logical topology?
3. What is a physical topology?
4. Why do fiber optic systems tend to be designed as a series of point-to-point links?
5. What two performance characteristics of optical fibers enable long-distance transmission?
6. Why are optical fibers immune to electromagnetic interference (EMI) and radio frequency interference (RFI)?
7. What advantage of fiber optic cables makes them attractive for use in shipboard, aircraft, and field tactical Army applications?
8. What characteristic of optical fibers makes them attractive for use in environments prone to lightning?

9. What is the optical power budget?
10. What is the significance of the optical power budget?
11. What are the four ways in which an FDDI network achieves reliability?
12. What is the maximum multimode distance between adjacent stations on an FDDI network?
13. What is the optical data rate at which an FDDI network functions?
14. At what wavelength does an FDDI network operate?
15. Why is an FDDI network not used as a LAN?
16. What does "FDDI" stand for?

CHAPTER

3

How To Install Cable and Prepare Ends

CHAPTER OBJECTIVES

From this chapter, you will be able to:

1. Install fiber optic cable successfully.
2. Avoid breakage and excess loss.
3. Handle fiber without breaking it and without being unnecessarily careful.
4. Prepare the ends of both tight tube (indoor) and loose tube (outdoor) cable designs.

INTRODUCTION

During cable installation and end preparation, the installer has two major concerns:

- To avoid fiber breakage
- To avoid reduction of the power level at the receiver

The installer can address both of these concerns by following the specific guidelines for successful installation and end preparation presented in this chapter. These techniques are applicable to most installation situations, but are focused on installation of cable in conduits, inner ducts, cable trays, and above suspended ceilings. Many of these techniques apply to direct burial, aerial self-support, and to aerial lashed installations. These last three installations have additional requirements specific to each of these situations.

TWENTY-FOUR GUIDELINES FOR SUCCESSFUL CABLE INSTALLATION AND END PREPARATION

Guideline 1

Plan the Installation Activities. Create a detailed, written plan of installation. Creation of this plan should eliminate over 95 percent of the problems installers can encounter. This plan should include the following ten elements:

- Equipment and supplies
- Information about cables to be installed
- People with appropriate types and levels of experience
- Location of equipment
- Installation methods
- Testing requirements
- Data forms for testing
- Assignment of activities
- Identification of potential problem areas
- Identification of safety issues

The equipment and supplies required vary from installation to installation, but include the following:

- Pulling equipment
- Tension limiting equipment
- Cable lubricant
- Cable cleaning chemicals
- Cable ties
- Cable pay-off stands
- Take-up reels and reel stands
- Cable pulleys (including pulley mounting equipment)
- Cones and signs to block off or segregate work area
- Communication equipment (walkie talkies, fiber talk sets)
- Testing equipment (optical time domain reflectometer, power meter loss test set)

The information required about the cables to be installed includes:

- Minimum long-term and short-term bend radii
- Maximum long-term and short-term loads
- Temperature installation and storage ranges
- Recommended methods for attaching pulling ropes to the cable

The number of people required for each cable pull depends on the specifics of the pull. The number of people will be discussed in Guideline 6.

The locations of equipment include the locations of the supply reel, pulling equipment, intermediate or assist pulling equipment, pulleys, and cones to segregate the work area.

There are two groups of installation methods to be used: methods of attaching pulling ropes to cables and methods of cable installation.

The method of attaching pulling ropes depends on the cable design. Many cables can be installed with a Kellems® grip, which is like a Chinese finger trap (Figure 3–1). This grip consists of a woven wire mesh, which is sized to the diameter of the cable being installed. The higher the installation load, the higher the compression load placed on the cable. A Kellems grip works well on cables with diameters larger than ⅜″.

However, three types of cables cannot be pulled with a Kellems grip. The first type is a small diameter cable (<⅜″), which will tend to pullout of such a grip. The second type is a larger diameter cable which allows stretching of the jacket, due to minimal bonding between the strength members and the jacket. With this type of cable, the jacket will stretch off the end of the cable and slip from the grip. The third type is a light duty, premise cable, in which the compression of the Kellems grip may crush and break the fibers.

Figure 3–1 Kellems Grip® Used for Attaching a Pull Rope to a Cable (Courtesy of Pearson Technologies Inc.)

For these types of cables, the pull rope must be tied to the cable strength members. However, not all strength member materials in a cable structure are used as strength members in the design. The installer must attach the pull rope to the strength members, as recommended by the cable manufacturer.

The second group of installation methods is methods of cable installation. There are two methods of pulling: end pulling and center, or back, pulling. In end pulls, the cable is pulled from one end to the opposite end. However, in some environments, the cable length will need to installed from a central location to each of the two ends. For center pulls, a large area needs to be identified and segregated for storage of the cable in a figure 8 pattern on the ground. For more details, see Guideline 10.

Both end and center pulls can be performed in a single pull or in multiple pulls. If the cable is to be installed in multiple pulls, a large area needs to be identified and segregated for storage of the cable at each intermediate location. For more details, see Guidelines 10 and 24.

The testing requirements include the number and sequence of tests to be performed. For more details, see Guideline 18.

The data forms for testing should be created before installation begins. These forms should indicate the type of test, location of test, direction of test, test parameters, and expected and allowable maximum test values. Types of tests are insertion, loss, OTDR trace, or continuity check. Test parameters are wavelength, spectral width, and core/clad diameters.

The identification of potential problem areas will enable supervisory personnel to assign extra personnel, to provide additional supervision, or to modify installation procedures for these areas.

The identification of safety issues will ensure a safe installation process. For more details, see Guidelines 21, 22, and 24.

Guideline 2

Do Not Exceed Maximum Recommended Installation Load. To avoid breakage due to excessive load on the cable, know and observe the maximum recommended installation load. Observing the load limitation requires:

- Knowledge of the load rating of the cable
- Estimation of the load likely to be imposed on the cable
- Use of tension monitoring or limiting equipment during installation

Knowledge of the installation load and estimation of the load likely to be imposed on the cable usually come from the system designer. However, not all system designers conduct their design to the level of the conduit, etc., in which the cable is to be installed. The load can be estimated from experience, or with the software package available from American Polywater Corporation (see Appendix 7).

Equipment for monitoring or limiting load is available in two forms: alarmed equipment and limiting equipment. Alarmed equipment (Figures 3–2 and 3–3) has a pulling mechanism, a load gauge, and an alarm that alerts the installer to violation of the installation load limit. The alarm can shut off the pulling motor. The load gauge can be connected to a chart recorder to provide evidence of compliance with the installation load. Such compliance during installation results in maximum cable reliability and lifetime. Exceeding the installation load can cause cracks to grow in the fibers. Such cracks can fail, either during installation or after installation.

Limiting equipment is available in two forms: a puller with a slip clutch, which is set to the installation load; and a break-away swivel, with a rating equal to the installation load. Use of a puller has the advantage of control of both the pull

Figure 3–2 Example of a Load Monitoring Puller (Courtesy of Greenlee Textron)

Figure 3–3 Example of a Load Monitoring Puller (Courtesy of Condux International)

speed and the load, but the disadvantage of relatively high cost and long time for set up of pulleys and/or slings to control the bend radius of the cable. However, use of a break-away swivel has a cost lower than that of pulling equipment and allows use of a vehicle as the puller.

Guideline 3

Comply with the Minimum Recommended Bend Radii. Know and comply with the two minimum recommended bend radii: the short-term, which is observed during installation; and the long-term, which is observed in the final installation. A change in direction or a change in height, due to routing the cable through different conduits, requires a pulley. Complying with the short-term installation bend radius requires use of pulleys sized to the cable being installed. Complying with this bend radius can be a problem when installing a large diameter cable in manholes. Large, high-fiber count cables can be 1 inch in diameter. If the 20× rule of thumb applies, this 1-inch diameter cable will require a pulley with a 20-inch radius or a 40-inch diameter. Most manholes are too small to accept a 40-inch pulley.

Complying with the long-term installation bend radius requires examination of the cable path, conduit elbow radii, and junction boxes, to verify compliance. Finally, complying with this bend radius requires careful use of cable ties. Excessively tight cable ties on small diameter, tight tube cables can cause excessive attenuation. Small diameter cables are one- and two-fiber, indoor, tight buffer tube cables and premise cables with twelve or fewer fibers.

Cable ties and/or clamps can be used to fasten the cable to the cable box as long as the ties or clamps are not excessively tightened. The box manufacturer will provide instructions on proper installation of a cable in the end box.

Finally, complying with the long-term installation bend radius requires care when installing cables in cable end boxes. This care is required when placing cable and buffer tubes inside the boxes. Loose buffer tubes tend to be stiff materials. Should you violate the bend radius of a loose buffer tube, the tube can kink and break the fibers.

Guideline 4

Pull, Do Not Push Cables. Pushing can result in violation of the bend radius. There are two exceptions: first, when sliding fibers into furcation tubing; and second, when pushing fibers into splices or connectors.

Guideline 5

Monitor the Supply Reel. Monitoring the supply reel during installation is necessary to prevent violation of the bend radius. Violation of the bend radius can occur in three situations:

- Improper winding of the cable on the reel
- Loosening of the cable on the reel
- Back wrapping

Improper winding occurs when an inner layer is on top of an outer layer. This situation can cause a violation when the outer layer resists being pulled.

Loosening of the cable occurs when the pull speed is reduced. When this happens, the momentum of the cable causes the cable to become loosely wrapped on the reel. Loosely wrapped cable can fall outside of the cable flange and wrap around the supporting shaft.

Back wrapping occurs because the supply reel has momentum. When the pull is stopped, momentum can cause back wrapping.

Guideline 6

Communicate Along the Path of Installation. When installing long lengths, communicate and monitor along the path of installation. This requires a crew of at least three people plus one person for each change in direction (or pulley location). Cables can, and do, jump from unmonitored pulleys. Off the pulley, the cable will be in violation of its short-term bend radius. In addition, a cable off its pulley can scrape the entry edge of metal conduit. Such scraping can damage the jacket and strength members immediately under the jacket. Damaged strength members reduce the installation load rating of the cable and can result in cable breakage. The minimum crew, with no pulley locations, consists of one person monitoring the pulling equipment, one monitoring the supply reel, and one coordinating all involved in the installation.

Guideline 7

Use Proper Tools and Techniques. A vehicle for pulling the cable is not a proper tool, unless a break-away swivel is also used. Proper techniques depend on the cable design and the location of the installation. For example, proper technique is pulling a cable into a conduit. Proper technique is placing or laying a cable in a cable tray or raceway. Kellems grips are proper tools for many, but not all, cables.

Guideline 8

Use Fiber Optic Cable Lubricant. Lubricate the cable when installing in conduits. Lubrication reduces the pulling load and the chance of breakage. Choose the lubricating fluid to be compatible with the jacket material.

Guideline 9

Train Installation Personnel. Properly train and instruct the people who are to do the installing. Installation personnel must be sensitive to the limitations of the products with which they are working. Proper training will reduce installation expense through reducing breakage and excess attenuation.

Guideline 10

Use the Figure 8 Technique. Divide long pulls into several shorter pulls, using the figure 8 technique (Figure 3–4) for storing cable at the intermediate locations. Reducing the length of the pull reduces the load on the cable and the probability of fiber breakage.

In this technique, the cable is placed on the ground in a figure 8 pattern. This pattern is large, at least 10–20 feet from top to bottom of the pattern. The size of this pattern depends on the length of cable to be placed in the pattern. The longer the length, the larger the pattern. When all the cable is placed in this pattern, the pattern is lifted and flipped over, so that the loose end is on top. This loose end is pulled into the next section of conduit or duct.

Guideline 11

Comply with Vertical Rise Limits. Know and observe the maximum vertical rise distance limit. Exceeding this limit can result in fiber breakage, excess attenuation, and, in loose tube cables, fibers sliding from cables. Cables in vertical installations that are longer than the vertical rise limit must be tied off at distances less than this limit. In addition, loose tube cables must be looped. For more details, see Guideline 17.

Figure 3–4 Figure 8 Technique for Storing Cable at Intermediate Locations During Multiple Pulls (Courtesy of Pearson Technologies Inc.)

Guideline 12

Allow for Thermal Expansion and Contraction. This guideline applies to outdoor applications. Failure to allow for thermal expansion can result in increased attenuation and/or breakage of fibers. A common practice is a 2.5-foot sag for a 150-foot span.

Guideline 13

For Tough Installations, Use Loose Buffer Tube Designs. Loose tube cables can be more forgiving of abuse than tight tube cables. The reason for this forgiveness is the excess length of the fiber in the buffer tube and the ability of the fiber to move from regions of lower stress to regions of higher stress. The fiber in the buffer tube can take a helical path. In addition, the fiber can take a helical path in the cable. Both of these conditions result in fiber length higher than cable length (excess fiber). Consider the fiber in the cable in Figure 3–5. Sections A and C are under no stress, but Section B is. The excess fiber length in the buffer tube allows the fiber in Sections A and C to move into Section B to equalize the stress on the fiber. After the load on Section B is removed, the fiber in Section B will move back into Sections A and C.

Guideline 14

Observe the Temperature Operating Range. Install a cable in locations in which the temperature range imposed is within the temperature operating range. An indoor fiber cable installed in a steam tunnel may experience a temperature above its maximum operating temperature, whereas an indoor climate may experience a temperature below its minimum. Violation of the temperature operating range can result in excessively high attenuation.

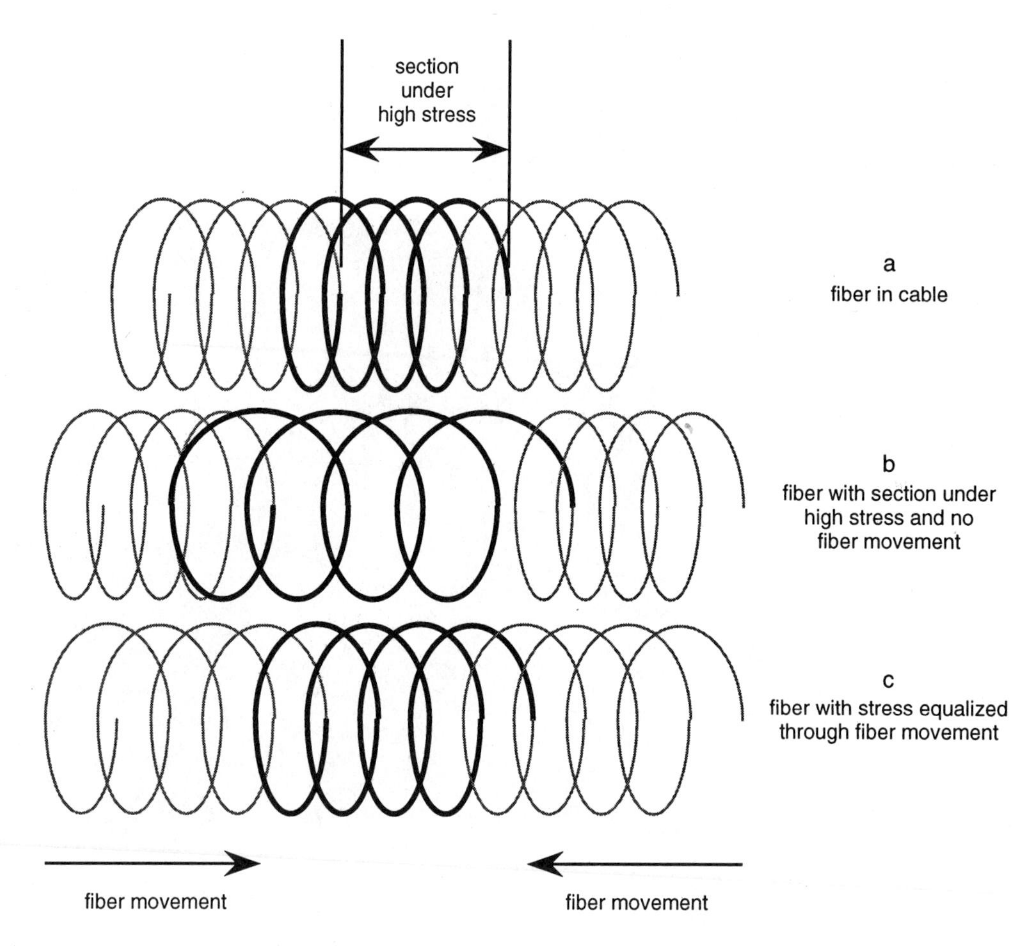

Figure 3–5 Fiber Movement in a Loose Tube Cable Under an Installation Load

Guideline 15

Cap or Seal Water Blocked Cables. Cap off or seal the ends of cables with water blocking gel or grease. These materials can flow out of the cable, causing maintenance problems in cable end boxes.

Guideline 16

Protect Fibers and Buffer Tubes. Confine fibers and buffer tubes in protective structures, such as splice trays and cable end boxes. Fibers and buffer tubes do not have sufficient strength to resist breakage due to the normal handling of copper cables.

Guideline 17

Loop Vertically Installed Loose Tube Cables. Install vertical, unfilled, loose tube cables with loops to prevent the fiber from slipping to the bottom of a vertical run. If this happens, attenuation can increase and fibers can eventually break.

Guideline 18

Check Continuity and Attenuation. Check the continuity and attenuation of the cable before each operation. These checks should be performed as-received/before installation, after installation, after splicing, and after connector installation.

You can check continuity with white light. You can check attenuation with temporary connectors (bare fiber adapters), which you can use to perform both insertion loss tests and optical time domain reflectometry tests (Chapter 5).

Guideline 19

Mark Cable as "Fiber Optic Cable". Mark cable as "fiber optic cable" in all locations in which it can be easily reached. (ACP International has snap-around cable markers.) Such marking will alert electricians to the nature of the cable.

Guideline 20

Make As-Built Data Logs. Make as-built data logs on all cables. Keep these data available to those who will perform maintenance and troubleshooting. Keep three copies: one for the person responsible for the installation; one for archives; and one in an open file available to those who will need to perform troubleshooting.

These data logs should include both insertion loss measurements and optical time domain reflectometer measurements. Data should identify wavelength, spectral width, equipment used, and the direction of test.

Guideline 21

Safety Precaution—No Food, Drink, or Smoke. Do not eat, drink, or smoke in any area in which bare glass fiber is created. Bare glass fibers can cause splinters which are very difficult to find and remove. A high-fiber diet does not mean ingesting optical fibers.

Safety glasses are recommended for end preparation activities. These glasses will protect eyes from glass fibers.

Guideline 22

Safety Precaution—Do Not Look into Fiber Without Checking Status. Never look into a fiber, cable, or connector unless you know, without any doubt, that there is no laser light in the fiber. A 850 nm and 1300 nm laser light is used in some test equipment and in some transmission systems. This light is invisible to the eye and can cause eye damage. For extensive work with fiber optics, safety glasses with IR filtering is advisable.

Guideline 23

Leave Service Loops. Leave cable and fiber service loops everywhere. You will regret not doing so. Service loops allow you to pull excess cable or fiber into a location in which you have experienced a problem. It is less expensive to pull in a service loop than to replace an entire length of cable.

Indoor service loops at cable ends should be at least 10 feet long. This length allows removal of the cable to a convenient location for installation of connectors or splices.

Common practice for outdoor service loops is 100 feet for each 1,000 feet of cable installed. In addition, cables that cross a street need to have 100 feet per crossing (up to 200 feet per 1,000 feet of installed cable) to allow for re-routing when poles are moved. A common practice is 20 feet of service loop every third manhole with a 50-foot service loop at each end of the link. Another common practice is a 50-foot service loop every third pole (pole spacing varies from 150 to 250 feet).

Guideline 24

Segregate or Isolate the Work Area. Rope off or otherwise isolate the area of cable installation, end preparation, and connector installation. By preventing access to unauthorized or untrained personnel, you will avoid safety problems, such as fiber splinters, and damage to partially completed work from unintentional abuse.

HANDS-ON ACTIVITIES: LEARN WHEN TO HANDLE FIBER CAREFULLY

Fiber Activity 1

Take a piece of all-glass fiber with buffer coating. Wrap the fiber around your fingers of both hands. Attempt to break the fiber by pulling. Was fiber stronger or weaker than you expected?

Fiber Activity 2

Bend the fiber into a loop near one end. Reduce the size of the loop until the fiber breaks. Estimate the diameter at which the fiber broke. Repeat this procedure one more time to verify your first result. Was the diameter of breakage larger or smaller than you expected?

Small diameter cables, such as sub cables from a break out cable design, have a typical diameter of 0.1 inch. Compare the diameter at which the fiber broke to the minimum recommended long-term bend radius of this 0.1-inch diameter sub cable. If you do not violate the minimum recommended bend radius, should you have any problems with breakage of fibers?

Fiber Activity 3

With a match, burn off the buffer coating from two inches of the end of a fiber. Wipe off the burnt buffer coating with your fingers. Push the fiber against any stiff object to bend it (Caution: do not push fiber against your finger. You may receive a fiber splinter, which can be difficult to find and remove). Continue pushing until the fiber breaks. Estimate the diameter at which the fiber broke. Repeat this procedure one more time to verify your first result.

The function of the buffer coating is to allow the fiber to retain its intrinsic high strength. How well does the buffer coating function?

To install splices or connectors, you will need to remove the buffer coating. With the buffer coating removed, you will push the fiber into a connector or splice with a fiber hole 1–4 μm larger than the fiber clad diameter. Based on your observations in this activity, at what step in your installation process will you need to be extremely careful?

END PREPARATION PROCEDURE FOR INDOOR CABLE DESIGNS

Most, but not all, indoor cables are tight buffer tube designs. Tight tube, single fiber cables are the simplest cables to prepare. The loose tube, single fiber per tube cables are almost as simple to prepare.

The procedure described in this section is to be used on a single fiber, tight tube cable design. This procedure, with minor modifications, can be used also on tight buffer tube premise cables, tight tube break out cables and two-fiber, tight tube zip cord duplex cables.

Tools and Supplies Required

tubing cutter (Ideal part number 45-162)

Kevlar® scissors (Fiskar)

single fiber cable with 125 μm clad

two Clauss No-Nik buffer tube/buffer coating strippers (one for spare) (part number NN200 or NN150)

small plastic bottle (for broken fiber)
Scotch™ tape
lens grade tissue
compressed air
95 percent isopropyl alcohol (without perfumes or oils)

PROCEDURE 1: END PREPARATION PROCEDURE OF SINGLE FIBER, TIGHT BUFFER TUBE CABLES

1. Obtain all tools and supplies.
2. Place the blade of the tubing cutter at a distance of approximately 2 inches from the end of the cable (Figure 3–6). Rotate the tubing cutter around the jacket one to three times. Remove the jacket with your fingers by pushing it from the cut towards the end of cable.
3. With the Kevlar® cutters or scissors, trim the Kevlar® or other strength members flush with the end of the jacket (in an actual in istallation, the Kevlar® would not be trimmed flush, but would extend beyond the end of the jacket). If the Kevlar® does not cut easily, twist or braid the Kevlar®, then cut. (Kevlar® dulls cutting edges very quickly. If cutting is consistently difficult, sharpen or replace scissors.)
4. Clean the buffer stripper. While holding the handles so that the heads are open, pull each of the plastic head covers so that the blade is exposed (Figure 3–7). Allow the head covers to snap back into place. Check the hole in the blade to ensure that it is free from debris. If the hole in the blade is not cleaned by snapping, use the compressed air to blow out the debris.
5. Wrap the cable around your fingers several times to keep the buffer tube from sliding out of the jacket.
6. Hold the buffer stripper so that the arrow on the head of the buffer stripper points towards the end of the cable.
7. Place the buffer tube between the teeth of the buffer stripping tool so that approximately ½ inch of buffer tube extends beyond the head (Figure 3–8).

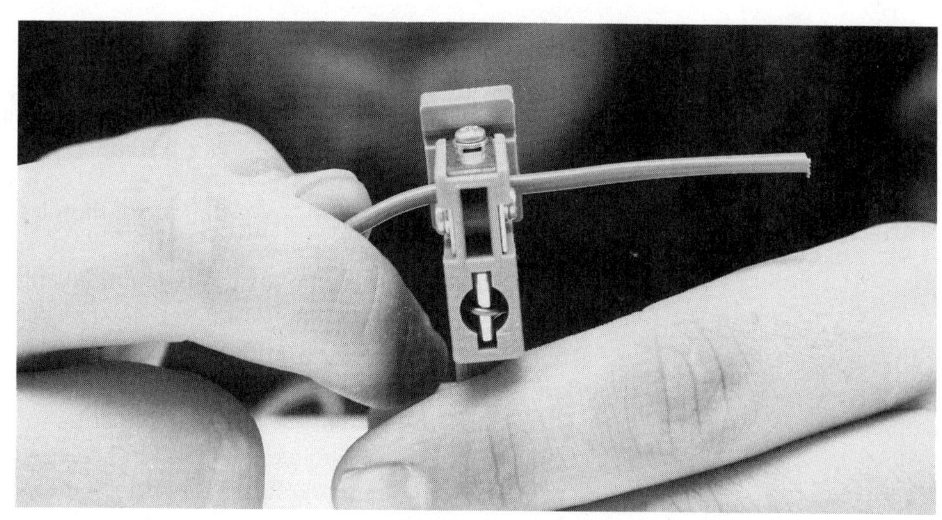

Figure 3–6 Tubing Cutter on Cable for Jacket Removal (Courtesy of Pearson Technologies Inc.)

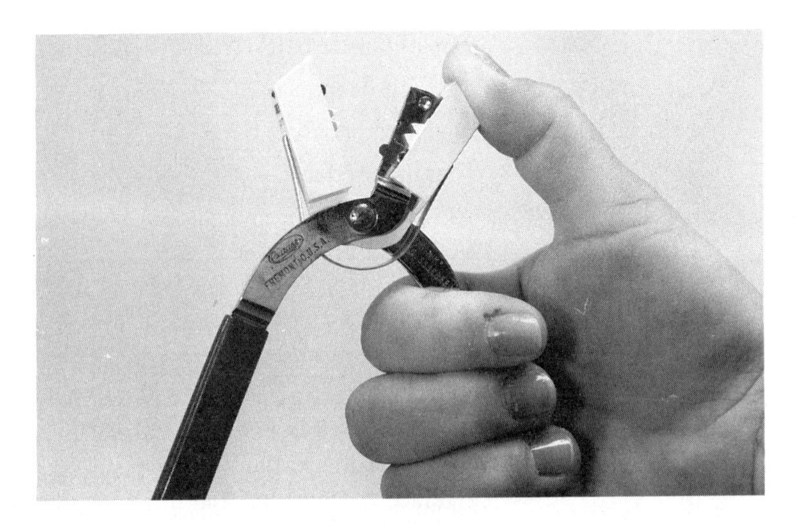

Figure 3–7 Head Covers of Stripper Open for Cleaning (Courtesy of Pearson Technologies Inc.)

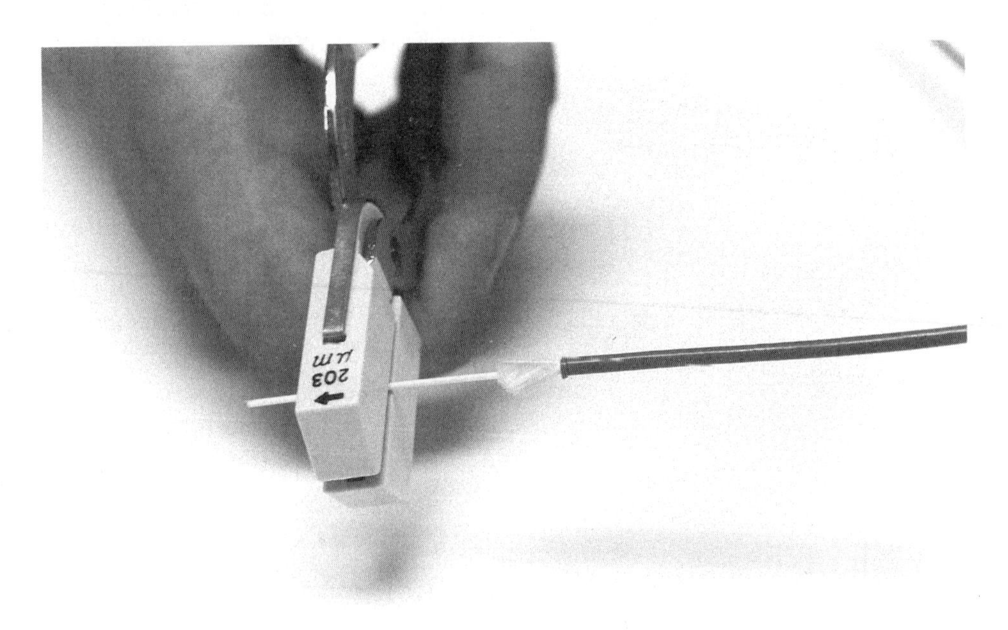

Figure 3–8 Correct Position of Buffer Tube in Stripper (Note arrow on head and position of buffer tube) (Courtesy of Pearson Technologies Inc.)

The ½-inch dimension is the rule of thumb for most cables, though some allow you to strip more in a single strip.

8. Slowly close the buffer stripper until you feel the blade cut into the buffer tube. While you need apply no additional pressure, do not release the pressure.
9. While holding the buffer stripper perpendicular to the axis of the fiber, slowly pull the stripper in the direction of the arrow. Use a low to moderate force to allow the buffer tube and buffer coating to slide off. After stripping the buffer tube, you may notice a thin clear plastic layer remaining on the fiber after you removed the buffer tube. This is the buffer *coating*, which can be removed by placing the fiber into the stripper and repeating this step.

10. Repeat Steps 6–8, until you have removed the buffer tube to within ¼ inch of the jacket. (In actual installation, this length will be determined by the connector manufacturer's instructions.)
11. Push the fiber against a stiff object so that it bends. Does it break? It should not. Cut off the exposed fiber, and dispose of it in a plastic container. (We recommend disposing of fiber pieces by wrapping them in aluminum foil. We do not recommend placing them in open wastebaskets, since people can get splinters.) If you drop fiber on your work surface, use the tape to pick up the fiber.
12. Repeat Steps 1–10 twelve times.

Troubleshooting Procedure 1

Novice installers experience two problems: breakage and removal of buffer tube without removal of the buffer coating. Most breakage results from five causes:

- Failure to clean the stripper before each strip (Step 4)
- Failure to hold the stripper perpendicular to the fiber (Step 9)
- Stripping excessively fast
- Using worn-out strippers
- Attempting to strip an excessive length in one strip

The cable is rarely the cause of breakage. If breakage persists after following all steps and replacing the stripper, reduce the length of buffer tube removed (Step 7) to ⅛ inch.

If breakage occurs before you strip the buffer tube, you have damaged the fiber during the jacket removal step. Reduce the number of turns you make with the tubing cutter until this problem ceases.

Failure to remove buffer coating is due to insufficient pressure on the handles of the stripper. Restrip with increased pressure to solve this problem.

PROCEDURE 2: END PREPARATION OF SINGLE FIBER, LOOSE BUFFER TUBE CABLES

1. Repeat Steps 1–3 of Procedure 1.
2. Place the tubing cutter on the buffer tube as close as possible to the jacket. (For actual installation, you will place the tube cutter at a distance from the end of the jacket that will be specified by the connector manufacturer.) Rotate the tubing cutter around the buffer tube once. Remove the tubing cutter. Quickly bend or snap the buffer tube at the score made by the tubing cutter.
3. Remove any gel, if present, with a rag or clean paper towel.
4. Repeat Steps 5 and 6 of Procedure 1.
5. Place the fiber between the teeth of the buffer stripper so that approximately 2 inches of buffer coating extends beyond the head.
6. Slowly close the buffer stripper until you feel the blade cut into the buffer coating. While you need apply no additional pressure, do not release the pressure. Repeat Step 9 of Procedure 1. If small particles of buffer coating remain on the clad, wipe the fiber with a lens grade tissue moistened with 91 percent isopropyl alcohol.
7. Repeat Step 11 of Procedure 1.
8. Repeat Steps 1–7 twelve times.

Results

Record your observations on the preparation of this type of cable.

Questions

What are the mistakes you can make in this procedure? How can you avoid such mistakes? Were the tools you used appropriate for the task?

END PREPARATION PROCEDURE FOR OUTDOOR CABLE DESIGNS

We present two procedures for end preparation of the MFPT (Figure 1–32) design. Both can be performed by teams of two or three. In Procedure 3, you will learn how to prepare the end of a gel filled and water blocked cable for connector installation. In Procedure 4, you will learn how to prepare the end of a gel filled and water blocked cable for splice installation. With minor modifications, this procedure is applicable to the central loose tube (CBT) design.

TOOLS AND SUPPLIES REQUIRED

tubing cutter (Ideal part number 45-162)

cable jacket slitter (Ideal part number 45-128)

needle nose pliers with protected tips (optional)

side cutters

Kevlar® cutters

fiber optic gel and grease remover (HydraSol® from American Polywater Corp., part number HS-1, or D'Gel™ Cable Gel Solvent, P-T Technologies Inc. part number GRDW)

95 percent isopropyl alcohol

paper towels or rags

dry fiber lubricant (such as unscented talcum powder)

length of MFPT cable, at least 6 feet long

splice tray with furcation cables attached

splice tray

Note: There are two methods for removing the jacket. Method 1 is described in Steps 2–8. Method 2 is described in Steps 9–14. If you are cutting the jacket with the rip cord, follow Steps 2–8. Then go to Step 15. If you are cutting the jacket with a jacket slitter, follow Steps 9–14. Then go to Step 15.

PROCEDURE 3: END PREPARATION FOR CONNECTOR INSTALLATION

1. Obtain all tools and supplies. Install the cable into the cable end box or the splice box. Pull the cable out the front of the box. Fasten the cable to the fastening locations inside the cable box with several cable ties. Tighten the ties so that the cable cannot be pulled from the box.
2. Find the rip cord in the cable end. It is often a yellow Kevlar® strand or a white or blue nylon strand.
3. Adjust the cable jacket slitter blade to a depth equal to the thickness of the cable jacket.
4. While holding the thumb rest down firmly to force the blade into the jacket (Figure 3–9), make a 2-inch long slit in the outer jacket slightly to one side of the rip cord.
5. Extract the rip cord from the cable.
6. Using the pliers with protected tips, pull the rip cord until you have "ripped"

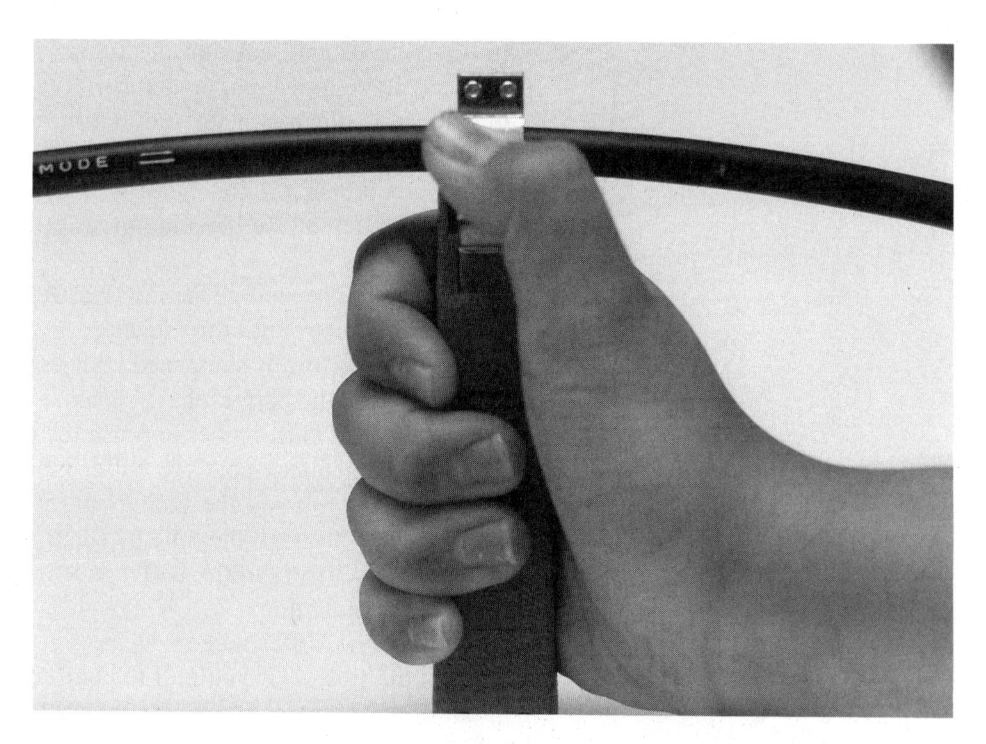

Figure 3–9 Correct Use of Jacket Slitter (Courtesy of Pearson Technologies Inc.)

the jacket to the length required for your application. (You could also wrap the rip cord around the handle of a large screwdriver to pull it.) For practice, rip the jacket to half the length of the cable. Common field practice is to rip the cable to a length of 6.5 feet.

7. Pull the core of the cable from the outer jacket.
8. Cut off the outer jacket with side cutters or a knife. Go to Step 15.
9. Adjust the cable jacket slitter blade to a depth equal to the thickness of the cable jacket.
10. Check the depth of slitter blade. Place the slitter on the jacket 2 inches from the end of the cable. While holding the thumb rest down to force the blade into the jacket, make a cut around the circumference of the cable.
11. Continue to hold the thumb rest down and pull the jacket slitter to the end of the cable. If the blade is set deeply enough, you will be able to remove the jacket. If the blade is not set deeply enough, increase the depth and repeat Steps 10 and 11.
12. Check the buffer tubes in the cable. Look for cuts in any of the buffer tubes. If you find cuts, reduce the depth of the blade and repeat Steps 10–12. If you find no cuts, proceed to Step 13.
13. Place the jacket slitter on the cable at the length of jacket to be removed. See Step 6 for guidance. While holding the thumb rest down to force the blade into the jacket, make a cut around the circumference of the cable.
14. Continue to hold the thumb rest down and pull the jacket slitter to the end of the cable. Remove the jacket.
15. Unwrap the wrapping tape or binder thread (if present) from the core of the cable.
16. Untwist the buffer tubes and fillers (black, solid pieces of plastic). Check the buffer tubes for damage from the jacket removal process. Cut off damaged sections and repeat Steps 2–8 or 9–14 so that the length of the cable core meets your requirements.

17. With the side cutters, cut out any fillers, any strength members under the jacket, and the central strength member. If the cable end box you are using has a location for attachment of the central strength member, leave enough strength member (2–6 inches) so that you can attach the strength member to the cable box properly.
18. Wipe off as much of the blocking grease (if present) as possible with paper towels or rags.
19. Using the pre-moistened gel remover towel, repeatedly wipe off the grease from the buffer tubes until they squeak.
20. Using fresh paper towels dampened with isopropyl alcohol, wipe off all of the gel remover until the buffer tubes squeak.
21. Choose one of the buffer tubes to finish the process of preparation.
22. With the tubing cutter, score the buffer tube at a distance of about 4 inches from the end by rotating the tubing cutter around the tube once—and only once! If you cut through the tube by rotating the cutter more than once, you may damage the fibers in the buffer tube and need the service loop you left during cable installation.
23. Quickly snap the buffer tube to break it at the score. Slowly pull the buffer tube from the fibers. (If you pull the buffer tube too quickly, the fibers may pull out of the short length of practice cable. This will not happen with long lengths of cable.)
24. Using paper towels or rags, wipe off as much of the filling gel as you can with one wipe.
25. Repeat Steps 22–24 at a distance of 4 inches from the end of the buffer tube until 6–12 inches of buffer tube remain extending beyond the end of the outer jacket.
26. Using the pre-moistened gel remover towel, repeatedly wipe the fibers until they squeak. Wipe the fibers two more times. Do not be concerned about untwisting or separating the fibers.
27. Using fresh paper towels damped with isopropyl alcohol, wipe off all of the gel remover. Wipe the fibers several times until the fibers squeak.
28. Place a small amount of dry fiber lubricant in the palm of one hand.
29. Starting at the buffer tube, drag the fibers through the dry fiber lubricant by holding the thumb from your other hand on top of the fiber.
30. Repeat Step 29 several times until the fibers separate. You should now see six color-coded fibers. (Some cables have 12 fibers per buffer tube.)
31. Fasten the buffer tube to the splice tray with one of the fastening ears at the end of the tray (Figure 3–10). (Or you could attach the buffer tube to a splice tray with a cable tie.)
32. Coil the fiber once around the inside of the splice tray. Slide the remaining length of fiber into one of the furcation cables attached to the tray.
33. Repeat Step 32 for each fiber in the buffer tube (Figure 3–10).
34. Place the cover on the splice tray. Place the splice tray on the retaining stud inside the cable box.
35. Repeat Steps 22–34 for each buffer tube in the cable.
36. Prepare the end of each furcation cable for connector installation by following the Procedure 2.
37. Install connectors.

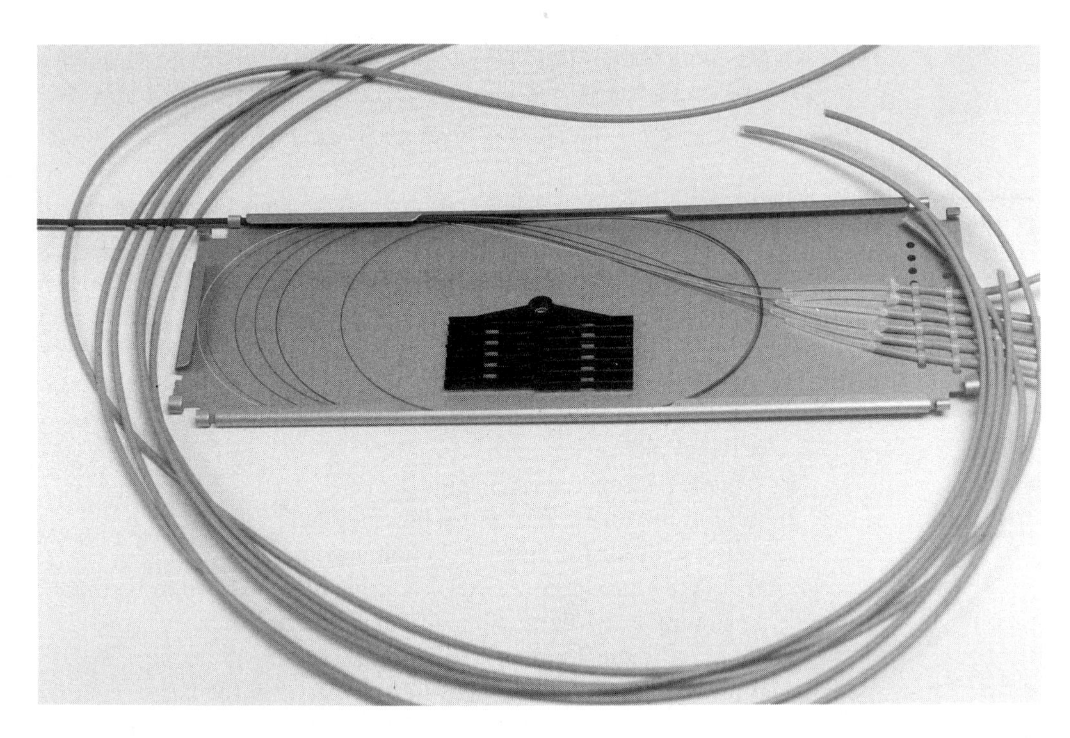

Figure 3–10 Fibers Installed into Furcation Cables on Splice Tray (Courtesy of Pearson Technologies Inc.)

PROCEDURE 4: END PREPARATION FOR SPLICING

1. Prepare all buffer tubes of the first cable to be spliced by following Steps 1–31 of Procedure 3.
2. Coil all fibers from the first buffer tube into the first splice tray.
3. Follow Step 1 for the first buffer tube of the second cable to be spliced, but attach the first buffer tube to the same tray as was used in Step 2. Both buffer tubes usually enter the splice tray from the same end (Figure 3–11).
4. Coil all fibers from the first buffer tube of the second cable into the first splice tray.
5. Place the cover on the splice tray. Place the splice tray on the retaining stud inside the cable box or in the splice enclosure.
6. Repeat Steps 2–5 for each of the buffer tubes of each of the two cables. Use a separate splice tray for each pair of buffer tubes.
7. Perform splicing.

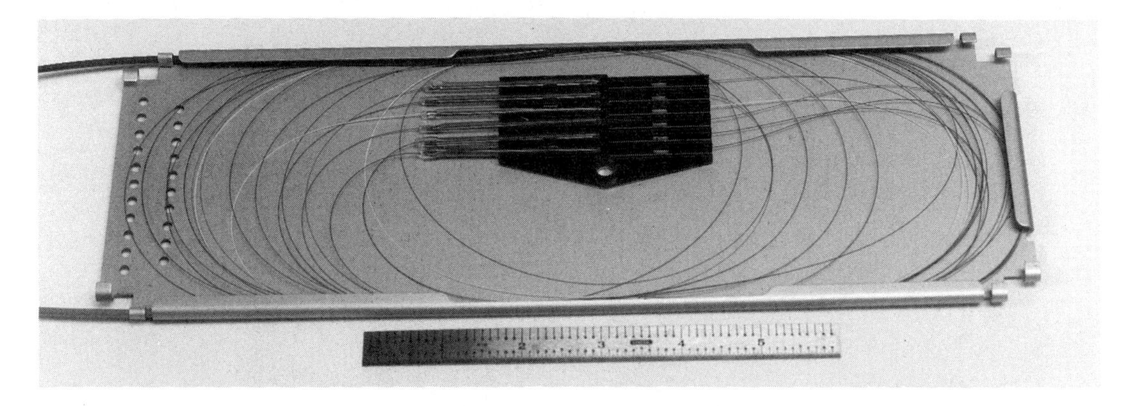

Figure 3–11 Two Buffer Tubes in Splice Tray for Splicing (Courtesy of Pearson Technologies Inc.)

Results

Record your observations on the preparation of this type of cable. Compare the preparation of this type of cable to the preparation of the other types of cables.

REVIEW QUESTIONS

1. What are the two possible problems that can occur during installation?
2. What are the ways to avoid both of these problems?
3. List the tools and supplies required for preparation of the ends of tight buffer tube cables.
4. List the tools and supplies required for preparation of the ends of gel-filled, loose buffer tube cables.
5. What are the five possible causes of breakage during removal of the tight buffer tube?
6. What is the function of a splice tray?
7. What is the function of water blocking gel?
8. Is it easier to break fibers in bending or in tension?
9. From your answer to Question 8, when do you need to be more careful, when tensioning or when bending a fiber optic cable?
10. What is the function of the furcation tube?
11. What happens to the strength of the fiber after the buffer coating is removed?
12. What happens to the strength of the fiber after the clad is abraded?

CHAPTER

4

Connector Installation: Four Methods and Two Styles

CHAPTER OBJECTIVES

From this chapter, you will be able to:

1. Install ST-compatible and SC style connectors.
2. Install connectors by four installation methods (epoxy, preloaded and preheated, epoxyless, and anaerobic adhesive).
3. Recognize the microscopic appearances of low-loss (good) and of high-loss (bad) connectors.
4. Improve your installation technique from your recognition of these appearances.
5. Determine the installation method best suited to your application.

INTRODUCTION

During connector installation, you will have two major objectives: low loss and high reliability. You will achieve low loss by following the installation instructions exactly. Following the instructions precisely is required to create an optical quality, lens grade surface on the end of a glass fiber approximately the size of the period at the end of this sentence. You will achieve high reliability by following the instructions exactly and by minimizing, and in some cases eliminating, the length of bare glass fiber inside the connector.

From this chapter, you will learn to install two styles of connectors: the ST-compatible style and the SC style. The ST-compatible style is the more commonly used connector in North America. Its features, advantages, and disadvantages are presented in Chapter 1. The SC style is expected to become the second most commonly used style in the near future.

Connector installation requires four steps:

1. Prepare the cable end.
2. Attach the fiber to the ferrule.
3. Attach the cable to the connector.
4. Finish the fiber end.

The second through fourth steps can be accomplished by eight different techniques, which we introduced in Chapter 1. From this chapter, you will learn six of these techniques. You will attach the fiber to the connector by three techniques:

- With epoxy or with anaerobic adhesive
- With preloaded and preheated hot melt adhesive
- With a crimp

You will attach the cable to the connector by two techniques, with a crimp and with hot melt adhesive. You will finish the end of the fiber by one technique, polishing. These six techniques are grouped into four installation methods:

- Method 1: Epoxy, crimp, and polish method
- Method 2: Preloaded, preheated, and polish method
- Method 3: Epoxyless crimp/crimp/polish method
- Method 4: Anaerobic adhesive, crimp, and polish method

These four methods are the most commonly used. The epoxy, crimp, and polish installation procedure is the method used for the first fiber optic connectors. This method is used on more connectors than on any others. This method is the least convenient, most time consuming, but the most forgiving of all methods. Pearson Technologies Inc. has found that novice installers have the greatest success with epoxy connectors (Table 4–1).

The preloaded, preheated, and polish installation method was pioneered by 3M in its Hot Melt connectors. These steps eliminate the need for epoxy, syringes, and relatively long cure times. Elimination of these needs results in reduced cost of consumables and reduced labor costs.

The crimp and polish method for installing connectors is the most convenient of those in this book. This method has one of the fastest installation rates and one of the lowest labor costs.

The anaerobic adhesive method is similar to the epoxy method, but offers faster installation and reduced labor cost. The installation time for this method can be slightly longer than that of the crimp and polish method. However, the anaerobic adhesive method is the least forgiving and most difficult for novices to learn (Table 4–1). If the installer inserts the fiber too slowly, the anaerobic adhesive hardens with excessive length of bare fiber in the connector. This bare fiber often breaks during polishing and use.

The procedures you learn will be applicable to other styles. The epoxy, etc., method is used on all connector styles. The preloaded, preheated, and polish method is used on ST, SC, and FC styles. The crimp and polish method is used on ST-compatible styles. Finally, the anaerobic method is used on ST-compatible and SC styles.

All methods presented in this chapter are a combination of those recommended by connector manufacturers and those Pearson Technologies Inc. has found to be highly successful in training novice installation personnel. As such, these methods are not exactly the same as those recommended by the manufacturers. Other methods will work.

These methods were written for the installation of connectors on short cables to create jumpers. The methods for installation of connectors on long lengths are the same as those herein, with the exception that you will install connectors on all fibers on one end of the cable before proceeding to the opposite end.

SAFETY PRECAUTIONS: *You should not smoke, eat, or drink during connector installation. In addition, you should not have cigarettes, food, or drinks in the work area. Bare glass fibers can be difficult to find and painful to remove. During connector installation, remove all pieces of broken fiber to a small plastic bottle (or wrap in aluminum foil) for disposal. You can use tape to pick up pieces of fiber. Finally, wash your hands before you rub your eyes during connector installation.*

Technique	%	Number of Connectors
epoxy and polish	86	159
preloaded and preheated	81	166
epoxyless	75	163
anaerobic adhesive	59	190

Table 4–1 Success Rates for Novice Installers with Three Installation Techniques (Data courtesy of Pearson Technologies Inc.)

METHOD 1: EPOXY, CRIMP, AND POLISH ST-COMPATIBLE CONNECTORS

Many companies manufacture ST-compatible connectors that require epoxy (Figure 1–51). The epoxy provides a reliable way to attach the fiber to the ferrule so that it will not move in or out of the ferrule (this motion is called "pistoning"). The crimping provides an acceptable level of strength between the connector and cable. A bead of epoxy placed on the tip of the ferrule can be controlled to provide a high degree of support for the fiber during polishing. This support increases the polishing time slightly, but reduces loss of connectors due to excessive pressure. The polishing provides a low-cost and reliable method of consistently producing low-loss fiber ends. Tooling costs are also low. Tooling has long life.

We present this method specifically for the AMP part number 503415-1, an ST-compatible connector with a liquid polymer ferrule. However, with modifications, this method can be used for any connector style from any manufacturer. Modifications include changes to the time, temperature, epoxy, fiber and ferrule material, crimper, and polishing tool. Consult the manufacturer's installation instructions to determine changes to this method appropriate for other connector styles or for ST-compatible styles from other manufacturers.

All parts listed below can be replaced by equivalent parts from other suppliers.

TOOLS AND SUPPLIES REQUIRED

tubing cutter (Ideal part number 45-162)

Kevlar® scissors (Fiskar)

at least 3 feet of single fiber, tight tube cable with 62.5 μm core and 125 μm clad

two Clauss No-Nik buffer tube/buffer coating strippers (one for spare) (part number NN200 or NN150)

crimper (Paladin part number 1331)[1]

ST polishing tool

glass or hard plastic polishing plate

wedge scriber (Fiber Optic Center, Inc., part number 58905-600)

12 μm polishing film (Fiber Optic Center, Inc., part number 9×11-12-A-3N-100)

1 μm polishing film (Fiber Optic Center, Inc. part number 9×11-1-A-3N-100)

0.3 μm or 0.5 μm polishing film (Fiber Optic Center, Inc., part number 9×11-.3-CA-2N-100)

two AMP connectors (part number 503415-1)

two ST-compatible curing shrouds[2]

electronic thermometer (Jensen Tools part number DT-205)

small plastic bottle (for bare fiber)

Tra-Con F230 epoxy

epoxy curing oven

syringe and needle for fiber optic connectors

Scotch™ tape

lens, or optical, grade compressed air

lens, or optical, grade tissue

toothpicks

pipe cleaners

95 percent Isopropyl alcohol

200× connector inspection microscope with ST fixture (Leica Fibre-Vue kit part number 31-22-70-98 or equivalent)

Alco pads (optional)

PROCEDURE 1: EPOXY, CRIMP, AND POLISH INSTALLATION

1. Obtain all tools and supplies.
2. Turn on oven to desired temperature. Unless otherwise instructed, use 100°C.[3]
3. Perform a white light test on the cable. Hold one end of the cable to be terminated to a bright light source. Look at the opposite end to see the light. If you cannot see the light, remove 1 inch of the jacket and cut the buffer tube with scissors to create a new fiber end. Repeat this step until you can see the light.
4. Install the boots on both ends of one cable. The large end of the boot should be near the end of cable, about 6 inches from end.
5. Install the crimp rings on both ends of the cable. Push the crimp rings slightly under the boots so that the crimp rings do not slide off the cable.
6. Prepare one end of the cable to the dimensions and by the sequence in Figure 4–1. Using tubing cutter, strip approximately 1.25–1.50 inches of jacket. This is done according to Step 2 of Procedure 1, Chapter 3.
7. Use Kevlar® scissors to cut Kevlar® so that ≥5⁄16 inch extends beyond jacket.
8. With buffer tube stripper, strip buffer tube so that no more than 15⁄16 inch of buffer tube remains.
9. While most connector manufacturers recommend cleaning the fiber, Pearson Technologies has found that this step is necessary for singlemode connectors, but optional for multimode connectors. Wipe the fiber with a lens grade tissue moistened with isopropyl alcohol. Should some buffer coating remain on the fiber, remove this coating from the fiber by wiping with the moistened lens grade tissue.
10. With the tape on the jacket, tape down the first end of the cable onto your work surface.
11. Repeat Steps 5–9 on second end.
12. Check fiber integrity (optional). Gently push one end of the bare fiber at a time against a hard surface until the fiber starts to bend. It should not break. If the fiber breaks, you have damaged the clad.
13. Perform a white light check of both connectors. Remove and save the connector caps. Hold the connector up to a bright light. Look into the backshell for a sparkle, which indicates a hole is in the ferrule. If you cannot see the sparkle,

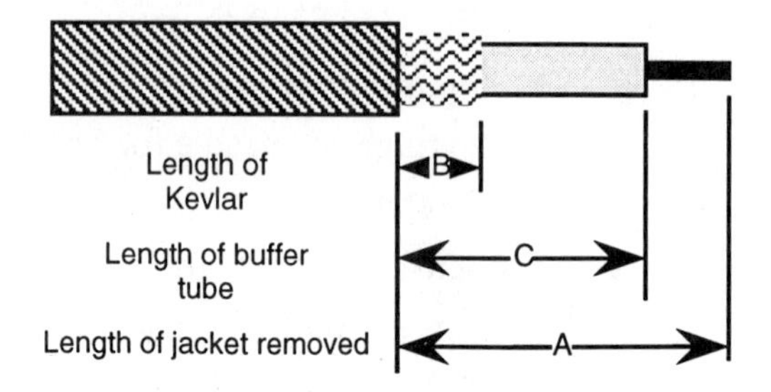

Figure 4–1 Stripping Dimensions for Epoxy Connector from AMP

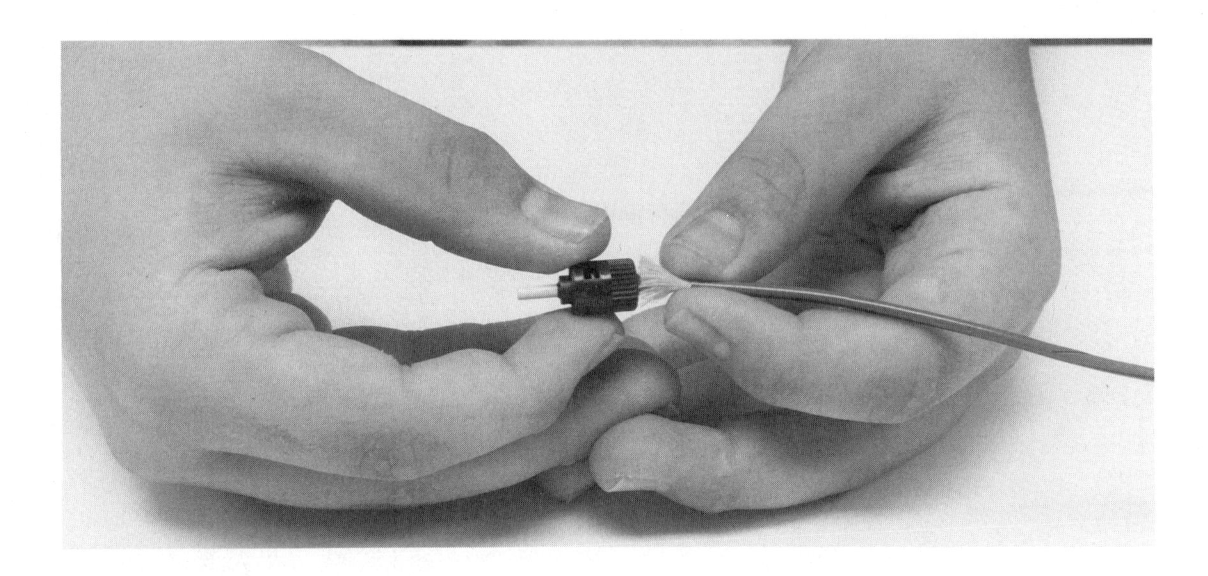

Figure 4–2 Positions of Hands, Fiber, Cable, and Connector During Dry Fit (Courtesy of Pearson Technologies Inc.)

the hole is missing or blocked with dust. Blow out the backshell of the connector with compressed air. Repeat the white light test. If you cannot see the sparkle, replace the connector with another connector and perform the white test on the replacement connector.

14. Perform a dry fit of both connectors. Rest your hands against one another. Slowly slide the fiber into the backshell. As soon as the fiber enters the backshell, start rotating the connector back and forth to allow the fiber to slip past any obstruction or non-uniformity inside the backshell. (Note that a rotating retaining nut does not ensure that the connector is rotating; be certain that the connector ferrule and backshell are rotating.) Insert the cable until the jacket butts against the backshell. You must see the fiber protruding beyond tip of ferrule (Figure 4–2). If you do not, the length of bare fiber is too short. Repeat Steps 6–14. Remove the fiber. Place the connector you checked near this end of cable. With tape on the jacket, tape this end of the cable onto your work surface. Perform a dry fit with the second connector. Place the second connector near the second end of the cable. Tape this end of cable onto your work surface.
15. Check the expiration date of the epoxy. Use the epoxy if the expiration date has not passed. Remove the epoxy from the outer package.
16. Mix the epoxy. Remove the separator (Figure 4–3). Mix by rubbing the epoxy inner package over the edge of the table or work surface 16 times, each time moving all the epoxy from one end of the package to the other until the epoxy is uniform in color. Squeeze the epoxy into the larger section of the epoxy package.
17. Prepare the syringe by removing the plunger from the barrel, needle cover, and end cap. Screw the needle tightly onto the base of the barrel. Cut off a small corner from the smaller section of the epoxy package. Slowly squeeze all of the epoxy into the barrel of the syringe. Insert the plunger into the barrel until the barrel holds the plunger in place. Invert the syringe. Allow the epoxy to flow onto the plunger. Remove the air from the syringe by depressing the plunger. Collect the epoxy from the needle onto the epoxy outer package or some other non-absorbent surface. You will have 40–50 minutes of useful pot life. (A 4-gram epoxy package can be used for 12 to 25 connectors.)

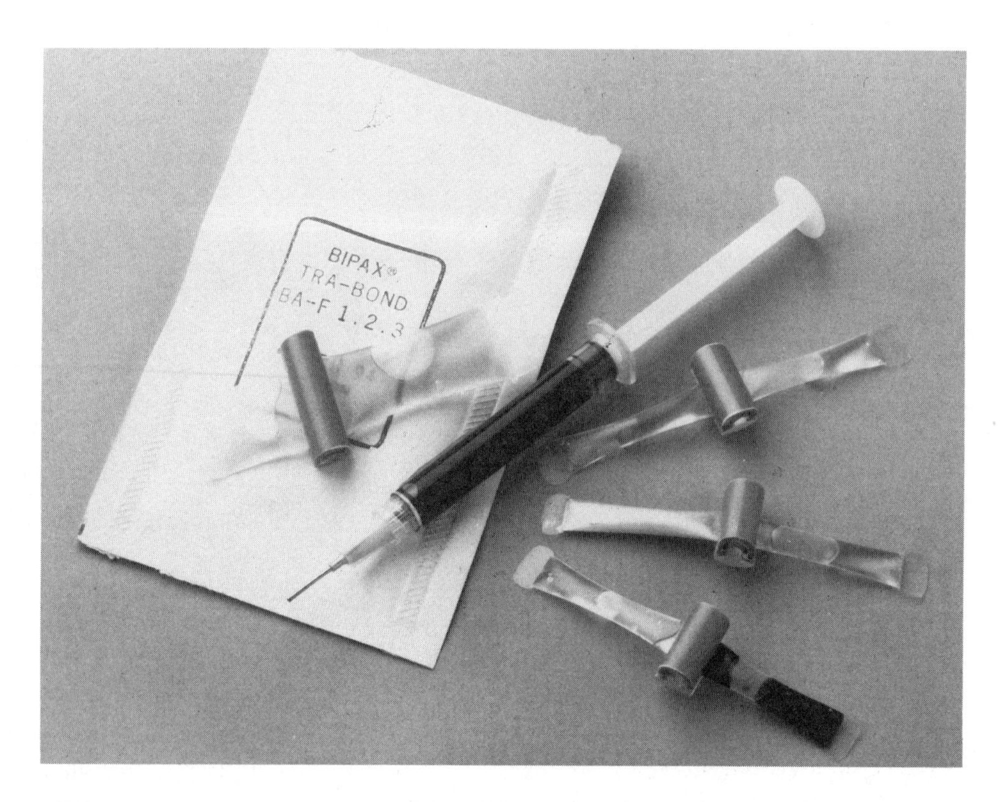

Figure 4–3 Epoxy Package with Separator and Assembled Barrel of Syringe (Courtesy of Tra-Con Inc.)

18. Check the epoxy viscosity. Push the plunger. If a rounded bead or drop appears, the epoxy is fluid enough to use. If a string of epoxy appears, the epoxy is too thick to use. Remove this epoxy from the needle.
19. Insert the needle all the way into the backshell. Inject epoxy until a small drop appears on the tip of the ferrule. Immediately begin to withdraw the needle from the backshell slowly. If epoxy flows out of the backshell before you withdraw the needle completely, you are withdrawing the needle too slowly. Your goal is to fill the backshell approximately half full. You may fill all the connectors you intend to install at this step. However, fill only as many as you can install within 30 minutes—approximately twelve for a beginner and twenty for a more experienced installer.
20. While rotating the connector back and forth, insert the fiber slowly into the backshell until the cable jacket butts against the backshell. If you see no epoxy forced out of the backshell, you removed the needle too quickly. If you see one or two drops forced out, you injected the proper amount of epoxy and withdrew the needle correctly. If you see more than two drops, you removed the needle too slowly and have excess epoxy in the connector. Remove this excess epoxy with a pipe cleaner. Correct your epoxy filling technique on the next connector. When the cable is completely in the backshell, you must see fiber protruding beyond the end of the ferrule. If you do not see fiber protruding beyond tip of ferrule, remove the cable from the connector, cut off this end of the cable, and prepare this end again (Steps 6–12 and 19–20).

> ***Note:*** *From here on, you must be careful not to break the fiber protruding beyond the tip of the ferrule. Handle the connector so that you do not touch this fiber. If you break the fiber, this connector is likely to exhibit high loss.*

21. Install the crimp ring. While rotating the crimp ring back and forth (to distribute the Kevlar® uniformly), slide it over the backshell. Hold the crimper with the flat surface up and the recessed area towards the floor. With the fiber pointing to the ceiling, install the crimp ring in the smallest nest of the

Figure 4–4 Connector in Crimper. (Note that flat surface is up.) (Courtesy of Pearson Technologies Inc.)

Note: Do not allow the cable to slide out of the connector during crimping. You must be able to see the fiber protruding beyond the tip of the ferrule.

crimper. Align the crimper with the top of the crimp sleeve (Figure 4–4). Squeeze the crimper handles together until they release. Move the crimper to the bottom of the crimp sleeve so that the uncrimped length of the crimp sleeve is in the smallest nest of the crimper. Squeeze the handles together until they release. Align the crimper with the top of the crimp sleeve. Rotate the connector 90 degrees from its first position. Squeeze the handles together until they release.

22. Tape down the first end with tape on the jacket in a location in which you will not touch or bump the fiber.
23. Repeat Steps 19–21 with the second cable end and the second connector.
24. Pick up a small drop of epoxy with a toothpick or a needle. Hold a connector with the fiber pointing up. Rest your hands together. Touch the fiber near the tip of the ferrule with the drop of epoxy, not with the toothpick. The epoxy will move to the fiber and flow down to the tip of the ferrule. The drop on the tip of ferrule should be $\frac{1}{32}$ inch high, approximately the width of a paper clip (Figure 4–5). If the bead is less than $\frac{1}{32}$ inch, repeat this step. If the bead is more than $\frac{1}{32}$ inch, remove the excess epoxy by wiping the bead, not the fiber, with a pipe cleaner. Wipe the pipe cleaner up and away from the fiber.
25. Pick up the connector. Rest your hands together. With a pipe cleaner, carefully remove all epoxy from the side of the ferrule. Wipe the pipe cleaner up the ferrule and away from the fiber.

Note: Do not touch the fiber with the toothpick or the pipe cleaner.

Figure 4–5 Proper Size of Epoxy Bead (Courtesy of Pearson Technologies Inc.)

> *Note: When cutting fiber, do not bend fiber.*

> *Note: Do not place the shroud flat on work surface, since epoxy may flow off the tip of the connector, epoxying the connector to the shroud.*

26. Install a curing shroud on the connector without touching the bead of epoxy. Inspect the shroud to be certain that the fiber does not protrude beyond the end of shroud. If the fiber protrudes beyond the end of the shroud, remove the shroud, trim the fiber with scissors, and reinstall the shroud. Tape the cable to a table with the shroud hanging over edge of table.
27. Repeat Steps 24–26 with the second connector.
28. Perform a white light test: hold one shroud to a bright light source. Look into the second shroud. You should see a sparkle indicating continuity of the cable.
29. Place a label on the cable (optional). The label can include an identification of the installer, a serial number, and the time at which the connector is installed in the curing oven.
30. Check the oven with an electronic thermometer. If the temperature is correct, install the shrouds in the curing oven. If the temperature is not correct, adjust the thermostat and allow the oven to achieve the correct temperature. The time in the oven must be more than the minimum specified by the connector manufacturer and the epoxy manufacturer. If appropriate, proceed to the next connectors.
31. When the curing time has expired, remove the shrouds from the oven.
32. Perform a white light test as in Step 28. Allow the connectors and shrouds to cool. Do not force cool. Remove the shrouds.
33. Check the epoxy to ensure that it has completely cured. There are two techniques for checking the epoxy. The first technique is color change: if the epoxy changes color when it is cured, check the color. The Tra-Con F230 changes from an uncured dark blue to a cured amber. The second technique for testing the epoxy is placing a razor blade or Exacto knife blade against the epoxy bead. If the epoxy is not completely cured, the knife will stick. If the blade sticks, return the connectors to the oven.
34. Remove the fiber protruding beyond the bead of epoxy (excess fiber). With a wedge scriber, lightly scratch the fiber at the surface of the epoxy bead (Figure 4–6). Do not break the fiber while scratching with the scriber. Break

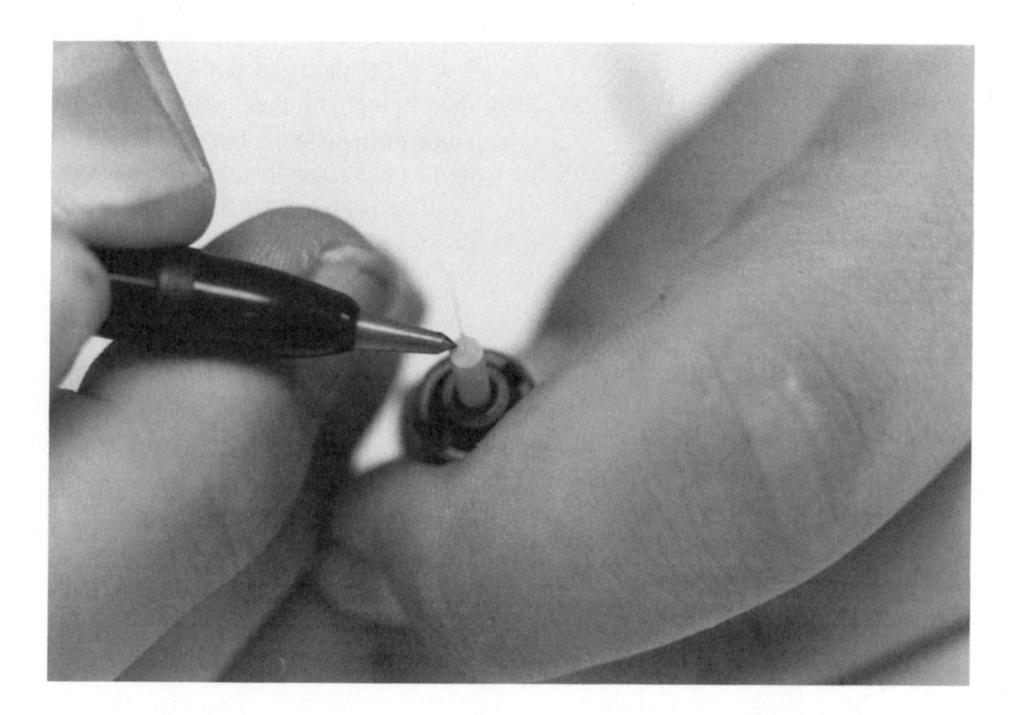

Figure 4–6 Location of Scribe (Courtesy of Pearson Technologies Inc.)

off the excess fiber by pulling the fiber away from the tip of the ferrule. If the fiber does not break, it may be covered with a thin layer of epoxy. Scratch the fiber slightly above the bead and pull. Remove the excess fiber from all the connectors you intend to polish.

35. Clean the polishing equipment. With compressed air, clean both sides of the polishing films, the polishing plate, and the polishing tool (also called a "fixture" or "puck").
36. Air polish the fiber stub. Hold the connector with the fiber up. Hold a piece of 12 or 15 μm polishing film (approximately 3 × 5 inches, dull side down) at one side. Rub the other side of the film against the fiber stub until it is flush with the bead of epoxy (Figure 4–7). Do not apply heavy pressure to the film. Do not place your fingers directly opposite the stub or near the stub. To check the stub, gently place your finger directly down on the end of the fiber, not from the side. The fiber is flush when you cannot feel the fiber with the tip of your finger. This step should take no more than 10 seconds.

> ***Note:** Avoid excessive pressure, since it can result in snapping of the fiber stub.*

37. Place a 12 μm polishing film on the polishing plate, dull side up. Place the cleaned polishing tool on film. Gently place the connector into the polishing tool until the bead rests on the film.
38. Polish with the 12 μm film. Hold the connector and the tool. With almost no pressure, move the tool on the film in a figure eight motion until the bead of epoxy is gone. (When polishing connectors with ceramic ferrules, polish until the bead is almost gone, leaving a thin film of epoxy.) Periodically, you can stop polishing to check the bead of epoxy. Another way to tell that the bead is removed is that the friction of the tool on the film drops noticeably when the bead is gone. Do not continue polishing on this film after you have removed the epoxy.
39. With compressed air, clean off the fixture, the connector and both sides of the medium (1 μm) polishing film. Place the medium film on the plate, dull side up.

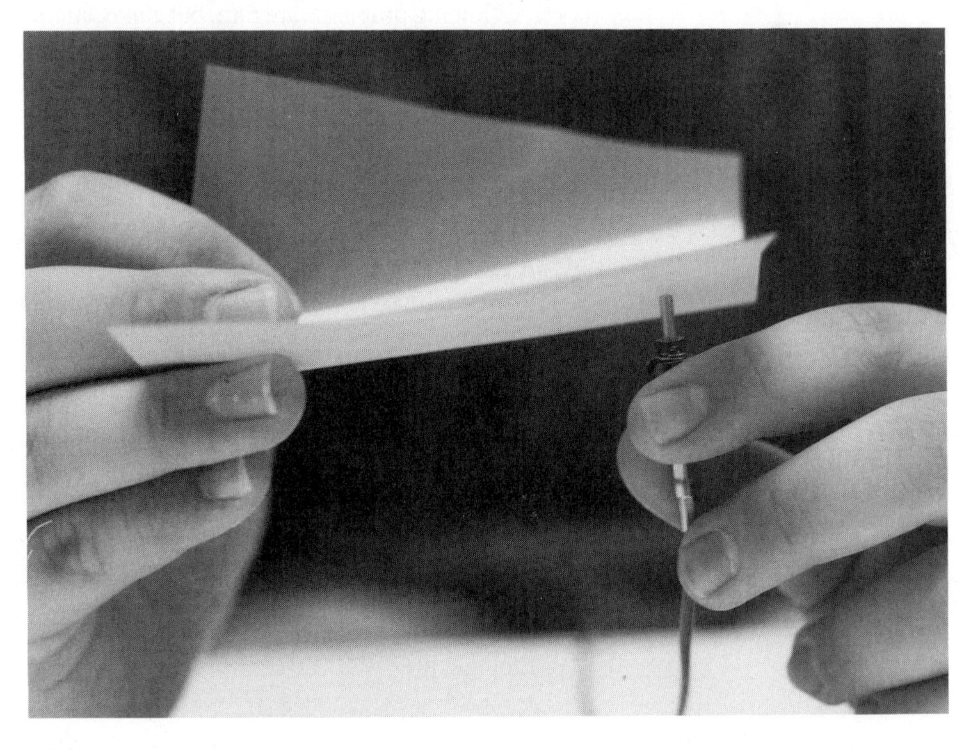

Figure 4–7 Air Polishing of Fiber (Courtesy of Pearson Technologies Inc.)

40. Hold the connector and the tool. With slightly more pressure than you used in step 38, move the tool on the film in a figure 8 motion for 8 to 12 strokes.
41. Clean the ferrule. Roll the side of the ferrule on the sticky side of Scotch Magic™ tape. Press the tip of the ferrule against the tape, never using the same area of tape twice.
42. Inspect the connector with a 200x microscope (inspection of singlemode fibers requires 400x). Follow Procedure 5 in this chapter. If this inspection reveals features, clean or repolish according to the instructions in Procedure 5.
43. With compressed air, blow out the connector cap. Cap the connector.
44. Repeat Steps 36–43 with the second connector.
45 Test the loss (Chapter 5).

Troubleshooting Method 1

For problems during end preparation of the cable, see Chapter 3, Procedure 1: End Preparation.

The problem that occurs during insertion of the cable into the connector is usually breakage of the fiber, which is infrequent. You can break the fiber during installation if you failed to remove all of the buffer coating from the fiber. This buffer coating has a translucent appearance. To solve the problem of breakage, review Procedure 1 in Chapter 3.

The fiber should always slide into the connector with minimum force, as it does during the dry fit. If you fail to rotate the connector while you are inserting the fiber into the connector, the fiber may break by catching on a machining imperfection in the backshell.

If the fiber moves very slowly into the connector, the epoxy may be too thick. You have allowed excessive time between the mixing of the epoxy and insertion of the fiber. The actual pot life, or usable time after mixing, depends on the epoxy, the age of the epoxy, and the temperature and humidity of the room.

Crimping problems are rare, but can occur. If the crimp ring is loose after crimping, the crimper may be out of adjustment. To solve this problem, adjust the crimper to a tighter setting or return the crimper to the manufacturer for adjustment.

Polishing problems are identified after microscopic inspection, described in Procedure 5 in this chapter.

METHOD 2: PRELOADED, PREHEATED, AND POLISH ST-COMPATIBLE CONNECTORS

This Hot Melt™ connector is available from 3M. Its advantages are the elimination of the use of epoxy, with all its time-related and product costs. Its disadvantages are limited applications (operation below), a boot that does not limit the bend radius as well as do other boots, and fixturing that can be slightly awkward to use. Overall, this product performs well and is well accepted in the marketplace.

Part numbers supplied in Method 1 are not repeated below. These instructions differ slightly from the 3M instructions. Review the manufacturer's instruction sheet before using this product.

TOOLS AND SUPPLIES REQUIRED

tubing cutter

Kevlar® scissors (Fiskar)

at least 3 feet of single fiber, tight tube cable with 62.5 μm core and 125 μm clad

two Clauss No-Nik buffer tube/buffer coating strippers
ST polishing tool (3M part number 08-00882)
3M hard rubber polishing pad
wedge scriber
2 µm 3M polishing film (3M part number 05-00015)
two 3M Hot Melt connectors (3M part number 6100)
two 3M connector holders (one per connector) (3M part number 05-00072)
small plastic bottle (for bare fiber)
3M preheating oven
3M cooling stand
Scotch™ tape
lens grade compressed air
lens grade tissue
95 percent isopropyl alcohol
200x connector inspection microscope with ST fixture
Alco pads (optional)

PROCEDURE 2: PRELOAD AND PREHEAT INSTALLATION

1. Obtain all tools and supplies. Open two connector packages. Assemble the connector cooling stand. Tighten the top of the connector holder to the bottom of the connector holder (these can loosen during use).
2. Turn on the oven. Proceed with the following instructions, but do not use oven for at least five minutes.
3. Align the connector key with the holder key and install two connectors into two connector holders (Figure 4–8).
4. Perform a white light test of the cable. Follow the instructions of Step 3 of Method 1.
5. Install a boot on one end of the cable.

> ***Note:*** *Do not touch the heating block, a hot connector, or a hot connector holder. All are hot enough to burn (400°F).*

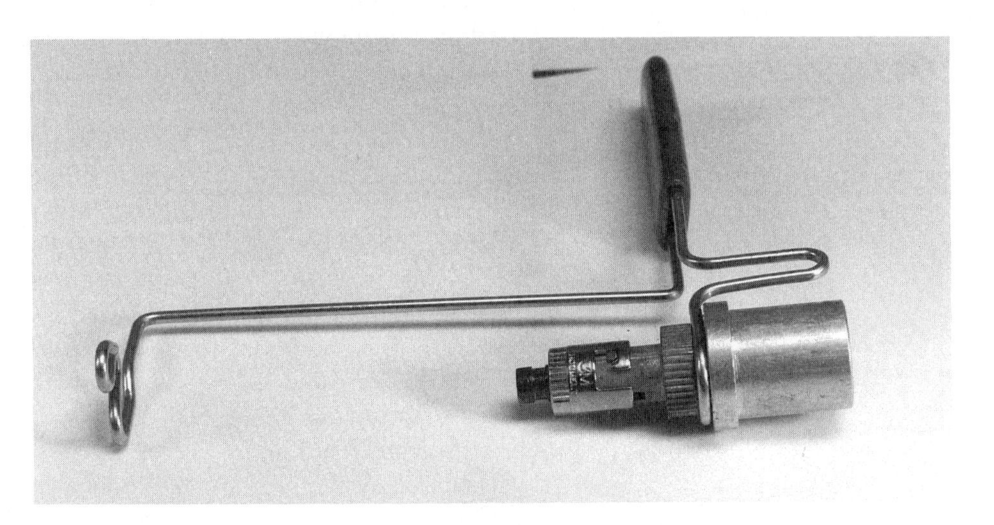

Figure 4–8 The 3M Connector in Holder (Courtesy of Pearson Technologies Inc.)

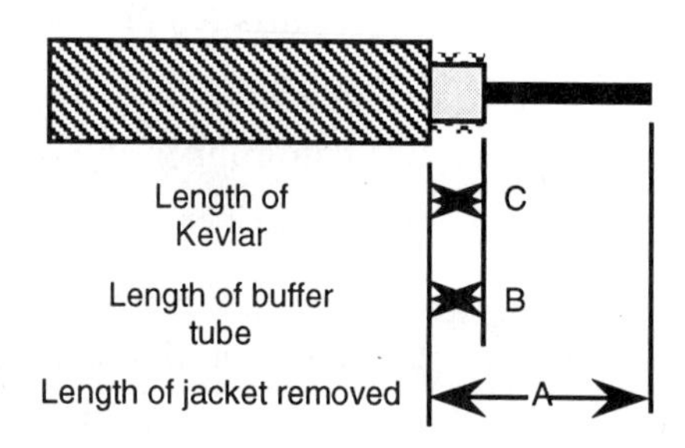

Figure 4–9 Stripping Dimensions for 3M Hot Melt Connectors

6. Prepare one end of the cable according to Figure 4–9. Using tubing cutter, remove ¾–⅞ inch of jacket. Remove ¾ inch for best results.
7. Hold the Kevlar® along the jacket. Use the buffer tube stripper to strip the buffer tube so that no more than 3⁄16 inch of the buffer tube remains.
8. Cut the Kevlar® so that less than 3⁄16 inch extends beyond the jacket. Align the Kevlar® strands so that they are parallel to and overlapping the buffer tube.
9. Check fiber integrity (optional). Gently push end of the fiber against a hard surface until the fiber starts to bend. It should not break. If the fiber breaks, you have damaged the clad during jacket removal. To solve this problem, reduce the number of turns you make with the tube cutter.
10. This step is necessary for singlemode connectors, but optional for multimode connectors. Wipe the fiber with a lens grade tissue moistened with rubbing alcohol. Should some buffer coating remain on the fiber, remove this coating from the fiber by wiping with the moistened lens grade tissue.
11. Place the connector holder in the oven. Wait for one minute. The connector should not be in the oven for more than 10 minutes.
12. Remove the connector holder from oven. While rotating either the cable or the connector holder, slide the Kevlar® and the cable into the connector backshell until the cable stops. The Kevlar® should fold back along the jacket. Most of the Kevlar® will be inside the backshell. A single drop of adhesive should appear at the top of the backshell.

Note: If the adhesive expands above the backshell during curing, discard the connector; it has been contaminated with moisture.

13. Without allowing the cable to slide out of the backshell, place the cable in the clip at top of connector holder (Figure 4–10). You may modify this step by withdrawing the fiber by a small amount, approximately 0.01 inch. This modification will allow for reheating and repolishing of a shattered end.
14. Place a label on cable (optional). On the label, write your initials and any other relevant information.
15. Perform a white light test. Follow the instructions of Step 28 of Procedure 1, but hold the connector holder instead of the shroud. If the connector fails the white light test, reheat the connector. Push the cable farther into the connector. Allow the connector to cool. If the fiber protrudes beyond the tip of the ferrule, follow Steps 18–29. If it does not, it is probably short or broken. In this case, reheat the connector (Step 11) and remove the cable. Inspect the length of fiber; if it is shorter than ¾ to ⅞ inch, cut off the end of the cable. Repeat Steps 6–15. Note that 3M does not guarantee the reliability of connectors that have been reheated. However, we have not noticed any degradation after the first or second reheating
16. Place the connector holder in the cooling stand.
17. While allowing the connector to cool for at least three to four minutes, repeat Steps 5–16 on the second end.

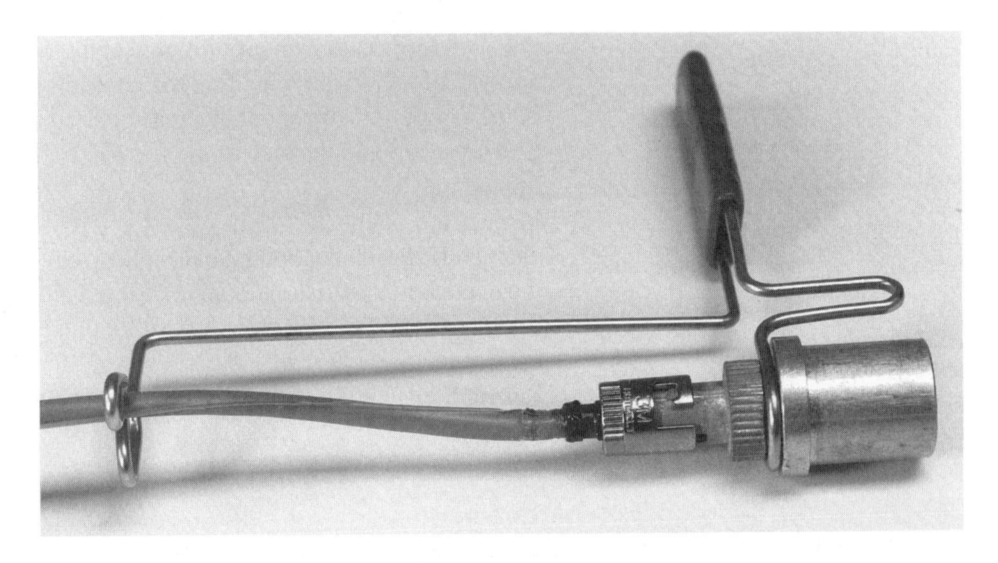

Figure 4–10 Cable in Clip of 3M Holder (Courtesy of Pearson Technologies Inc.)

Note: From here on, you must be careful of the fiber protruding beyond the tip of the ferrule. Handle the connector so that you do not touch this fiber. If you break the fiber, the connector may have high loss.

18. With compressed air, clean the rubber polishing pad, both sides of the 3M polishing film, and the polishing tool.
19. If you are using a 3M polishing tool, inspect it. If the three wear pads are not visible on the bottom, replace the tool.
20. Remove the cable from the clip of the connector holder. Remove the connector from the bottom of the holder.
21. Remove the fiber protruding beyond the bead of adhesive. With a wedge scriber, lightly scratch the fiber at the surface of the adhesive bead. Do not break the fiber while scratching with the scriber. Break off the excess fiber by pulling the fiber away from the tip of the ferrule. If the fiber does not break, it may be covered with a thin layer of adhesive. Scratch the fiber slightly above the bead and pull the fiber away from the tip of the ferrule.
22. Air polish the fiber stub. Hold the connector with the fiber up. Hold a piece of 12 μm or 15 μm polishing film (approximately 3 × 5 inches, dull side down) at one side. Rub the other end of the film against the fiber stub until it is flush with the bead of epoxy (Figure 4–7). Do not apply heavy pressure to the film. Do not place your fingers directly opposite the stub or near the stub. To check the stub, gently place your finger directly down on the fiber, not from side. The fiber is flush when you cannot feel the fiber with the tip of your finger. This step should take no more than 10 seconds. Avoid excessive pressure, since it can result in snapping of the fiber stub.
23. Place the polishing tool on the film. Place the connector into the tool slowly and gently until the fiber rests on the film.
24. Hold the connector and the polishing tool. While applying a very light pressure, polish with a figure 8 motion until the adhesive has been removed from the connector. You will recognize complete removal of the adhesive by looking at the tip of the ferrule: any blue or blue green material indicates adhesive. Any dull film that prevents tip of ferrule from having a glass-smooth surface indicates adhesive. If you feel or see the fiber scratch the polishing film, check the film for "tears," which are regions in which the abrasive has been removed from the polishing film. If you experience tears, replace the polishing film, since these tears may snag the fiber and break it.

25. Polish the connector for an additional eight to ten figure 8s. Check for complete removal of adhesive by holding the connector so that light reflects from the tip of the ferrule. The tip should be glass smooth, as is the side of the ferrule.
26. Clean the end and the side of the ferrule. Follow the instructions of Step 41 of Procedure 1.
27. Inspect the connector with a 200x microscope. Follow Procedure 5 in this chapter. If this inspection reveals adhesive remaining on the surface or polishing scratches, repolish with an additional ten figure 8 strokes.
28. Slide the boot over the backshell. With compressed air, blow out the cap. Cap the connector.
29. Repeat Steps 20–28 with the second connector.
30. Test the loss.

Troubleshooting Method 2

A shattered end is the most commonly occurring problem. Follow the instructions of Troubleshooting Procedure 1.

High loss measurements with no apparent reason may be due to over polishing. Over polishing of ceramic ferrule connectors results in the fiber surface being below the ferrule surface, resulting in an air gap. This air gap occurs because the fiber is slightly softer than the polishing film while the ceramic ferrule is harder than the polishing film.

The Salvage Procedure

Frequently, shattered ends can be salvaged through reheating the connector. Place the connector into the holder. Place the cable into the clip of the holder with a slight bow. Place the holder into the oven for one minute. After reheating, push the cable further into the connector. This procedure can allow additional fiber to protrude beyond the tip of the ferrule.

Since this fiber is not supported by a bead of adhesive, it must be carefully air polished (Step 22) and repolished (Steps 23–26). In Step 24, slowly insert the connector into the polishing tool until you feel the fiber touch the film. During your initial polishing strokes, use a very light pressure as long as you can feel the fiber scratch the polishing film. After the scratching has stopped, you can increase the pressure on the connector.

METHOD 3: EPOXYLESS, CRIMP/CRIMP/POLISH ST-COMPATIBLE CONNECTORS

The epoxyless, crimp, and polish connectors have extremely short installation times. However, the absence of a bead of epoxy or adhesive to support the fiber during scribing and polishing increases the difficulty of achieving low loss. However, you can achieve low loss with practice.

Part numbers supplied in Methods 1 and 2 are not repeated below. These instructions are an abbreviated form of those supplied by the manufacturer. Review and follow the manufacturer's instructions when using this product.

TOOLS AND SUPPLIES REQUIRED

tubing cutter

Kevlar® scissors

at least 3 feet of single fiber, tight tube cable with 62.5 μm core and 125 μm clad[4]

two Clauss No-Nik buffer tube/buffer coating strippers
Automatic Tool and Connector crimper (part number ATCC-NC)
connector cleaver (AMP part number 504032-1)
ST polishing tool
glass or hard plastic polishing plate
12 µm polishing film
1 µm polishing film
0.3 µm or 0.5 µm polishing film
two Automatic Tool and Connector Co. Inc. connectors, (part number STA-01-CT)
Scotch™ tape
lens grade compressed air
lens grade tissue
95 percent isopropyl alcohol
200x connector inspection microscope with ST fixture
Alco pads (optional)

PROCEDURE 3: EPOXYLESS, CRIMP/CRIMP/POLISH INSTALLATION

1. Obtain all tools and supplies.
2. Perform a white light test of the cable.
3. Install a boot on one end of the cable. The small end of the boot slides on first. Position the boot approximately 6 inches from the end of the cable.
4. Install the crimp ring on one end of the cable. The small end of the crimp ring slides on first. Slide the crimp ring slightly under the boot.
5. Prepare one end of the cable to the dimensions in Figure 4–11. Using the tubing cutter, strip approximately 1.75 inches of the jacket.
6. Use the Kevlar® scissors to cut the Kevlar® so that ≥⅜ inch extends beyond the end of the jacket.
7. Use the buffer tube stripper to strip the buffer tube so that no more than 1.25 inches of buffer tube remains.
8. This step is necessary for this connector. Wipe the fiber with a lens grade tissue moistened with isopropyl alcohol. Should some buffer coating remain on the fiber, remove this coating from the fiber by wiping with the moistened lens grade tissue, or, if that fails, restrip the fiber with the buffer tube strippers.
9. Remove and save the caps.

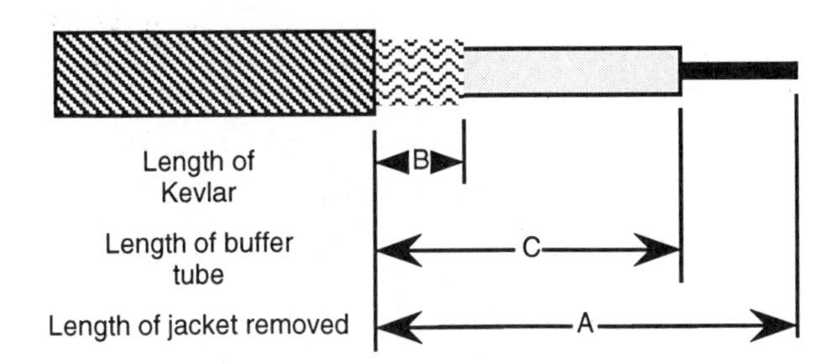

Figure 4–11 Stripping Dimensions for the Crimp and Polish Connector

Figure 4–12 Connector in Cavity 1. (Note the large side of the crimper accepts the large diameter of the backshell.)

10. Perform a white light check of both connectors according to Step 13 of Procedure 1.
11. Check fiber integrity (optional). Gently push the end of the fiber against a hard surface until the fiber starts to bend. It should not break. If the fiber breaks, you have damaged the clad
12. While rotating the connector back and forth, insert the fiber slowly into the connector until the cable jacket butts against the backshell. Insert the fiber into this connector more slowly than with the connectors in Methods 1 and 2, because the fiber hole tolerance seems to be tighter than those of the epoxy connectors and of the preloaded and preheated connectors. If you feel resistance to fiber insertion, stop. Do not force the fiber into the connector. Try another connector.
13. Crimp the backshell of the connector to the buffer tube of the cable. Hold the cable and the connector with the fiber pointing up. Hold the crimper so that the large diameter of Cavity 1 is on top of the crimper. Place the connector into Cavity 1 (Figure 4–12). Crimp area A of the connector backshell (Figure 4–13) with Cavity 1 (the smallest) of the crimp tool. Then crimp area B (Figure 4–13) with Cavity 4 (the second smallest). The Kevlar® trapped between the jaws of the crimper will cause no problem.
14. While rotating the crimp sleeve back and forth to uniformly distribute the Kevlar® strength members around the backshell, slide the crimp sleeve over the backshell. Crimp over area C of the crimp sleeve (Figure 4–12) with Cavity 2 (the second largest) of the crimp tool. Crimp over area D of the crimp sleeve (Figure 4–12) with Cavity 3 (the largest) of the crimp tool.
15. Cleave the fiber. Slide the fiber into the side slot of the AMP connector cleaving tool. Slide the ferrule into the ferrule hole of the AMP connector cleaving tool until the tip of the ferrule rests against the stop screw (Figure 4–14). Very

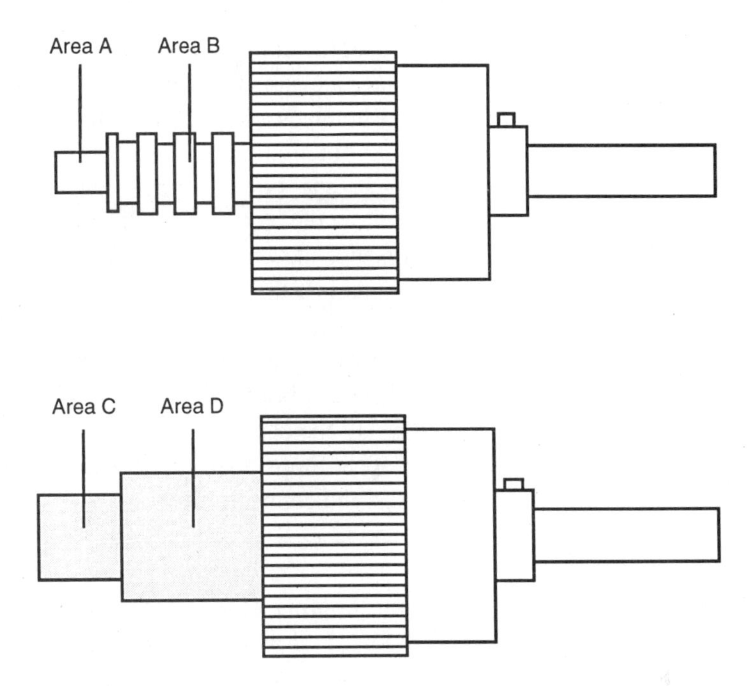

Figure 4–13 Crimp Areas in the Crimp and Polish Connector (Courtesy of Pearson Technologies Inc.)

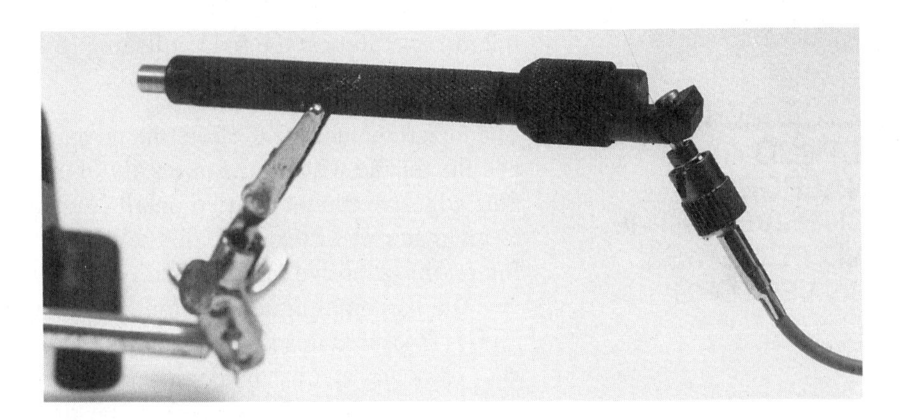

Figure 4–14 Connector in Cleaving Tool (Courtesy of Pearson Technologies Inc.)

slowly (so as not to shatter the fiber end) press the cleaving button on the end of the AMP tool until the cleaving blade lightly touches the fiber. The fiber will fall out of the AMP tool. Perform a white light test. If the connector fails the test, the fiber is broken inside it. Replace this connector. Using a 12 µm film, air polish the fiber stub until it is flush with the tip of the ferrule. Follow the instructions in Step 36 of Method 1.

16. With compressed air, clean both sides of the three polishing films, the polishing plate, and the polishing tool.
17. With the dull side up, place a 1 µm film and the polishing tool on the plate. Place the connector into the polishing tool slowly and gently until the fiber rests on the film.
18. With almost no pressure, move the tool on the 1 µm film in a figure 8 motion for eight to twelve figure 8s until you see dark traces on the polishing film. When you see traces, move the tool on the film in a figure 8 motion twelve more times.

19. Clean the connector and the polishing tool with compressed air.
20. Place a cleaned 0.3 μm or 0.5 μm film on the polishing plate (optional). Install the connector into the polishing tool. With light pressure, move the tool on the film in a figure 8 motion for five strokes.
21. Clean the side of ferrule by rolling the connector on Scotch Magic® tape. Follow the instructions of Step 41 of Method 1.
22. Inspect the connector with a 200x microscope. Follow Procedure 5 in this chapter. If this inspection reveals features that may be removed by polishing, repolish.
23. Blow out the cap with the compressed air. Cap the connector.
24. Slide the boot over the backshell. With compressed air, blow out the cap. Install the cap.
25. Repeat Steps 3–24 on the second connector.
26. Test the connector loss.

Troubleshooting Method 3

Troubleshooting of Method 3 is similar to that described in Troubleshooting of Method 1. Follow the instructions of that section.

The Salvage Procedure

Shattered ends in stainless steel and liquid crystal polymer ferrules may be recovered by polishing on a coarse film (12 μm to 15 μm) and on a 1 μm film. Polish on the coarse film until the core is apparent when you view the connector under a microscope. Repeat the final polishing (Steps 18–23 of Method 3).

METHOD 4: ANAEROBIC ADHESIVE, CRIMP, AND POLISH SC CONNECTORS

The anaerobic adhesive offers the advantage of a strong bond between the fiber and the ferrule without the necessity of epoxy and curing ovens. However, anaerobic adhesive cannot be used on all connectors, since this adhesive will not bond to all materials. Presently, this adhesive is used with ceramic ferrules. In the future, this adhesive may be used with metal and liquid crystal polymer ferrules.

The hardening time is both an advantage and a disadvantage. The hardening time is measured in seconds, so installation rates are high and labor costs are low.

However, the fast hardening time forces installers to work quickly, or risk hardening of the adhesive before the fiber is fully inserted into the connector. This fast hardening can result in high connector loss and low success rates with novice installers.

The SC connector has a number of advantages. From an installation viewpoint, the pull-proof and wiggle proof performance complicates installation, since excessive adhesive in the backshell can glue the ferrule to the backshell, destroying these advantages.

Part numbers supplied in Methods 1, 2, and 3 are not repeated below. These instructions are a modified form of those supplied by the manufacturer. Review the manufacturer's instructions when using this product.

Tools and Supplies Required

tubing cutter

Kevlar® scissors

at least 3 feet of single fiber, tight tube cable with 62.5 μm core and 125 μm clad

two Clauss No-Nik buffer tube/buffer coating strippers

Automatic Tool and Connector crimper (part number ATCC-SC2)
SC polishing tool
glass or hard plastic polishing plate
12 μm polishing film
1 μm polishing film
0.3 μm or 0.5 μm polishing film
two Automatic Tool and Connector Co. Inc. connectors, (part number SCA-02-T2)
Scotch™ tape
anaerobic adhesive in syringe, accelerator, and extra needles for syringe (Automatic Tool and Connector Co. part number FRP-QT)
lens grade compressed air
lens grade tissue
95 percent isopropyl alcohol
200x connector inspection microscope with SC fixture
Alco pads (optional)

PROCEDURE 4: ANAEROBIC ADHESIVE INSTALLATION

1. Obtain all tools and supplies.
2. Perform a white light test of the cable.
3. Install a boot on one end of the cable. Slide the small end of the boot on first.
4. Install the crimp ring on one end of the cable. Slide the small end of the crimp ring on first.
5. Prepare one end of the cable (Figure 4–15). The sequence will be the same as for the epoxy connectors, but the dimensions will be those in Figure 4–15. Follow the instructions of Steps 6–8 of Method 1.
6. Moisten a lens grade tissue with 95 percent isopropyl alcohol. Wipe the fiber at least twice with the folded tissue.
7. Wipe the fiber at least twice with a dry folded tissue. Moisture left on the fiber may act as an accelerator for the adhesive.
8. Remove and save the caps.
9. Perform a white light check of both connectors.
10. Check fiber integrity (optional). Gently push the end of the fiber against a hard surface until the fiber starts to bend. The fiber should not break. If the fiber breaks, you have damaged the clad.
11. Perform a dry fit. Follow the instructions in Step 14 of Procedure 1.

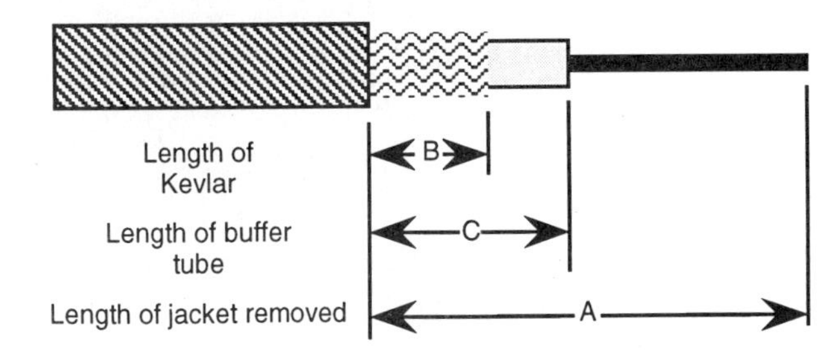

Figure 4–15 Stripping Dimensions for Anaerobic Adhesive, SC Connectors

> *CAUTION: Avoid excessive adhesive on the tip of the ferrule. Excessive adhesive on the tip significantly increases polishing time. In addition, excessive adhesive on the tip can flow onto the side of the ferrule, from which it is difficult to remove. If the adhesive cannot be removed, the connector will not fit into the adapter or the receptacle.*

> *CAUTION: During the next step, you must proceed rapidly: if you insert the fiber too slowly, the adhesive will set before the fiber is completely installed. You will lose the connector.*

> *CAUTION: Do not spray accelerator on hands. Some people are allergic to the accelerator.*

12. Remove the cap from the syringe of adhesive. Install a needle on the syringe. Insert the needle into the backshell of the connector until the needle stops. Inject the adhesive until adhesive begins to flow from the fiber hole in the tip of the ferrule. As soon as adhesive appears through the fiber hole, quickly remove the needle from the connector.
13. While rotating the connector, insert the fiber quickly into the connector until the cable jacket butts against the backshell.
14. Crimp area A, the small diameter of the backshell in Cavity 1 of the crimper. Area A of this connector looks like area A of Figure 4–13. Hold the cable and the connector for two seconds. The adhesive will set. While rotating the crimp sleeve back and forth around the backshell, slide the crimp ring over the backshell. Crimp the smallest diameter of the crimp sleeve with Cavity 2 (middle size) of the crimper. Crimp the largest diameter of the crimp sleeve with Cavity 3 (largest) of the crimper.
15. Cure the adhesive. Shake the bottle of accelerator. Install the spray cap on the bottle of accelerator. Hold the bottle 4 to 6 inches away from the tip of the ferrule. Spray once. The adhesive will change color, indicating complete hardening.
16. Remove the fiber protruding beyond the tip of the ferrule (excess fiber). With a wedge scriber, lightly scratch the fiber just above the tip of the ferrule. Do not break the fiber while scratching with the scriber. Break off the excess fiber by pulling the fiber away from the tip of the ferrule. If the fiber does not break, it may be covered with a thin layer of epoxy. Scratch the fiber slightly above the bead and pull. Remove the excess fiber from both connectors.
17. Air polish the fiber (Step 36 of Procedure 1).
18. With compressed air, clean both sides of the two polishing films, polishing plate, and polishing tool. Place the polishing film on the plate. Place the polishing tool on the film.
19. Install the connector slowly into the polishing tool until the connector rests against the film.
20. With almost no pressure, move the tool on the medium (1 μm) film in a figure 8 motion three times or until a thin film of adhesive remains.
21. Clean the connector and the polishing tool with the compressed air.
22. Place a 0.3 μm polishing film on the polishing plate. Polish with a light pressure in a figure 8 path until the adhesive is completely removed. When the adhesive is completely removed, the tip of the ferrule will be glass smooth.
23. Install the outer housing (or plug frame). Align the filled-in corners of the outer housing with the missing corners of the ferrule (or plug) (Figure 4–16).

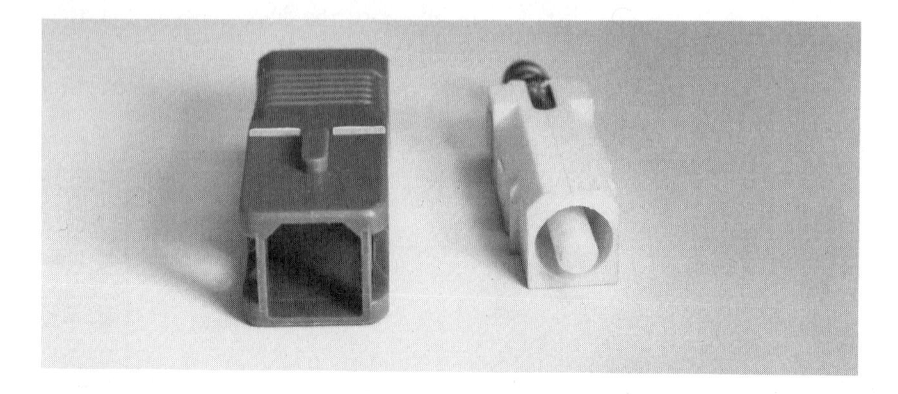

Figure 4–16 End View of Unassembled SC Connector (Courtesy of Pearson Technologies Inc.)

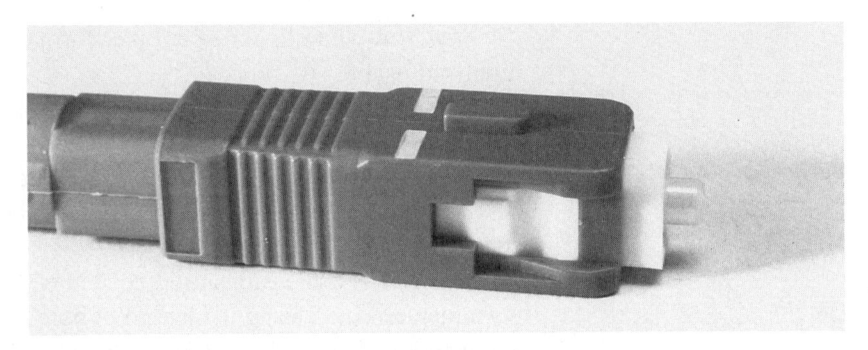

Figure 4–17 Assembled View of SC Connector (Courtesy of Pearson Technologies Inc.)

Slide the outer housing onto the ferrule until the housing clicks and slides beyond the retaining tabs. The tip of the ferrule must protrude beyond the end of outer housing (Figure 4–17). Slide the boot under the back end of the outer housing.

24. Clean the ferrule tip with Scotch™ tape. Clean the ferrule sides with compressed air.
25. Inspect the connector with a connector inspection microscope with an SC fixture or adapter.
26. With compressed air, blow out the cap. Install the cap. Cap the connector.
27. Repeat Steps 5–24 on the second connector.
28. Test the connector loss.

Troubleshooting Method 4

The most common problem with the anaerobic adhesive is curing of the adhesive before the fiber is completely inserted into the connector. To avoid this problem, practice dry fitting until you can insert the fiber without any delay.

The second most common problem is a shattered end. Follow the instructions of Procedure 5 of this chapter for details on causes.

TWENTY-THREE ALTERNATE STEPS

The procedures presented in this chapter evolved from field installation and training activities. As such, they are not identical to those recommended by connector manufacturers. This is not to say that the procedures in this chapter are better than those recommended by manufacturers. Rather, these procedures work well in both field and training situations. Other procedures, including those recommended by manufacturers, can work equally well. In this section, we will present twenty-three alternate steps and the conditions under which such steps will work.

Unless otherwise indicated, these alternative steps correlate to the steps in Procedure 1. Some of these alternative steps can be used in Procedures 2–4.

Step 2: An oven may not be required, since some epoxies cure at room temperature. Room temperature epoxies can be slow cure (12 hours) or fast cure (5 to 20 minutes). Slow cure epoxies are easy to use and have long pot lives (30 to 50 minutes). They are ideal for the installation of a large number of connectors. Fast cure epoxies are somewhat difficult to use, and are better suited for installation of a small number (one to four) of connectors.

The difficulty in using fast cure epoxies occurs during insertion of the cable into the backshell: if the epoxy hardens too rapidly, the fiber will break during insertion. If this happens, you will discard the connector.

Note that the time and temperature will be different for each epoxy used.

Step 3: A white light test or other form of continuity test is not required but is highly advisable.

Step 4: Heat shrink tubing may be supplied instead of boots. Note that the use of boots increases the reliability of the connector.

Step 5: Some connectors do not use crimp rings. Instead, the strength members are epoxied or glued to the inside of the backshell, as in the 3M Hot-Melt connector.

Step 6: The dimensions used depend on the connector style and the connector manufacturer. Increasing the length of bare fiber that protrudes beyond the end of the ferrule makes the fiber easier to grasp after scribing. However, increasing the length of bare fiber increases the chance of breaking the fiber during handling and during insertion of the connector into a curing oven. If the fiber is broken before the epoxy has cured, the connector is often high loss. Note that the length of bare fiber should not extend beyond the end of the shroud.

Step 7: The length of Kevlar® in Procedure 1 is longer than that recommended by manufacturers. We recommend this excess length in order to have a visual confirmation of adequate Kevlar® length. While this excess length might result in a cosmetic concern, it will be concealed under the boot.

Step 9: All connector manufacturers recommend that the fiber be cleaned. However, Pearson Technologies Inc. has observed that the fiber does not need to be cleaned in order to fit into the fiber hole, except for the QT™ (Quick Term) products from Automatic Tool and Connector, Co. Inc. In addition, as long as you do not wipe the bare fiber with your fingers (which can leave oil on the fiber), cleaning is not required to ensure proper adhesive of the epoxy to the fiber.

Step 12: This optional check of fiber integrity is a useful technique for novices. As you gain experience and confidence, you will not need this step.

Step 13: The white light check is useful for novices. As you gain confidence with your supplier, you can eliminate this step. However, if you experience breakage problems during insertion of the fiber into the connector, you should include this step in an effort to identify the cause of the problem. Blocked or missing fiber holes are extremely rare.

Step 14: The dry fit is useful for novices. As you gain confidence with your abilities and your supplier, this step can be eliminated. However, if you experience breakage problems during insertion of the fiber into the connector, you should include this step in an effort to identify the cause of the problem. Undersize fiber holes are rare.

Step 16: Mixing the epoxy by rubbing the package over the edge of a table is not the only mixing technique. Some epoxy manufacturers recommend use of a roller to ensure complete mixing of the resin and hardener.

Step 17: Some professional installers transfer three drops of epoxy to the top of the backshell with a toothpick. Use of this technique requires that you make one of two assumptions: the epoxy will flow to the fiber hole or the epoxy will be dragged to the fiber hole during cable insertion. If both of these assumptions are not true, bare glass will be left inside the connector. This bare glass can cause low reliability.

While some professional installers achieve high reliability with this technique, use of the syringe places the epoxy exactly in the correct location, ensuring high reliability independent of the installation personnel. In addition, the deliberate overfilling of the backshell with the syringe provides visual indication that sufficient epoxy has been injected. Finally, overfilling the backshell allows faster installation times by allowing looser control of the length of the buffer tube dimension (Step 8, Procedure 1). Looser control is possible without a reduction in reliability because the bare glass will be buried in the epoxy.

Step 20: Most connector manufacturers do not recommend that one to two drops be forced out of the backshell during cable insertion. We do recommend this as a positive indication that sufficient epoxy has been injected, and because the addition of epoxy to the strength members results in additional strength between the cable and the connector.

Step 21: Some connector manufacturers recommend one or two crimps. The third crimp is an "insurance crimp." Occasionally, but rarely, the crimp rings will be loose after the first crimp. The third crimp on top of the first has eliminated this problem completely.

Step 24: Some connector manufacturers recommend removal of all epoxy from the tip of the ferrule before curing. These manufacturers accurately state that a bead on the tip increases polishing time. While this is true, the bead also supports the fiber during scribing and polishing. An unsupported fiber can break below the surface of the tip of the ferrule. With such a break, a connector will have either high loss or increased polishing time (and cost). We find the bead to be useful and often necessary to achieve low loss and high connector yield.

Some professional installers add a bead to the tip of the ferrule. This bead provides a visual indication that polishing is complete.

A reasonable strategy is to use a large bead until you become confident in your ability to scribe and polish the fiber successfully. When confident, you can use a reduced bead size, until you are confident with this reduced size. You can repeatedly reduce the size of the bead and establish your ability to scribe and polish. Eventually, you can eliminate the bead.

When you eliminate the bead, you will benefit from the use of the AMP connector cleaving tool to remove the excess fiber. The use of this tool is described in Procedure 3. If you use this cleaving tool, always air polish (Step 36 of Procedure 1). Do not use this tool when there is a bead of epoxy or adhesive on the tip of the ferrule. The scribing blade will be damaged from being embedded in the bead.

Step 26: Use of shrouds is not universally recommended by connector manufacturers. Use of a shroud on each connector protects the fiber from being broken prior to curing of the epoxy. If you decide not to use a shroud, you must handle the connector carefully prior to curing. You must insert the connector carefully into the curing oven.

Step 36: Air polishing is not universally recommended. This polishing usually brings the fiber flush with the bead. When the fiber is flush, it is difficult to break when the connector is inserted into the polishing tool. If the fiber is broken in this manner, the break may extend below the surface of the ferrule. In this case, the connector can be lost, or the polishing time can be increased significantly.

Steps 38 and 40, alternative A: Connectors can be automatically polished in polishing machines (available from Buehler and Seiko). Use of machines results in reduced training costs and increased uniformity of results. Use of machines for polishing singlemode connectors to achieve low back reflection is common.

Steps 38 and 40, alternative B: Polishing can be done dry or wet. Wet polishing requires distilled or deionized water or isopropyl alcohol as a lubricant. This liquid extends the life of the polishing film by flushing the fiber and epoxy from the film. We have found that wet and dry polishing produce equal loss values for multimode connectors.

Steps 38 and 40, alternative C: Polishing can be done on films with or without an adhesive backing (if you use film with an adhesive, choose a film that is easily removed from the plate). Films without an adhesive backing can slide on the polishing plate. You can reduce this slipping by placing one drop of water on the plate under the film. Surface tension reduces slipping.

Step 39: Some manufacturers recommend three-step polishing, instead of the two-step polishing procedure in Procedure 1. The third film is usually 0.3 μm or 0.5 μm. We have found no improvement through use of the third film on multimode connectors, as long as the final film used is at least as fine as 1 μm.

Step 41: Ferrules can be cleaned with Alco pads or with alcohol soaked, lens grade tissues.

Step 15, Procedure 4: Some connectors using anaerobic adhesive require application of the accelerator by brushing the accelerator onto the fiber or dipping the fiber into the accelerator. This step is called priming. Priming is done prior to insertion of the fiber into the connector. Based on experience in training with anaerobic adhesives, we suggest priming only for skilled and experienced installers.

HOW TO RECOGNIZE LOW LOSS AND HIGH LOSS CONNECTORS THROUGH MICROSCOPIC INSPECTION

Microscope inspection of connectors by is a fast and convenient method by which to estimate the expected loss of the connector. This expected loss will be low, questionable, or high.

This inspection takes a few seconds, which is much less time than it takes to make an insertion loss test. Microscopic inspection will also reveal the characteristic(s) of the high loss. The characteristics will indicate the corrective actions you need to take in your connector installation and/or handling techniques.

Low loss connectors require no special treatment. If the connector looks slightly bad (as determined from experience), its loss will be higher than the average, but may or may not, be in excess of the specifications of the connector manufacturer. A useful procedure is tagging those connectors with questionable microscopic appearance. You will replace high loss connectors without performing an insertion loss test.

Whenever possible, perform microscopic inspection at a magnification of at least 200 (400 for singlemode fibers) both with and without backlight. Inspection should be performed both ways, since backlight can both reveal and conceal features in the core of the connector.[5]

Microscopic inspection indicates whether a connector will have acceptable loss or unacceptable loss. A connector looks "good" if it has a round, clear, featureless core (Figure 4–18) and if the fiber is flush with the ferrule. (Note that a fiber is epoxyless, adhesiveness connectors can be above the ferrule if the connector is underpolished. The solution is additional polishing.) If the connector looks

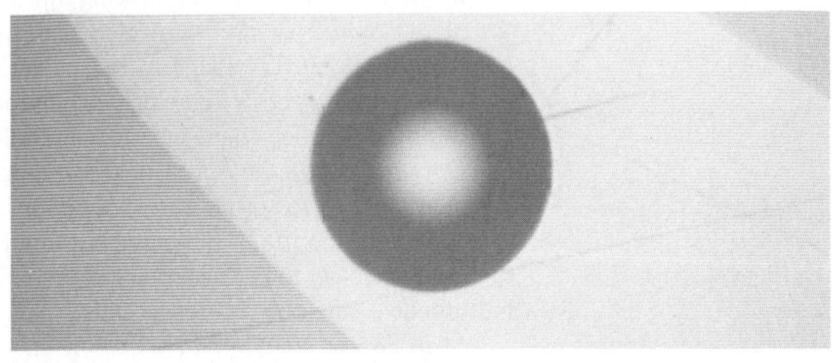

Figure 4–18 A Low Loss Connector Has a Round, Clear, Featureless Core (Courtesy of Automatic Tool and Connector Co. Inc.)

good, its loss will be within the specifications of the manufacturer most of the time.[6] If the connector does not look good, it may exhibit high loss.

If a connector does not look good, the core will not be round, it will not be clear, it will not be featureless, or it will not be flush. A core will not be round (Figure 4–19) if the fiber was broken below the surface of the ferrule (Figure 4–20). This breakage can occur during insertion of the fiber into the backshell, during scribing, during air polishing, and during polishing. Excessive pressure on the wedge scriber can break the fiber during scratching. Excessive pressure during air polishing can break the fiber. Incomplete air polishing can leave a fiber stub

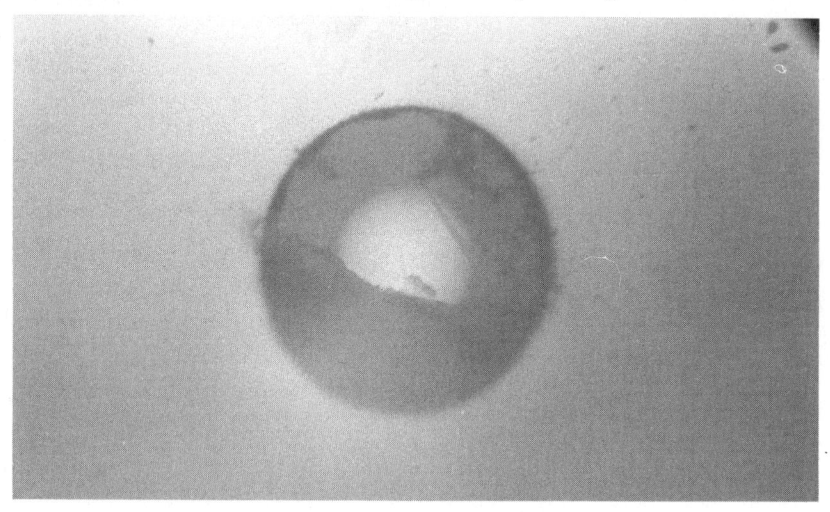

Figure 4–19 Example of a Non-Round Core (Courtesy of Pearson Technologies Inc.)

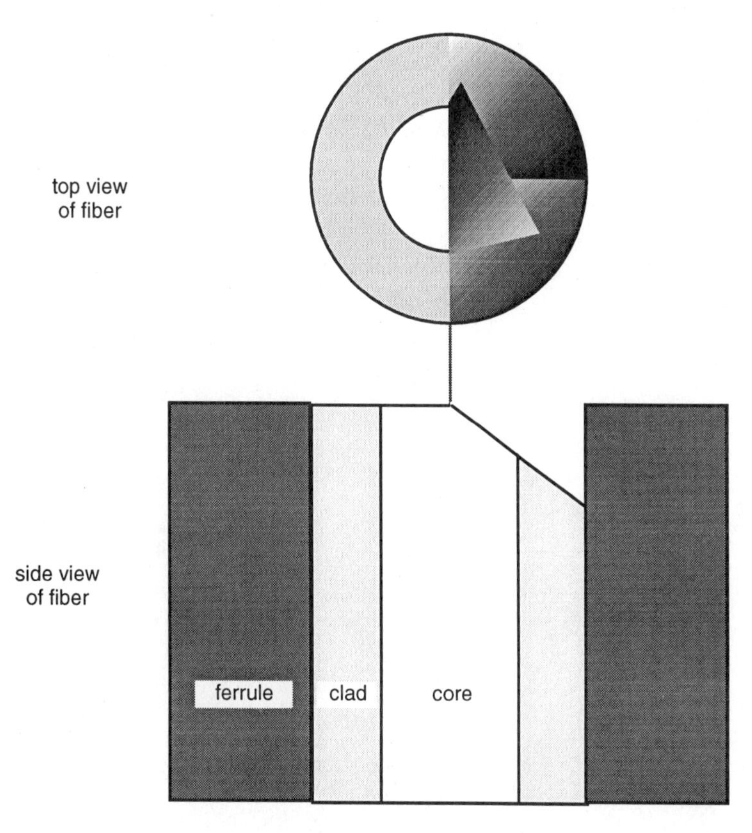

Figure 4–20 Fiber End That Results in a Non-Round Core

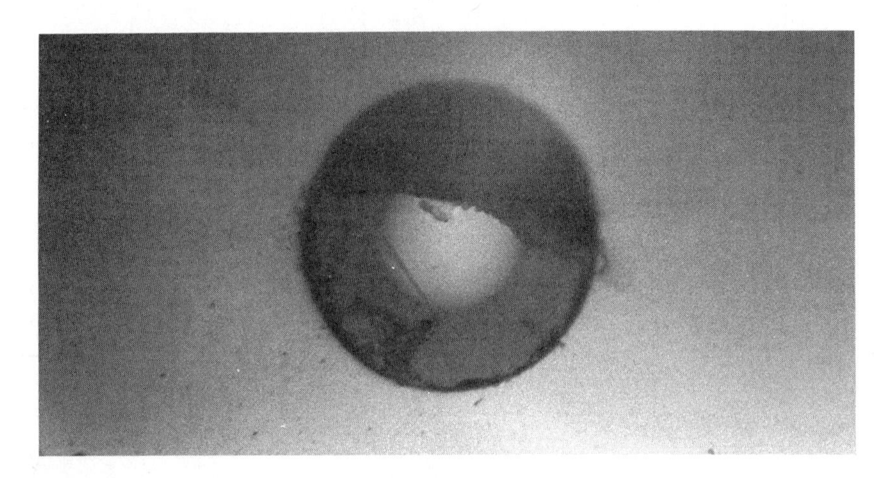

Figure 4–21 Shattered End (Courtesy of Pearson Technologies Inc.)

above the bead of epoxy or adhesive. This stub can be broken during the insertion of the connector in the polishing tool or by excessive pressure during the initial polishing strokes. Punching the scribing button instead of slowly pressing the button of the AMP cleaving tool (Step 15 of Procedure 3) can break the fiber below the surface of the ferrule. In the worst case, the fiber will have no clearly definable core, which is called a shattered end (Figure 4-21).

Shattered ends in stainless steel and liquid crystal polymer ferrules may be recovered by polishing on a coarse film (12 μm to 15 μm) and on a 1 μm film. Polish on the coarse film until the core is apparent when the connector is viewed under a microscope. Repeat the final polishing (Step 38).

A core will not be clear for three reasons:

- Incompletely cured epoxy (Figure 4–22)
- Adhesive that is smeared across the core during polishing
- Water stains from water or 70 percent alcohol used to clean the ferrule

You will be able to remove water stains with 91 percent isopropyl alcohol and lens grade tissue or with three to five strokes on a 0.3 μm or 0.5 μm polishing film.

Incompletely cured epoxy will have irregular shapes and look like flakes. Depending on the epoxy used, the epoxy will have a color. For the Tra-Con

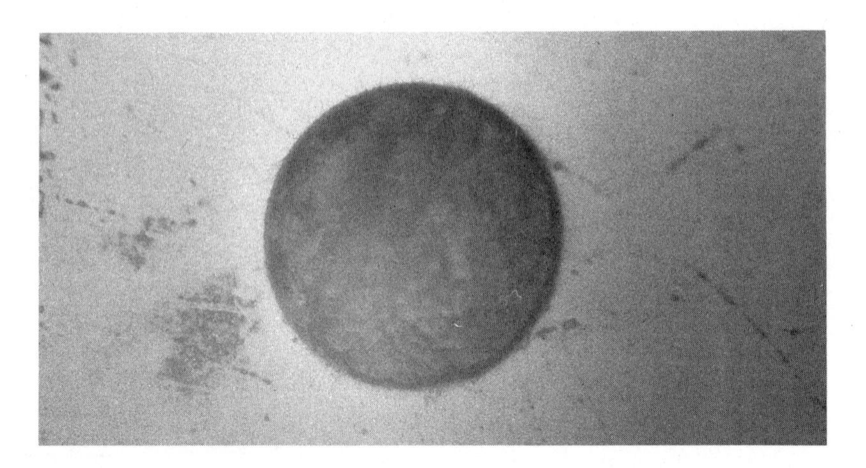

Figure 4–22 Smeared Epoxy on a Connector (Courtesy of Pearson Technologies Inc.)

F113SC, the color will be blue if the epoxy fills in depressions in the fiber. If this same epoxy is thin, it may have a pink tinge. Thick epoxy on core is due to insufficient polishing.

Incomplete curing of epoxy is a rare problem. Epoxies will not cure completely if the time is too short, the temperature is too low, the epoxy has been allowed to freeze, or the epoxy has passed its expiration date.

A core will not be featureless for the following reasons:

- If there are scratches because the connector has been incompletely polished
- If the core has been scratched by particles that have contaminated the polishing film
- If the core has been scratched from use
- If particles from the polishing film remain
- If there are cracks in the core
- If dirt from the environment remains on the core
- If the cleave was bad (for multimode connectors only)

Incomplete polishing leaves many parallel, fine scratches that continue across both core and clad (Figure 4–23). These scratches can usually be removed from contact connectors (ST-compatible, SC, FC, D4, FDDI, and ESCON) with additional polishing on a 1 μm or finer film.

These scratches may or may not be removable from non-contact connectors. (SMA, biconic, and mini-BNC). If the polishing of the non-contact connector was not completed, so that ferrule material remains to be removed to bring the ferrule to the proper length, additional polishing will usually remove polishing scratches. However, if the non-contact connector was over polished on a coarse film so that the ferrule length is proper, additional polishing will either not be possible (SMA) or result in short ferrule length (biconic). Short ferrule length results in an excessively large air gap and high loss.

Scratches from contaminated polishing films are different from polishing scratches. Contamination scratches are usually singular and thicker than those from incomplete polishing (Figure 4–24). Such scratches usually continue across both core and clad. Contact connectors with metal or liquid polymer ferrules can often be repolished to remove such scratches. Contact connectors with ceramic ferrules are difficult to repolish.

Figure 4–23 Scratches Due to Incomplete Polishing (Courtesy of Pearson Technologies Inc.)

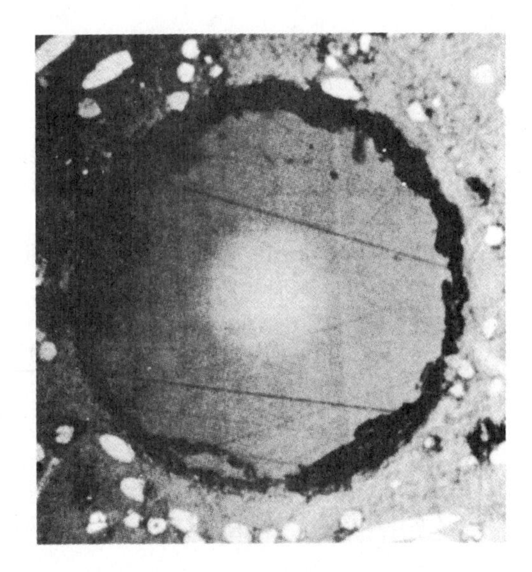

Figure 4–24 Scratches Due to Contamination of Polishing Film (Courtesy of AMP Inc.)

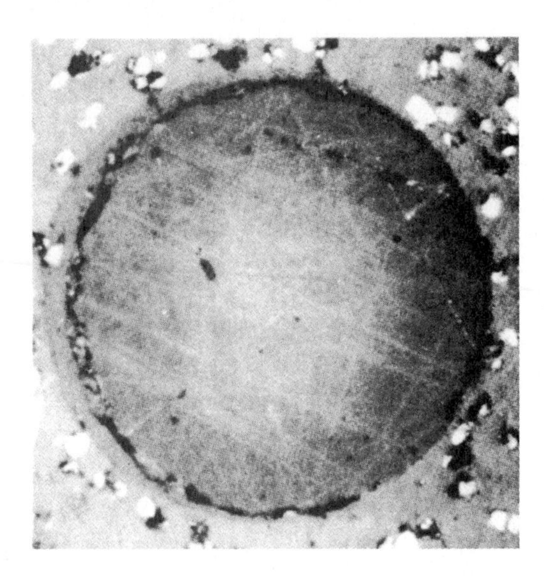

Figure 4–25 Scratches from Use (Courtesy of AMP Inc.)

Use scratches can have length on unkeyed connectors. Use scratches will be points (have no length) on keyed connectors (Figure 4–25).

Certain polishing films can leave polishing particles on the core (Figure 4–26). These particles can be removed with Alco pads, isopropyl alcohol and lens grade tissues, or with additional polishing on a 0.3 or 0.5 μm film.

A bad cleave has many features, with alternating featured and featureless regions. Bad cleaves occur when cleaving tools are out of adjustment (Figure 4–27).

Cracks in the core differ from scratches in thickness; cracks are usually thicker than scratches. Cracks are either mechanical or thermal. Mechanical cracks are due to use of excessive force. Thermal cracks are due to the different thermal contraction rates of fiber, ferrule material, and epoxy during cooling of heat cured connectors.

Dirt from the environment (Figure 4–28) is obvious, since such dirt is significantly larger than any other feature on the ferrule.

If a fiber is not flush with the surface of the ferrule, the fiber stub can be damaged, creating a shattered end. Careful, additional polishing of this stub is necessary.

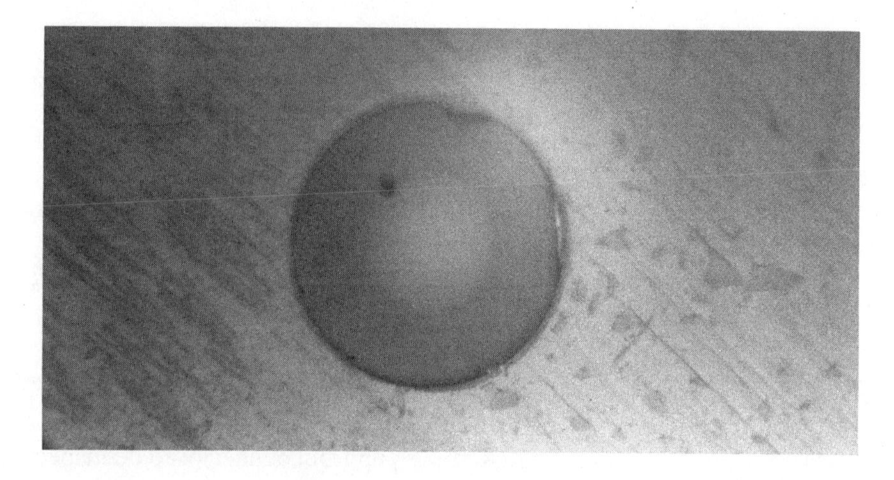

Figure 4–26 Polishing Grit on Core (Courtesy of Pearson Technologies Inc.)

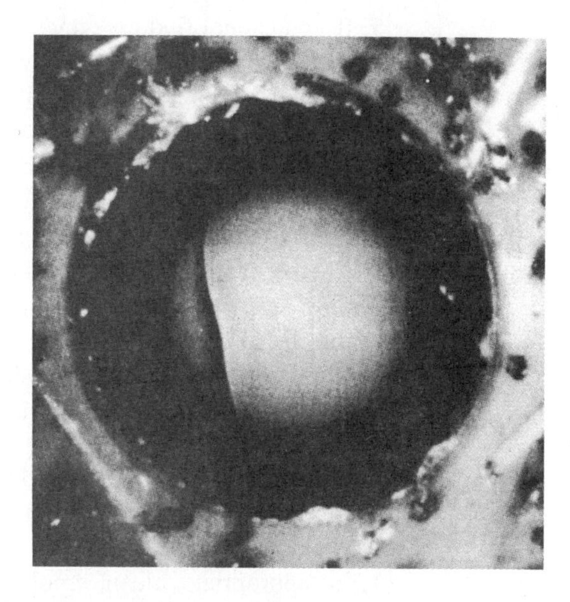

Figure 4–27 Bad Cleave (Courtesy of AMP Inc.)

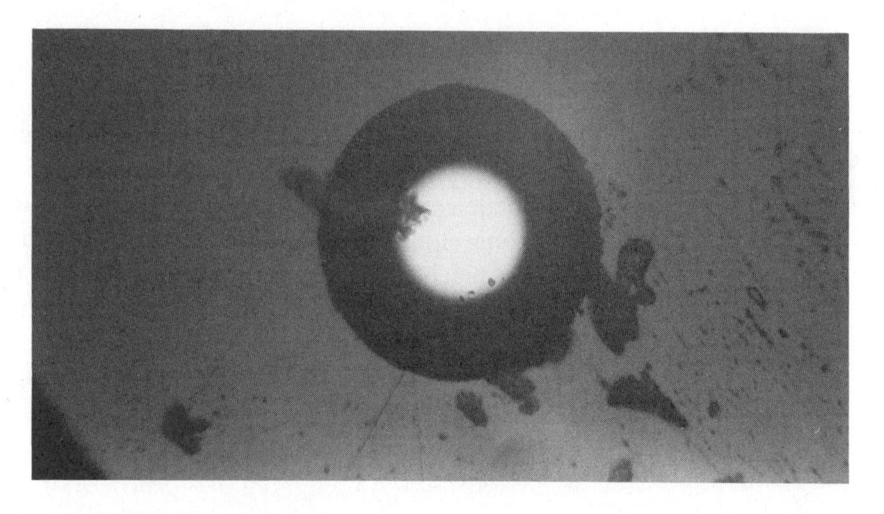

Figure 4–28 Dirt from Environment on Core (Courtesy of Pearson Technologies Inc.)

Tools and Supplies Required

200x or 400x microscope with adapter for connector style to be inspected

Scotch™ tape

Alco pads or 95 percent isopropyl alcohol with optical lens grade tissues

source of light

Procedure 5: Microscope Installation

1. Remove cap from connector to be inspected.
2. Insert connector into adapter of microscope.
3. Turn on microscope.
4. Focus microscope.
5. If you cannot find core and cladding in field of view, adjust the adapter until the fiber is in the field of view.
6. Inspect the core of the fiber. If properly installed, the core will be round, clear, featureless, and flush with surface of ferrule.
7. Inspect the clad of the fiber. If properly installed, the clad will be round, clear, featureless, and flush. If dirt or debris appears on the clad, clean and reinspect the connector. If the core is good (as defined above) but the clad is not good (by the same definition), the connector will test good, but the installation technique was improper. Features in the clad have the same causes as features in the core.
8. Remove the cap from the connector on the opposite end of the cable. Position the opposite end so that light can be seen in the core of the first end. Inspect the first core again.
9. Repeat steps 1–7 for the second end of the cable.
10. Place a tag or label on an end that does not have a round, clear, featureless core that is flush with the ferrule.
11. As appropriate, clean, repolish, or replace any end that is not good. Repeat steps 1–10.

SINGLEMODE POLISHING

With the exception of polishing, the installation procedures for singlemode connectors are the same as those for multimode connectors. The polishing of singlemode connectors to achieve low back reflectance is similar to that for multimode connectors. However, four grades of polishing films are required to create the extremely smooth surface that results in low back reflection. In addition, cleaning after polishing requires more care for singlemode connectors than for multimode connectors. Most singlemode connectors are machine polished for low back reflectance and low cost. Only pre-radiussed singlemode connectors are hand polished. Procedure 6 is presented for hand polishing of connectors with pre-radiussed ceramic ferrules and begins after the scribing and air polishing of the fiber stub.

Tools and Supplies Required

wedge scriber

metal ST polishing tool

nylon ST polishing tool

four 3M hard rubber polishing pads

5 μm diamond polishing film (Fiber Optic Center, Inc., part number 5-5-D-3N-1)

3 μm alumina polishing film (Fiber Optic Center, Inc., part number 9x11-3-CA-2N-100)

1 μm diamond polishing film (Fiber Optic Center, Inc., part number 5-1-D-3N-1)

0.3 μm alumina polishing film (Fiber Optic Center, Inc., part number 9x11-.3-CA-2N-100)

acetone (do not use if cyanoacrylate adhesives are used)

lens grade compressed air

lens grade tissues

95 percent isopropyl alcohol

400x or 200x connector inspection microscope with ST fixture

Alco pads (optional)

Procedure 6: Achieving Low Reflectance

1. Clean the polishing pads with alcohol. Clean both sides of the polishing films with compressed air. Place the four polishing films on the hard rubber polishing pads. Clean the films with rubbing alcohol and lens grade tissues.
2. Clean the polishing tool with alcohol and tissues. Install the connector in the metal polishing tool.
3. With a figure 8 motion, polish the connector with light pressure on the 3 μm alumina film until you have removed the adhesive or epoxy. Clean the connector and the polishing tool.
4. With a figure 8 motion, polish the connector with light pressure on the 0.3 μm film for five seconds. Inspect the connector at a magnification of at least 200x. If necessary, continue polishing to remove all scratches.
5. Remove the connector from the steel polishing tool. Clean the connector and the nylon polishing tool with alcohol and tissues. Insert the connector into the nylon polishing tool.
6. Polish the connector on the 5 μm film for five seconds. Use light pressure and a figure 8 motion. Clean the connector and the nylon polishing tool with alcohol and tissues.
7. On the 1 μm film, polish the connector with light pressure and a figure 8 motion for five seconds. Clean the connector and the polishing tool with alcohol and tissues.
8. On the 0.3 μm film, polish the connector with light pressure and a figure 8 motion for five seconds. Be careful: excessive polishing on this film can result in undercutting, in which the fiber surface ends up below the ferrule surface. This creates an air gap, which results in high reflectance. If reflectance measurements are high, undercutting may be the problem. Repeat Step 3 with

heavy pressure to polish the ferrule flush with the connector. Then repeat Steps 4–11. Use more pressure on this film than on previous films. Remove the connector from the polishing tool.

9. Clean the connector with acetone and tissues.
10. Clean all films after every fifth connector. Remember that polishing must be performed in a clean environment. Dust particles from the air can create scratches, which will result in high reflectance.
11. With a microscope, inspect the connector for any form of core damage or surface contamination. Core damage can be a shattered end or scratches. Surface contamination can be dust, water stains, or oils.

    If you observe a shattered end, repeat Steps 6 and 7 until you see the core and clad defined. Then repeat Steps 8–10. If you see fine polishing scratches, repeat Step 7 until you have removed such scratches. Then repeat Steps 8-11. If you see scratches from contamination, clean all films with alcohol and tissues. Repeat Step 7 you have removed the scratches. Then repeat Steps 8–11. If you see contamination, repeat Steps 8 and 9. If you are unable to remove the contamination with Step 8, repeat Steps 6–11.
12. Test insertion loss and reflectance.

REVIEW QUESTIONS

1. What is the advantage of a bead of adhesive or epoxy on the tip of the ferrule?
2. What is the most common cause of a shattered end?
3. Describe the microscopic appearance of a good connector. Try to use only five words.
4. What is a white light test of a cable? What does it demonstrate?
5. What is a dry fit of a connector? What does it demonstrate?
6. What is the difference between polishing scratches and scratches that result from contaminates on the polishing film?
7. What safety precautions should you take during connector installation?
8. Why should you rotate a connector while inserting fiber?
9. If the curing temperature is too high, what would you expect to see when you view a connector under the microscope?
10. What is the advantage of using a syringe to inject epoxy?
11. What is an advantage of the Hot Melt connector over an epoxy or epoxyless connector?

CHAPTER

5

How To Make Loss Measurements

CHAPTER OBJECTIVES

From this chapter, you will be able to:

1. Perform insertion loss measurements.
2. Verify the six Insertion Loss Rules of Thumb.
3. Identify the seven basic features of the OTDR trace.
4. Interpret each of the ten basic OTDR traces.
5. Make OTDR length and loss measurements.
6. Identify the test equipment best suited to your fiber network.

INTRODUCTION

The proper functioning of fiber optic systems is determined by two mechanisms: pulse spreading and signal attenuation. Based on proper functioning, one might assume that an installed fiber optic cable and connector system (also called "cable plant") would require testing of both types of performance. However, errors during installation will not increase pulse spreading and degrade the bandwidth or bit rate capacity of the system. To understand the reason that pulse spreading cannot be increased by such errors, we must reexamine the mechanism of pulse spreading.

Pulse spreading in multimode fibers is a result of three mechanisms:

- Modal dispersion
- Chromatic dispersion
- Material dispersion

We will not consider chromatic and material dispersions, since the basic chemistry of the fiber determines such dispersions. No errors during installation will change this basic chemistry.

However, modal dispersion (also called "multimode" and "intermodal" dispersion and "multimode distortion") can be changed by such errors. In multimode fibers, rays of light travel paths parallel to the axis and at all angles up to and including the critical angle (Figure 1–9). As long as the fiber remains reasonably straight, the light will continue to experience total internal reflection.

If the fiber is bent to a very small radius, the critical angle in the fiber is rotated relative to its normal position (Figure 5–1). When the critical angle is rotated, rays of light that were within the critical angle may not be within the critical angle of the bent region. When these rays of light strike the clad, they will escape the core into the clad (Figure 1–9). These rays are lost, creating excess attenuation in the fiber.

The normal radii to which the fiber is bent in any stranded cable are large enough to avoid all excess attenuation by this rotation. However, installation errors can result in excessive rotation of this angle.

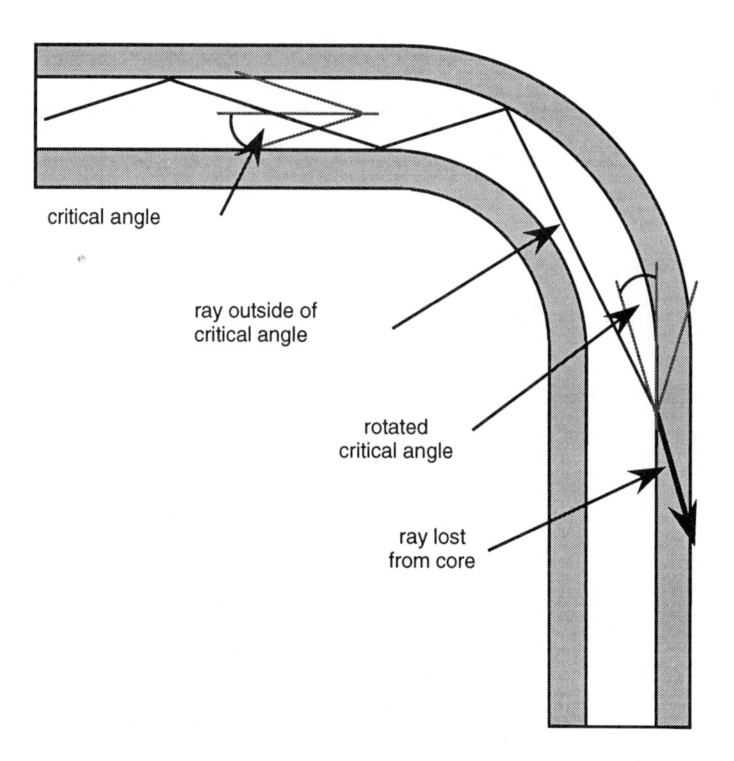

Figure 5–1 Rotation of Critical Angle During Bending of Fiber

Installation errors are violations of one of the performance limits of the cable. For example, violation of the bend radius, crush load, long term use load, and temperature operating range all result in a rotation of the critical angle of the fiber.

This change is some form of bending. For example, when the bend radius is violated, the fiber is bent on a macroscopic scale. This form of bending is commonly called microbending, but is more accurately a combination of microbending and macrobending (Frederick C. Allard, *Fiber Optics Handbook for Engineers and Scientists* [McGraw Hill, 1990], 1.32–1.34, 2.10–2.13).

The rays of light that may escape are those rays at or near the critical angle. Such rays are the last rays to arrive at any point along the fiber (Figure 1–15). Since these rays are the last to arrive, loss of these rays into the clad during bending results in a reduction in pulse spreading (Figure 5–2).

In singlemode fibers, installation errors result in microbending and/or macrobending, which, in turn, allow energy to escape from the core. Since installation

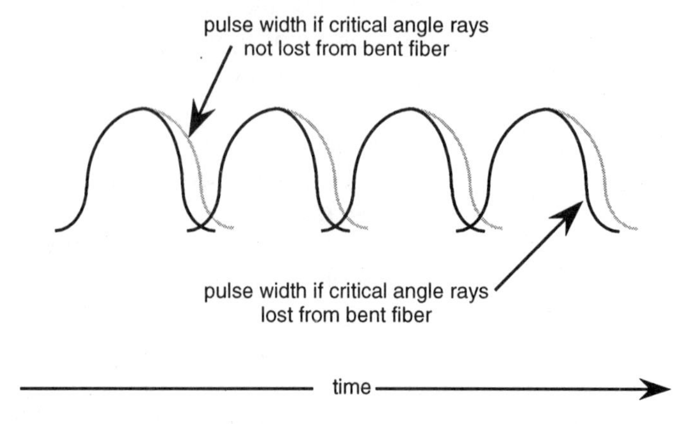

Figure 5–2 Reduction in Pulse Spreading Due to Loss of Rays Near Critical Angle

errors can cause increased attenuation in both multimode and singlemode cables, and reduced pulse spreading in multimode cables, attenuation of installed cables must be checked, but pulse spreading need not be.

There are two forms of loss: *insertion loss* and *reflectance* or return loss. Insertion loss is a measurement that simulates the loss in power a transmitter-receiver pair would experience. Insertion loss is a result of connection loss and fiber attenuation. Reflectance is a measurement of the power reflected back towards a singlemode light source. Reflectance is determined by the smoothness of the surface of the connector.

There are two forms of loss or attenuation testing, insertion loss testing and optical time domain reflectometry (OTDR). In this chapter, you will learn how to perform both forms of testing and interpret the data obtained.

THE INSERTION LOSS MEASUREMENT PROCEDURE

Requirements and Features of Test Equipment

Attenuation testing of installed cable and connector systems is performed to simulate, as closely as possible, the operating conditions of the transmitting and receiving optoelectronics. Attenuation testing measures the end to end loss of the optical path. This loss includes attenuation by the fiber, loss from splices and connectors, and loss that occurs from any passive devices in the optical path (Figures 1–2 to 1–6). This testing is performed using a stabilized light source, an optical power meter, barrels, and low-loss reference test leads (several sets for each test, since test leads become damaged in use).

The stabilized light source (Figure 5–3) has five important characteristics:

- Feedback circuitry to maintain constant output
- A light source (LED or laser) with a wavelength close to that of the transmitter light source that will be used
- A light source with a spectral width close to that of the transmitter light source that will be used
- Optical launch conditions as close as possible to those of the optoelectronics
- A connector receptacle

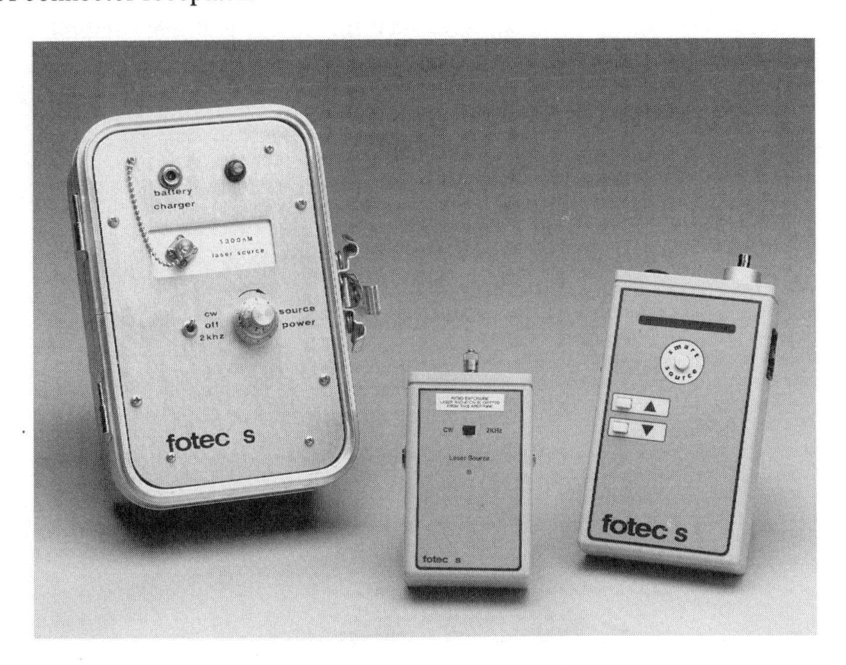

Figure 5–3 Stabilized Light Sources (Courtesy of Fotec Inc.)

These launch conditions are spot size and the angular distribution of the output of the light source. For simplicity, we will call this the NA of the light source. Most, but not all, commercially available test sets have launch conditions that simulate typical optoelectronic light sources.

The connector receptacle accepts the connectors on the test leads. The connectors on these leads are usually the same style as those on the cables being tested.

Stabilized light sources are available with different capabilities. The test light source can have a single wavelength, or two wavelengths, for multiple wavelength testing.[1] If the source has two wavelengths, each wavelength may have a separate receptacle, or both wavelengths may be fed through a coupler (Figure 1–2) to a single receptacle. A single receptacle is convenient, since you can perform testing of both wavelengths without changing the test lead at the light source.

Stabilized light sources have different features. A light source can have a fixed power output or a variable power output. A fixed output light source (which most test set manufacturers use) has the advantage of reduced cost. A variable power output is convenient to use, since it allows you to adjust the output power to a convenient reference value, such as 0 dBμ or 0 dBm.[2] Use of a 0 dB reference value means that the loss reading on the power meter will be the insertion loss of the cable system under test. If you cannot set a 0 dB reference value, you will need to record two values, the reference value and the test value. You will subtract these two values to determine the insertion loss. Some light sources offer the option of cycled output (1 kHz, 5 kHz, 10 kHz, etc.) in addition to continuous wave output.

The optical power meter (Figures 5–4 and 5–5) has five important characteristics:

- The range of power level that can be accurately measured
- The wavelengths at which the meter is calibrated
- The accuracy of the measurement
- The amount of drift that can occur during extended measurement times
- A connector receptacle

The range of power level is determined by the application. For telephone applications, this range is large, typically +3 dBm to -45 dBm (-45 dBm is 45

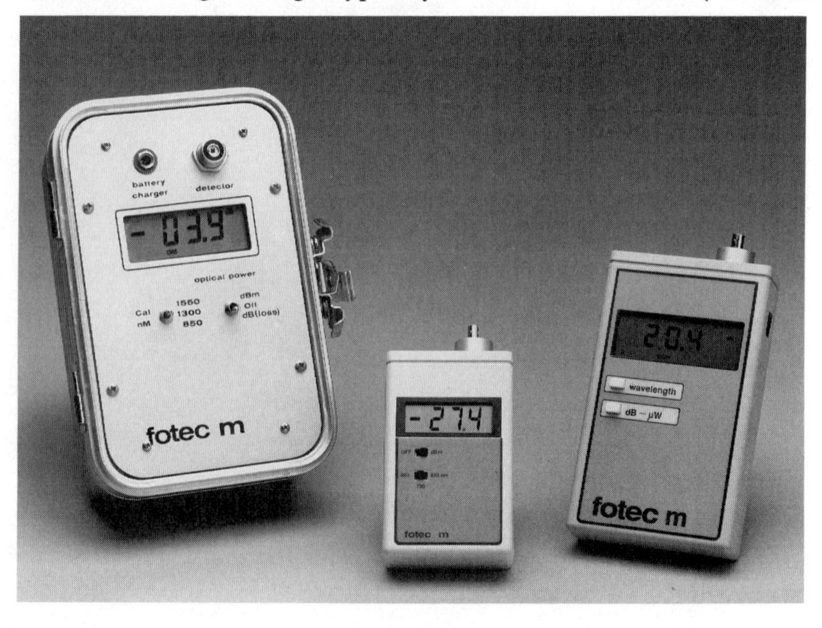

Figure 5–4 Power Meters (Courtesy of Fotec Inc.)

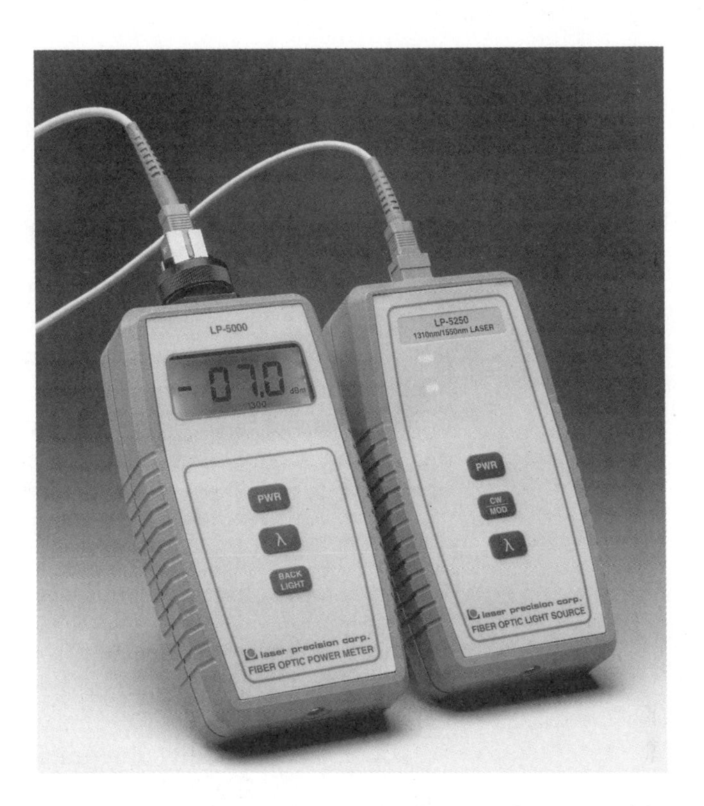

Figure 5–5 Power Meter and Stabilized Light Source (Courtesy of Laser Precision Corp.)

dBm below 1 milliwatt). For cable TV applications, this range is large, typically +20 dBm to -50 dBm. For local area networks and data communication applications, this range is smaller, typically 10 dBm to -60 dBm.

Power meters are available with different capabilities. Some power meters allow an offset to be applied to the power measurement in order to create a 0 dB reference signal. This feature is highly convenient, as long as the linearity and accuracy of the meter are not reduced.

Some power meters remember or store a reference signal level for future use. Some power meters have low battery indicators. The latest products available can store the power meter measurements for downloading to a computer or printer (Laser Precision Corp.) or directly transfer the measurements to a preformatted data file in an attached computer (Fotec).

Both power meters and light sources need to have long battery life: short battery life will result in interruption of the testing in order to replace batteries and establish a new reference signal. Both power meters and light sources need to be stable (i.e., not drift) with time in order to ensure measurement accuracy. Both power meters and light sources can have AC power options, which eliminate concern for battery life. The barrels are standard barrels used in patch panels.

The test leads must be 1–4 meters long in order to comply with the standard test technique. The leads must have low loss connectors with the same fiber type (singlemode or multimode), core diameter (9 μm, 50 μm, 62.5 μm, etc.), and NA (0.2, 0.275, 0.3, etc.) as the fiber in the cable being tested. Test leads have the same connector style (ST-compatible, SC, etc.) as on the cable being tested.

The insertion loss test technique contained herein is Method A of ANSI/EIA/TIA-526-14-1990. This method is commonly used in field testing of cables. Method A is applicable to testing of installed multimode cable plant, but is frequently used for testing singlemode cable plant. Method A of ANSI/EIA/

TIA-526-14-1990 is used because its results are realistic or slightly conservative measurements. The results from Method A can be conservative, since Method A overestimates the loss experienced by the optoelectronics if they are directly connected to the ends of the cable being tested. The results from Method B of ANSI/EIA/TIA-526-14-1990 are always conservative. (See Appendix 2 for an analysis of Methods A and B.)

If the cable being tested will be connected to optoelectronics via jumper cables from the cable ends to the optoelectronics, such jumpers should be included in the test. However, such jumpers are often not available. Fortunately, this lack of availability rarely causes problems, since the optical power budget of most optoelectronics is higher than needed. However, such jumpers need to included in the loss test when the maximum loss of the cable under test is expected to be close to the optical power budget available from the optoelectronics.

The Five Step Procedure

The five steps of the insertion loss measurement procedure are as follows.

1. Set an initial, or reference, power level.
2. Measure the insertion loss in both directions.
3. Compare the measurements to expected loss.
4. Determine variability or repeatability of the loss measurement.
5. Check the reference power level.

We present Steps 1, 2, 3, and 5 in Procedure 1 in this chapter. We present Step 4 in Procedure 2. Procedure 1 is used for testing installed cable plants. Procedure 2 is used for practicing testing and learning about the variabilities and problems that can occur during testing.

Normal practice requires making a loss measurement in both directions. Multiple measurements in both directions are advisable for some, but not all, links in the system in order to determine variability. Knowledge of variability will be required during maintenance testing and troubleshooting, as presented later in this chapter.

Tools and Supplies Required

light source with receptacle of the same connector style as on the test lead (Fotec FS370 or Laser Precision LP 5150)

power meter with receptacles of the same connector style as on the test lead (Fotec FM310 or Laser Precision LP 5025)

two test leads 1–5 meters long with the same connector style as on the cable to be tested (we will call them "lead in" and "lead out" cables to distinguish them from the cables under test)

two barrels or bulkheads for connector style under test

power supplies and/or new batteries for light source and power meter

Scotch™ tape or other connector cleaning supplies

lens, or optical, grade compressed air

cable assemblies (lengths of cable with connectors at both ends) to be tested

PROCEDURE 1: MEASURE LOSS AND DIRECTIONAL DIFFERENCES

1. Obtain all equipment. Clean all test lead connectors. Inspect all test lead connectors with a microscope to ensure that they look good (Procedure 5, Chapter 4).
2. Make appropriate entries at the top of the test data form (Table 5–1). Use a separate form for each combination of fiber type and connector style you test. Enter the optical appearance of both connectors for each fiber you intend to test. Use G for a good connector and NG for a connector that does not look good.
3. Inspect the barrel for dirt. If necessary, clean out with lens grade compressed air.
4. Connect lead in and lead out cables, barrel, power meter and source as shown in Figure 5–6. After the lead in and lead out cables are attached, do not disconnect the connectors directly connected to the power meter and light source, unless you start this procedure at Step 1.
5. Turn on the source; if necessary, set the wavelength switch to 850 nm (or to the wavelength at which you are testing).
6. Turn on the power meter; set the wavelength switch to the same value as on the light source. Set the readout to dBm or dBμ.
7. Set and record the initial value. There are four methods for setting the initial value. Choose the method suitable to the equipment you are using.

Connector type:__________ Manufacturer:__________

Fiber type: _____/_____ Source/meter vendor: _____

Wavelength: _____

Note: If initial value is not zero, record initial value here: _____

	Appearances		Measurements		Calculated Loss	
Cable ID	End 1	End 2	First	Opposite	Eq. 5–1	Eq. 5–2
_____	_____	_____	_____	_____	_____	_____
_____	_____	_____	_____	_____	_____	_____
_____	_____	_____	_____	_____	_____	_____
_____	_____	_____	_____	_____	_____	_____

Check of initial value: _____ (Figure 5–6)

Table 5–1 Insertion Loss Data Form

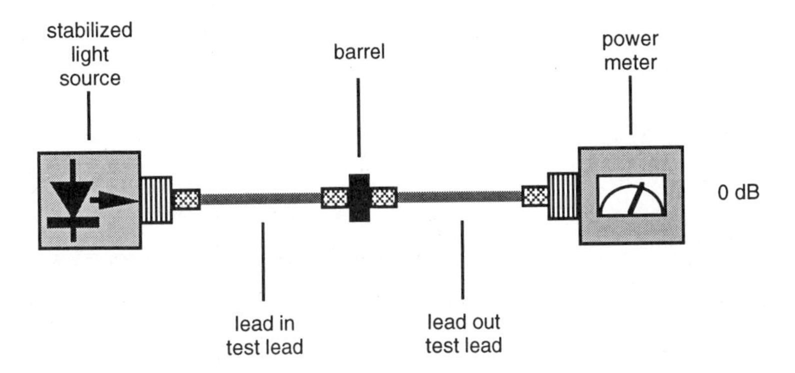

Figure 5–6 The Test Procedure for Establishing a Reference Power Level

7a. If your equipment has a fixed output light source, record the initial value.

7b. If your light source has an adjustable light source, adjust the output power so that the meter reads a convenient value. For the Fotec S series, the value is 0 dBμ.

7c. If your power meter has a reference feature, push the reference button or key according to the instructions from the manufacturer so that the meter reads 0 dB. (Note the reference power level in case you lose the reference level offset.)

7d. If your light source will adjust itself to a fixed value when connected to the power meter with a copper cable, follow the instructions of the manufacturer to set the initial reference signal.

8. Disconnect the end of the lead in cable from the bulkhead; connect a barrel to the lead in cable; connect both ends of the cable under test to the barrels. The equipment will be set up as shown in Figure 5–7.
9. Read and record the power meter reading on the data form (Table 5–1) in the column labeled Measurement/First.
10. Disconnect your first cable from the barrels.
11. Reverse the ends of your cable. Reconnect your first cable to the barrels as shown in Figure 5–8.
12. Record the power meter reading on the data form (Table 5–1) in the location labeled Measurement/Opposite.
13. Calculate the maximum and expected losses (Equations 5–1 and 5–2). To perform these calculations, you calculate the maximum loss and the expected loss with Equations 5–1 and 5–2.

(Equation 5–1)

Maximum loss = (maximum loss/km of the cable) × (length of the cable) + (maximum loss/connector pair) × (number of connector pairs)

(Equation 5–2)

Expected loss = (actual loss/km [or typical loss, if actual is unknown] of cable) × (length of cable) + (typical loss/connector pair) × (number of connector pairs)

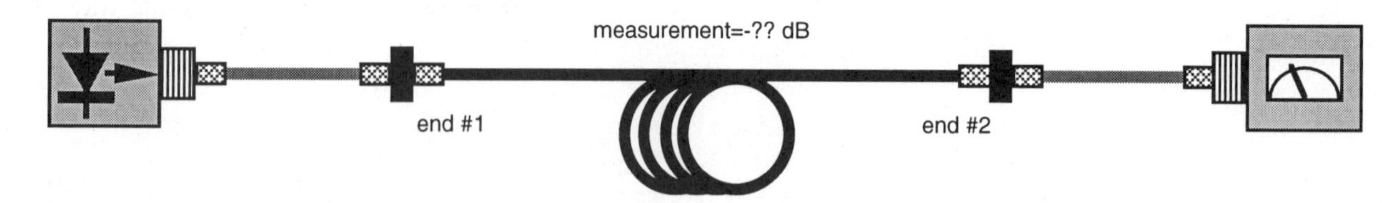

Figure 5–7 The Test Procedure for an Insertion Loss Measurement

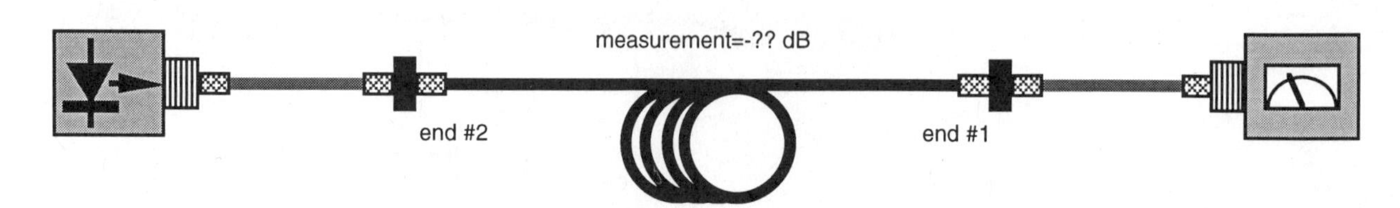

Figure 5–8 The Test Procedure Measuring Directional Effects on Insertion Loss

If splices are also present, add the product of the number of splices and the maximum acceptable splice loss to Equation 5–1 and the product of the number of splices and the average splice loss to Equation 5–2. When using the value dB/pair, be certain that the connector manufacturer states the performance of the product as db/pair, not dB/connector. If the manufacturer uses the dB/connector rating, estimate the dB/pair by doubling the dB/connector value.

The maximum loss calculation (Equation 5–1) determines the maximum loss you can expect if the cable and connectors are installed properly. The expected loss calculation (Equation 5–2) determines the average loss you can expect if the cable and connectors are installed properly. If the measured loss is closer to the expected loss than to the maximum loss, the cable and connectors are likely to be properly installed. (If a cable has a low attenuation rate, low loss connectors, and excess loss due to improper installation, the measured loss can be less than the maximum loss.) Enter these values on the data form (Table 5-1).

14. If you followed Step 7a, subtract the initial value from each of your loss readings to determine the loss measurements. Compare your insertion loss measurements to the maximum and expected values. How do your measurements compare to the maximum and expected measurements specified by the manufacturer?
15. Repeat Steps 8–14 on any additional cable assemblies you wish to test.
16. Disconnect the cable under test from the barrels. Remove one barrel. Reconnect the lead in and lead out cables as shown in Figure 5–6. Record your result as "Check of initial value" (Table 5–1).
17. Repeat the entire procedure at a different wavelength (optional).

Questions about Procedure 1

Is the insertion loss the same in both directions? Are the initial value and check of initial value the same? If not, how large is the difference? Did you obtain positive insertion loss measurements?

In most cases, the insertion loss will be different in both directions. There are three causes of this directional effect. The first cause of directional effect is mechanical: the same two connectors will not align exactly the same way when removed from and inserted into the same barrel. This difference between successive measurements of the same connectors is called repeatability or reproducibility.

The difference due to this cause is small (0.1–0.3 dB) with keyed connectors, but can be large (0.5 dB) with unkeyed connectors (Table 1–7). Repeatability can result in the difference you observe between the initial value and the check of the initial value. (Dirt on or damage to the connectors of the lead in and lead out of cables and weakened batteries in the light source and power meter are other potential causes.)

The second cause of the directional effect is a lack of perfect concentricity of the connectors on the lead in and lead out cables. All connectors are slightly imperfect. The most common form of imperfection is offset (Figure 1–38 a). There are three causes of this offset:

- The fiber hole can be slightly off center in the ferrule
- The fiber can be slightly off center in the fiber hole
- The core can be slightly off center in the clad

As a result of this offset, a connector under test can align with the connector on the lead in cable better or worse than with the connector on the lead out cable. A directional difference will result.

This second cause (lack of perfect concentricity of the lead in and lead out cables) can result in positive insertion loss measurements of short cables. If the connectors on the cable under test align better to the connectors of the lead in and lead out cables than the connectors of the lead in and lead out cables align with each other, you will measure a positive insertion loss.

Positive insertion losses indicate a potential problem with the lead in and lead out cables. If you frequently measure positive insertion losses, you should replace the connectors of the lead in and lead out cables with the lowest insertion loss connectors you can obtain. (Pearson Technologies uses 3M Hot Melt connectors for multimode lead in and lead out cables.) Occasional positive insertion loss measurements are inevitable.

The third cause of directional effects is found in long lengths of multimode fibers. This direction effect is caused by *differential modal attenuation* (DMA). DMA occurs because different rays of light travel different paths.

To understand DMA, consider axial and critical angle rays (Figure 1–8). The axial rays travel the shortest distance in a cable; the critical angle rays, the longest. Axial rays will be attenuated the least; critical angle rays, the most. The critical angle rays are most likely to be at or near the core-clad boundary at a connector pair and at a splice.

Most connector insertion loss is due to core offset (Figure 1–38 a). If the critical angle rays are relatively weak when they reach a connector pair with an offset, there will be little light energy that can be lost; if they are relatively strong, there will be some light energy that can be lost. In other words, if the critical angle rays are relatively weak, the insertion loss of a connector pair will be low; if relatively strong, the insertion loss will be higher.

Consider the cable in Figure 5–9. It consists of two lengths of cable connected at a patch panel. The patch panel is much closer to End 2 than to End 1. When the light source is connected to End 2, the critical angle rays reaching the patch panel connectors are stronger than they will be when the light source is connected to End 1. Thus, the insertion loss measurement with the light source at End 1 will be lower than that with the light source at End 2.

In summary, directional effects require measurement of insertion loss in both directions. In addition, the first cause of directional difference can result in differences between successive measurements of the same cable with the same lead in and lead out cables in the same direction. In order to interpret insertion loss measurements during periodic maintenance checks of the insertion loss of installed cable plants and during troubleshooting activities, you will need to quantify these changes.

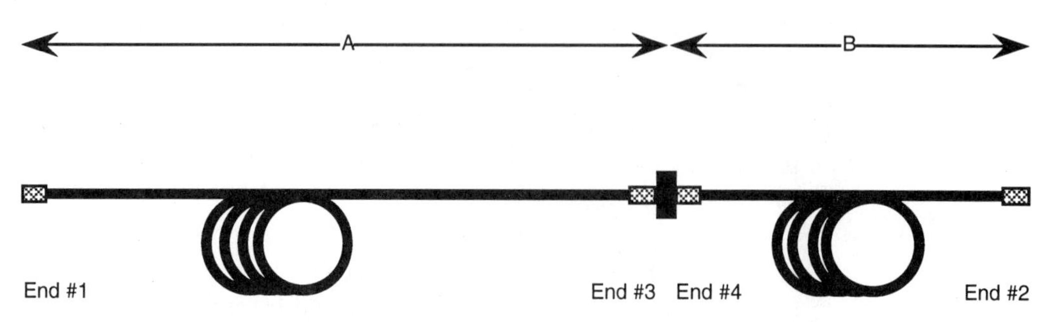

Figure 5–9 A Cable Plant That Will Exhibit Directional Differences Due to DMA

For example, if you are performing a maintenance insertion loss measurement of the cable plant in Figure 5–9, you may determine that the current insertion loss is 1.5 dB higher than the value after the cable plant was initially installed. Is this increased insertion loss indicative of a problem that requires troubleshooting? The answer is: it depends on the range of measurements you can expect from the connectors under test. In Procedure 2, we present a test technique with which you can determine this range. We will return to Figure 5–9 after you have performed Procedure 2.

Procedure 2: Measure Range of Insertion Loss

1. Use Table 5–2 instead of Table 5–1. Repeat Steps 1–9 of Procedure 1. Record the first measurement in Table 5–2 on the first line of the column labeled "Measurements/First."
2. Disconnect both ends of the cable under test from the barrels. Reconnect both ends of the cable under test to the barrels. Record the result in Table 5–2 on the second line of the column labeled "Measurements/First" of the data form. (Whenever you disconnect and reconnect unkeyed connectors, such as the SMA style, randomly rotate the connectors relative to the barrels. This stimulates a worst case situation. You will not be able to do this with the ST-compatible connectors.)
3. Repeat Step 2 two more times. Record these two additional measurements in the third and fourth lines of the column labeled "Measurements/First."
4. Disconnect your first cable from the barrels.
5. Reverse the ends of your cable; reconnect your first cable to barrels as shown in Figure 5–8.
6. Record the meter reading in Table 5–2 on the first line of the column labeled "Measurements/Opposite" of the data form.
7. Disconnect both ends of the cable under test from the barrels. Reconnect both ends of the cable under test to the barrels. Record the meter reading in the second line of the column labeled "Measurements/Opposite" of the data form (Table 5–2).
8. Repeat Step 7 two more times. Record these two additional measurements in the third and fourth lines of the column labeled "Measurements/Opposite."
9. Disconnect the cable under test. Reconnect the lead in and lead out cables as shown in Figure 5–6. Record your result as "Check of initial value."

Connector type:________________ Manufacturer:________________

Fiber type: _____/_____ Source/meter vendor: _________

Wavelength: _________ Cable ID_________

Note: If initial value is not zero, record initial value here: _____

Measurement	Appearances		Measurements		Calculated Loss	
Number	End 1	End 2	First	Opposite	Eq. 5–1	Eq. 5–2
1	_____	_____	_____	_____	_____	_____
2	_____	_____	_____	_____		
3	_____	_____	_____	_____		
4	_____	_____	_____	_____		

Check of initial value: _______ (Figure 5–6)

Table 5–2 Range Measurement Data Form

10. Calculate the ranges in both directions. The range in one direction is the difference between the maximum and minimum values in the column labeled "Measurements/First." The range in the opposite direction is the difference between the maximum and minimum losses in the column labeled "Measurements/Opposite" of the data form (Table 5–2).

Questions about Procedure 2

Were all measurements in both directions the same? In most cases, the measurements are not the same. If the connectors are keyed connectors (e.g., ST-compatible styles and SC style), the range can be from 0.0-0.3 dB. If the connectors are unkeyed (e.g., 906 SMA), the range can be 1.0 dB.

We can now use this measurement of range to determine whether the cable plant in Figure 5–9 needs further troubleshooting. You determined that the current insertion loss measurement is 1.5 dB higher than the initial measurement. We will assume that the connectors in Figure 5–9 are keyed ST-compatible connectors with a range of 0.3 dB/pair (Table 1–7). Since there are two pairs in Figure 5–9, the maximum range you can expect is 0.6 dB (a worst case assumption). With this information, you can conclude that the increase of 1.5 dB is outside of the range you can expect from normal behavior of the connectors. Therefore, troubleshooting is required to determine the location of the increased loss.

Let us examine Figure 5–9 with a different assumption: the connectors in this cable plant are unkeyed, 906 SMAs. These products can have an average range of 1 dB/pair (Table 1–7). Since there are two pairs in Figure 5–9, the expected range is 2.0 dB. With this information, you can conclude that the increase of 1.5 dB is less than the range you can expect from normal behavior of the connectors. Therefore, troubleshooting is not required.

Verify the Six Insertion Loss Rules of Thumb

You can verify the six rules of thumb after you have performed Procedure 2 on at least ten cables. For each cable assembly tested, calculate the average insertion loss measurement in both directions. For each cable assembly tested, calculate the range in insertion loss measurements in both directions. Enter the data on Table 5–3: enter data from cables with two good connectors from the top; enter data from cables with one or two bad connectors from the bottom. Review all the data in Table 5–3. Then answer the questions that follow.

Cable #	First Direction		Opposite Direction		Both Good?
	Average	Range	Average	Range	
1	____	____	____	____	____
2	____	____	____	____	____
3	____	____	____	____	____
4	____	____	____	____	____
5	____	____	____	____	____
6	____	____	____	____	____
7	____	____	____	____	____
8	____	____	____	____	____
9	____	____	____	____	____
10	____	____	____	____	____

Table 5–3 Summary Insertion Loss Data for All Cables Tested

Rule of Thumb 1. If the connector looks good, does it test good? If a connector looks good, it will usually test good. Testing good means an insertion loss measurement less than the maximum insertion loss (Equation 5–1). If a connector looks good but tests bad, the problem is usually, but not always, in the test procedure. Most measurement problems will be revealed by the final "Check of the initial value." If this check reveals a difference of >0.3 dB (for keyed connectors), there is likely to be a measurement problem. See "How to Recognize Measurement Problems" in this chapter for guidelines on recognizing and indentifying the causes of measurement problems.

Rule of Thumb 2. If the connector looks good, does it test closer to the average for the product or closer to the maximum for the product? Calculate the average insertion loss of those cables with two good connectors. If you have performed the installation and insertion loss procedures correctly, the average insertion loss for connectors should be closer to the average than to the maximum. In Table 5–4, we present typical values obtained during training using tests performed according to Method A of FOSTD-526-14. The crimp connector was from Automatic Tool and Connector Co., Inc.

Rule of Thumb 3. You must develop and use a rule. The rule will be either: the insertion loss is always the same in both directions; or, the insertion loss will not always be the same in both directions. Based on your answer, what measurement step is advisable?

Rule of Thumb 4. If the connector looks bad, does it always test bad? Some connectors can have features that classify them as not good. However, not all features will cause a connector to exhibit excessive insertion loss. Since all features do not cause excessive loss, you will not replace such connectors until after you have performed insertion loss testing. A common practice is tagging connectors that do not look good. This enables you to distinguish between high loss measurements due to high loss connectors and high loss measurements due to measurement problems.

Through experience, you will learn to classify connectors by microscopic appearance into three categories: good, not good, and terrible. You will replace terrible connectors before insertion loss testing, because they always test high loss.

Rule of Thumb 5. Is each insertion loss measurement exactly the same as the previous measurement? It there some measurement variability? How large is this variability? When is variability important?

Rule of Thumb 6. Did you experience any measurement problems? Did you obtain any positive insertion loss measurements? If so, what might cause such measurements?

Product	Average dB/Pair	Stated Average dB/Pair	Number of Connectors
3M Hot Melt	0.33	0.30	286
AMP 503415	0.42	0.30	204
Crimp	0.63	0.50	234

Table 5–4 Average dB/Pair Obtained During Training with ST-Compatible Connectors on 62.5/125 Fiber

How to Recognize Measurement Problems

There are two types of bad measurements: positive losses and excessively high losses. Positive losses occur when the connectors on the cable under test align better with the lead in and lead out cables than the lead in and lead out cables align with each other. Positive losses can be occasional or frequent.

Occasional positive losses result from the normal measurement procedure. Unless you obtain the highest quality connectors for your test leads, you will experience occasional positive measurements. These positive measurements are a result of the relatively small imperfections, and/or variations in core diameter, NA, and eccentricity of the core in the ferrule. In addition, should the fiber in the test leads have either a relatively small core diameter or a relatively small NA, you will experience occasional positive losses.

Frequent positive losses have different causes in measurements of keyed and unkeyed connectors. In measurements of keyed connectors (FC/PC, D4, ST-compatible, and SC styles), inferior quality test leads can result in frequent positive losses. In addition, an initial reference measurement of contact connectors taken before the springs have completely set will result in frequent positive losses. If you take the initial reading immediately after inserting the connector into the barrel, the spring may take a few seconds to set completely. This delay in ferrule setting will result in a lower initial or reference power level. The low initial reading can result in frequent positive losses.

This delay in setting may be a result of dirty barrels. Dirty barrels can result from extensive use. You can clean dirty barrels with lens grade compressed air and/or lint free pipe cleaners moistened with 95 percent isopropyl alcohol.

In measurements of unkeyed connectors (such as 905 and 906 SMAs), frequent positive losses can reflect the normal performance of such connectors. In this case, the initial reference value may be obtained with the ferrules at close to the maximum loss position. To minimize positive losses with unkeyed connectors, you can use the average of several initial values as the initial value. Consistent positive losses are unusual and have the same causes as do frequent positive losses (above).

High losses can be increasing, occasional, or constant. Losses that increase with each measurement have three causes:

- Damage to the test leads
- Dirt or contamination on the core
- Low batteries

Damage can occur when excessive pressure is used to seat the connector in the barrel. Damage can also occur when non–pull-proof connectors are pulled by the cable and allowed to snap back into the barrel. Low batteries in either or both the light source and power meter can result in increasing loss measurements with increasing use.

Occasional high losses have four causes:

- Dirt either on the connectors being tested or on the connectors of the test leads
- Lateral pressure on the backshells of the connectors under test
- Axial tension on the cables near those connectors
- Incorrect identification of a connectors being good

If dirt or contamination is the cause of the high losses, cleaning and retesting all connectors should result in reduced losses. Barrels can become dirty after many tests.

If the connector style is not wiggle proof, lateral pressure will cause the ferrules to cock slightly. This cocking will result in increased loss. You can avoid this with careful attention to the cables used during testing.

If the connector style is not pull-proof, axial tension on the cable can cause the ferrules to separate. You can avoid this cause of occasional high loss with careful attention to the cables during testing.

Incorrect identification of a connector can result in an unexpected high loss measurement. If you obtain high loss measurements, check your connectors to ensure that they are good. Good connectors have a round, clear, featureless core at the surface of the ferrule. You will recognize a core that is below the surface from the shadow of the fiber hole that appears on the cladding.

High losses that remain constant are most frequently caused by damage to the test leads. This damage can be identified through microscopic inspection or by a test in both directions made according to Method B of FOSTD-526-14.

If the connectors look good, check the initial reference reading (Step 16 of Procedure 1). If the reference reading is close to the initial value (0–0 3.dB for keyed connectors; 1 dB for unkeyed connectors), your testing equipment is functioning properly. The problem is in the connectors or the cable.

Significant differences between initial and check of initial readings have three causes: low initial reading, low batteries, or contamination or damage to the test leads.

HOW TO MAKE AND INTERPRET OTDR MEASUREMENTS

After performing an insertion loss test, you may observe that the measured insertion loss is greater than the maximum loss (Equation 5–1). You will want to identify the location of the excess loss. How will you proceed? The high loss could be in either of the connectors or in the cable between the connectors. How will you find the location(s) of high loss? An insertion loss test will not provide you with this information.

However, an optical time domain reflectometer (OTDR, Figures 5–10 and 5–11) will provide this information with a picture of almost the entire length of the

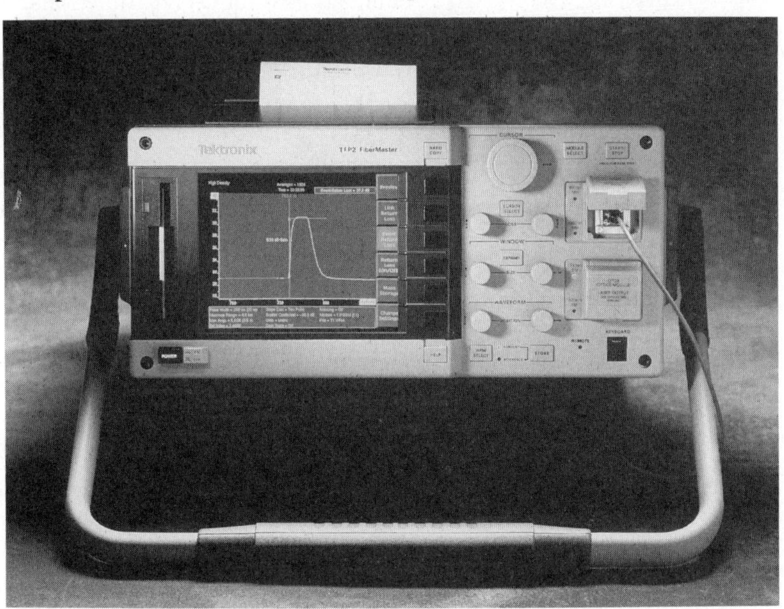

Figure 5–10 An Optical Time Domain Reflectometer (Courtesy of Tektronix Inc.)

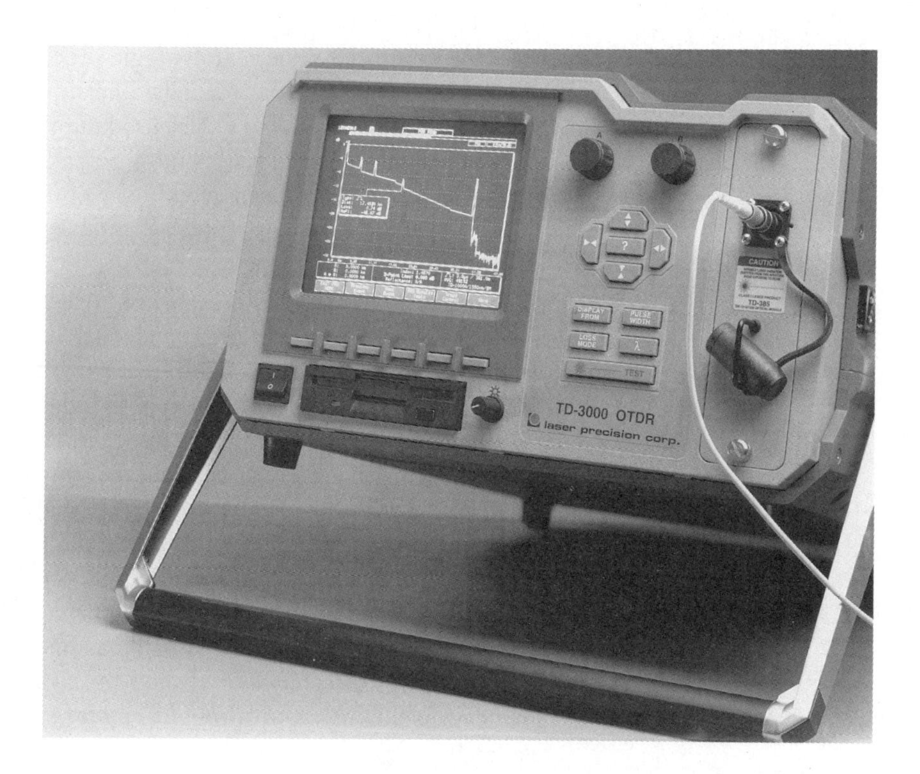

Figure 5-11 An Optical Time Domain Reflectometer (Courtesy of Laser Precision Corp.)

cable plant. With this picture, and the associated data, you will be able to determine the exact location of the high loss. This picture will include the data you need to measure distances and losses. Losses include: cable loss, splice loss, connector loss, and microbend loss. With some OTDRs, you can also measure reflectance, a so-called back reflection, or return loss.

How the OTDR Functions

To understand how an OTDR functions, we must return to the basics of the fiber. In Chapter 1, you learned that light is lost from the core when light strikes atoms and is scattered outside the critical angle (Rayleigh scattering). As shown in Figure 1–11, some of the light is scattered back towards the input end within the critical angle of the fiber. This light will experience total internal reflection as it travels back towards the input end of the fiber. An OTDR launches light into a fiber and measures this back scattered light.

An OTDR consists of five basic components (Figure 5–12):

- A high-power laser light source
- A coupler-splitter
- A detector
- Signal averaging electronics
- A connector interface

The laser creates a brief pulse of light which travels through the coupler to the cable under test. At each point along the fiber, a very small portion of this light experiences back scattering. This back scattered light travels back to the OTDR and is directed through the coupler to the detector. The detector creates an electrical signal of intensity vs. time, which is distance. The averaging electronics create an average of this signal along the length of the cable.

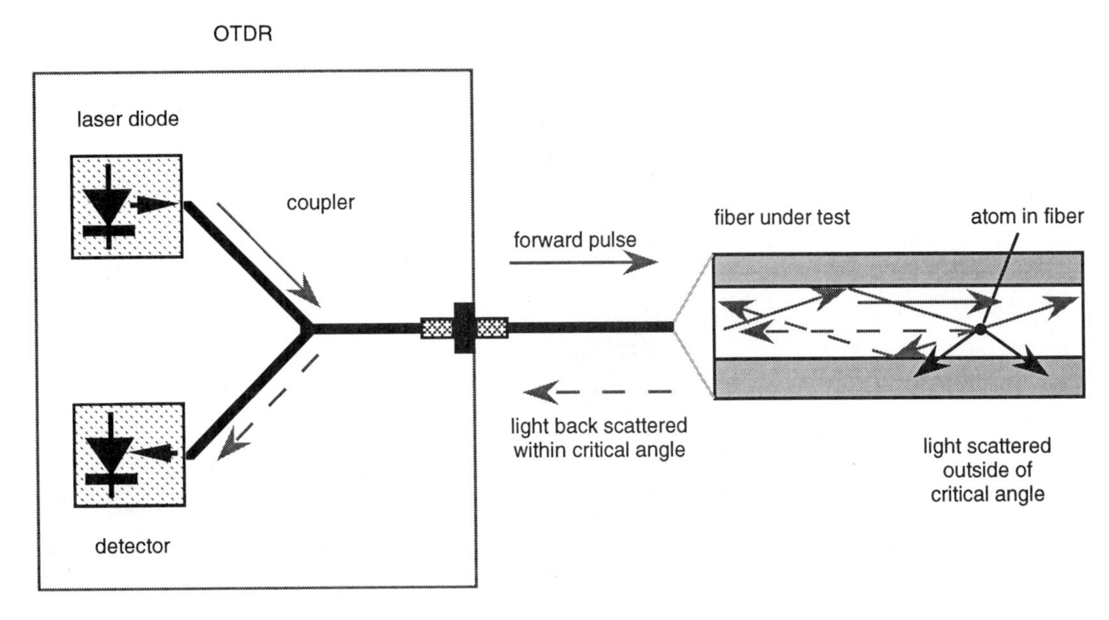

Figure 5-12 Basic Mechanisms by which an OTDR Functions

Assume that there is a cause of loss in the cable. The cause could be a connector, splice, microbend, macrobend, or attenuation of the signal by the fiber. The intensity of light returning to the OTDR from before the cause of loss will be higher than the intensity of light returning to the OTDR from after the cause of loss. By measuring the difference between these two intensities, we can determine the loss of the connector, etc.

The Four Basic Features of an OTDR Trace

The laser is pulsed repeatedly, thousands of times, creating thousands of values of back scattered light intensity for each location along the cable. These values are averaged to create an OTDR trace.

All OTDR traces have four features (Figure 5–13). The first feature is the initial peak followed by a rapidly decaying peak.

This pulse and decaying peak are a result of the Fresnel reflection from the connectors that connect the OTDR to the cable (Figure 1–38 h). Any portion of the cable within the initial peak cannot be examined with the OTDR, since the peak is characteristic of the OTDR and not the cable attached to the OTDR. This region is called the dead zone or blind zone. It is possible to inspect a cable by using a test or lead in cable longer than the width of this peak. It is also possible to "see" this region of a cable by using the OTDR from the opposite end of the cable. Both techniques are used, though attaching a lead in cable is advisable: connectors can be damaged during use; it is less expensive to replace one on the lead in cable then to repair an OTDR.

The second feature is a straight region, which is the back scattered signal from the cable under test. From this region, we obtain all of the useful information about the loss of the cable system.

The third feature is an end peak, which is a Fresnel reflection from the end of the cable. Note that this reflection has a width, or length. This width is similar to that of the dead zone. The width of this reflection places a limit on the ability of an OTDR to resolve closely spaced reflective features. If two connectors, two mechanical splices, or two breaks in tight buffer tube cables are closer than the width of this reflective event zone, the OTDR will reveal a single feature, not two separate features.

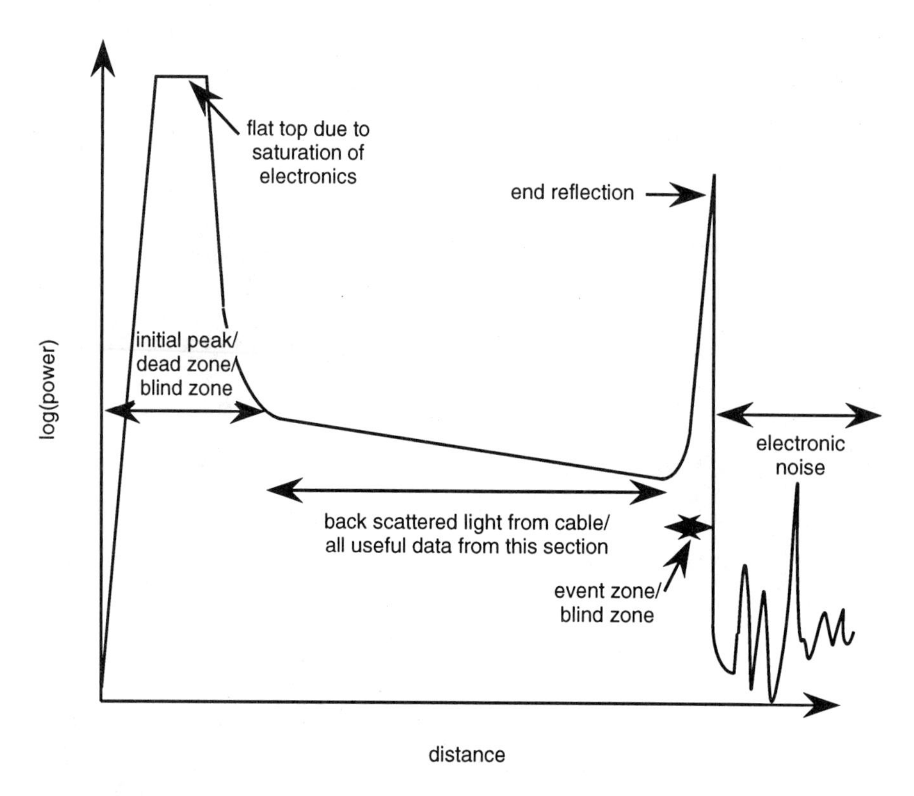

Figure 5–13 The Five Basic Features of the OTDR Trace

The fourth feature is the noise signal, which appears after the end reflection at the right end of the trace. This noise signal will depend upon the condition of the OTDR, the strength of the signal coupled into the fiber, and the total loss in the fiber under test.

Ten Typical Traces and Interpretations

An OTDR trace will rarely look like Figure 5–13. Instead, traces will exhibit specific features (Figures 5-14 to 5-22).

- No end reflection
- Accurate measurement
- Inaccurate measurement due to high loss connector at OTDR
- Inaccurate measurement due to high total loss
- Improper launch
- Low attenuation rate
- High attenuation rate
- Uniform loss
- Non-uniform loss without Fresnel reflection/peak
- Non-uniform loss with Fresnel reflection/peak

No End Reflection. Under certain circumstances, the end reflection may not appear (Figure 5–14). There are two possible causes of Figure 5–14: the far end of the fiber may have a badly shattered end. A second possibility is a cable end, in which the bend radius has been violated for a significant length. In both cases, all of the reflected light (a Fresnel reflection) will be directed outside the critical angle. No end peak will appear.

Accurate Measurement. When setting up an OTDR, it is important to launch as much optical power into the fiber as possible in order to obtain accurate and reproducible measurements. It is possible to identify the relative quality of loss measurements obtained from an OTDR by comparing the difference between the signal level at the end of the fiber under test to the level of noise, which occurs after the end reflection. The larger the difference, the more accurate and reproducible the measurements (Figure 5–15).

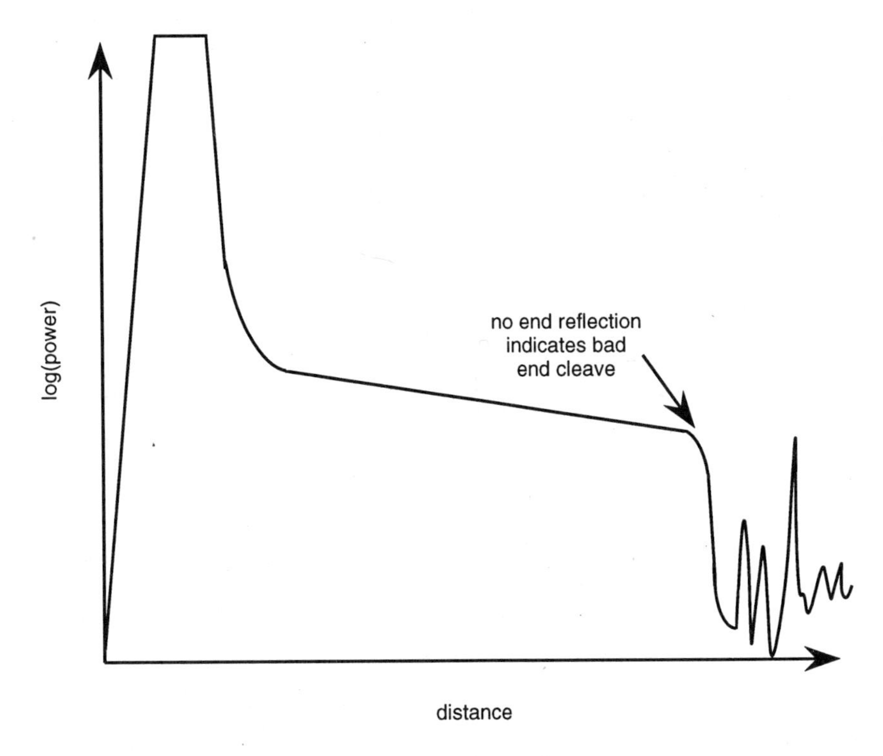

Figure 5–14 An Unusual Trace without an End Reflection

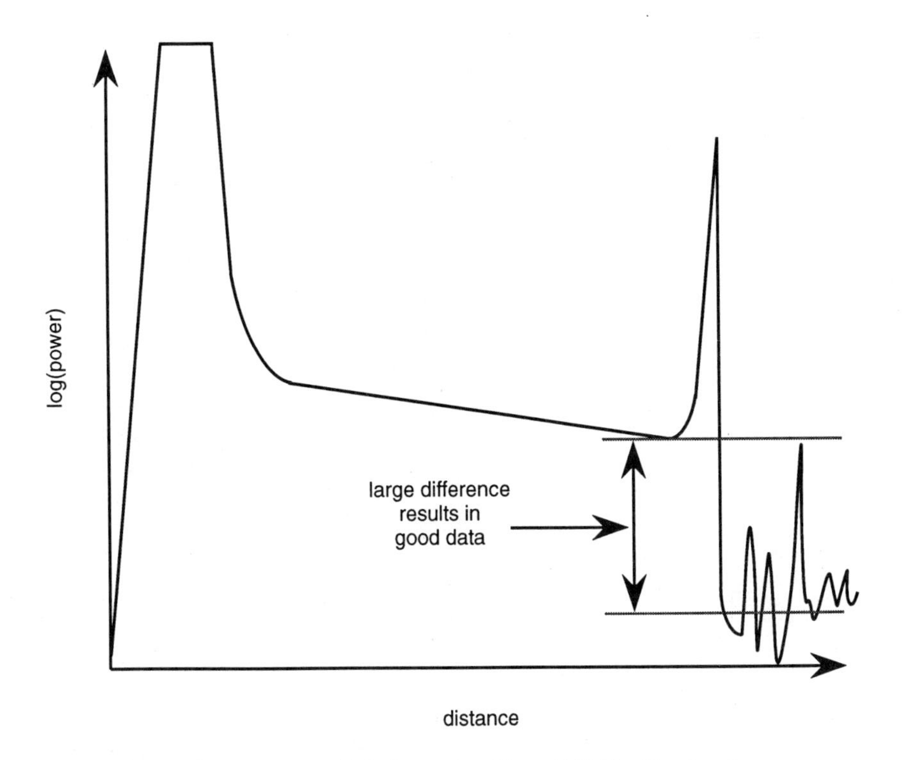

Figure 5–15 An Trace from which Accurate Data Will Be Obtained

Inaccurate Measurement. If the connector attached to the OTDR is high loss, the power level at the end of the fiber can be low enough to produce inaccurate measurements (Figure 5–16). In addition, if the total loss (the sum of the cable attenuation, connector losses, and splice losses; also known as "link loss") of the cable being measured is high, close to the loss limit of the OTDR, you will obtain inaccurate measurements (Figure 5–17).

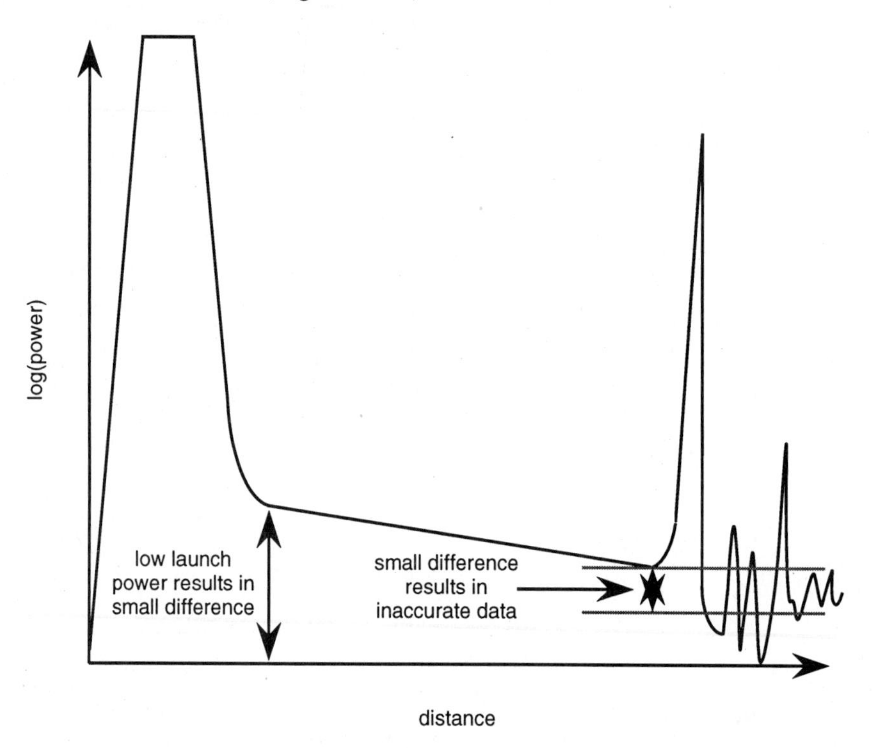

Figure 5–16 Inaccurate Data Will Be Obtained from this Trace Due to Low Launch Power

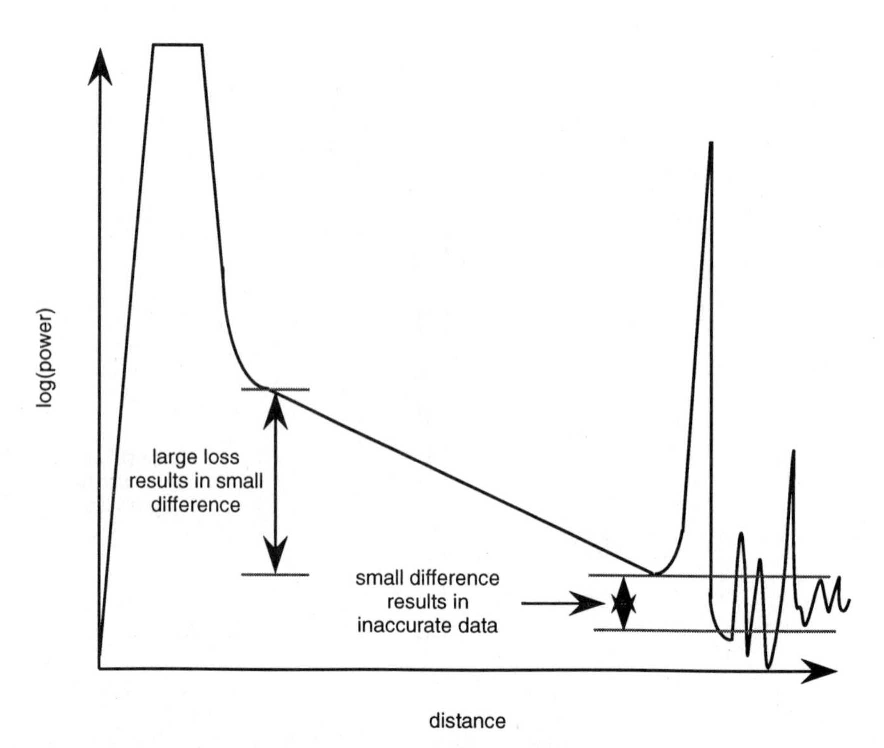

Figure 5–17 Inaccurate Data Will Be Obtained from this Trace Due to High Loss

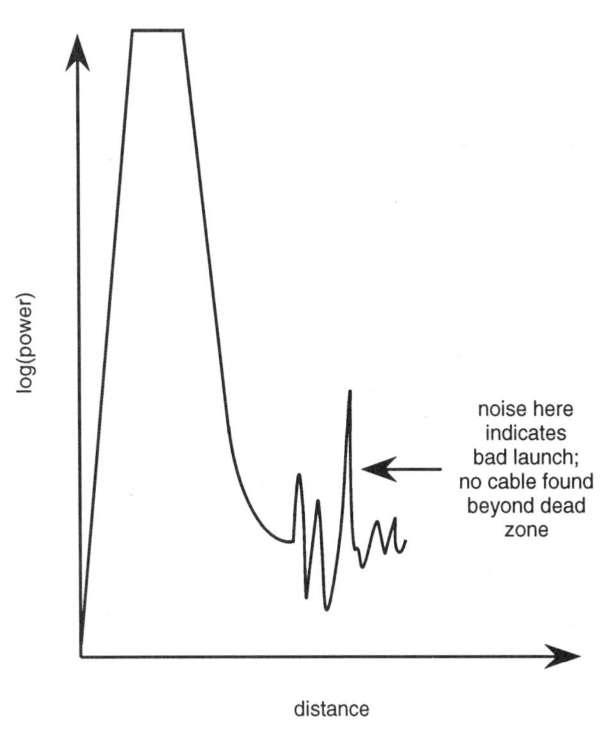

Figure 5–18 A Trace with No Backscatter

Improper Launch. It is possible to set up the OTDR incorrectly, such that light is not properly launched into the fiber. Then the trace will look like that in Figure 5–18. Note that there is no straight back scatter region. Lack of such a straight region indicates that no light is coming back to the OTDR from the fiber. This trace results from a broken connector connected to the OTDR or a break in the fiber within the optical dead zone.

Low and High Attenuation Rates. With a good launch, you can make qualitative determinations. Low attenuation rate fibers and high attenuation rate fibers are shown in Figures 5–19 and 5–20. Such a comparison requires the same horizontal and vertical scales for both traces.

> ***Rule of Thumb:*** *For OTDR Trace Interpretation, a properly designed, properly manufactured, properly installed cable will always have a straight line trace. Any deviation from linearity indicates a connection or installation problem.*

Uniform Losses. Once you have a proper launch, you can distinguish between uniform and non-uniform losses. If there are no splices or connectors in a cable system, you will always see a straight trace, which indicates uniform loss (Figure 5–13).

Non-Uniform Losses. Non-uniform losses may or may not have reflections (Figure 5–21). A non-reflective, non-uniform loss has at least four possible interpretations. Such a loss can result from a bend in the cable to below its minimum recommended bend radius. Such a loss can result from a localized crushing of the cable, from a cable tie that was excessively tightened on a tight tube, low fiber count (or small) cable. Such a trace can result from a short section of cable under excessive tension or excessively high or low temperature. In summary, any violation of the performance characteristics of the cable can result in a non-reflective, non-uniform loss.

Such a non-uniform trace can result from a fusion splice in the cable. Since a fusion splice does not have an air gap, there is no Fresnel reflection.

A reflective non-uniform loss (Figure 5–22) has five possible explanations. This trace can result from two sections of cable connected with connectors. The presence of connectors always results in a Fresnel reflection.

This trace can result from a mechanical splice. Mechanical splices have an index matching gel that approximates the index of refraction of the cores of the fibers, but does not exactly match this index. Because the match is not exact, there is some Fresnel reflection.

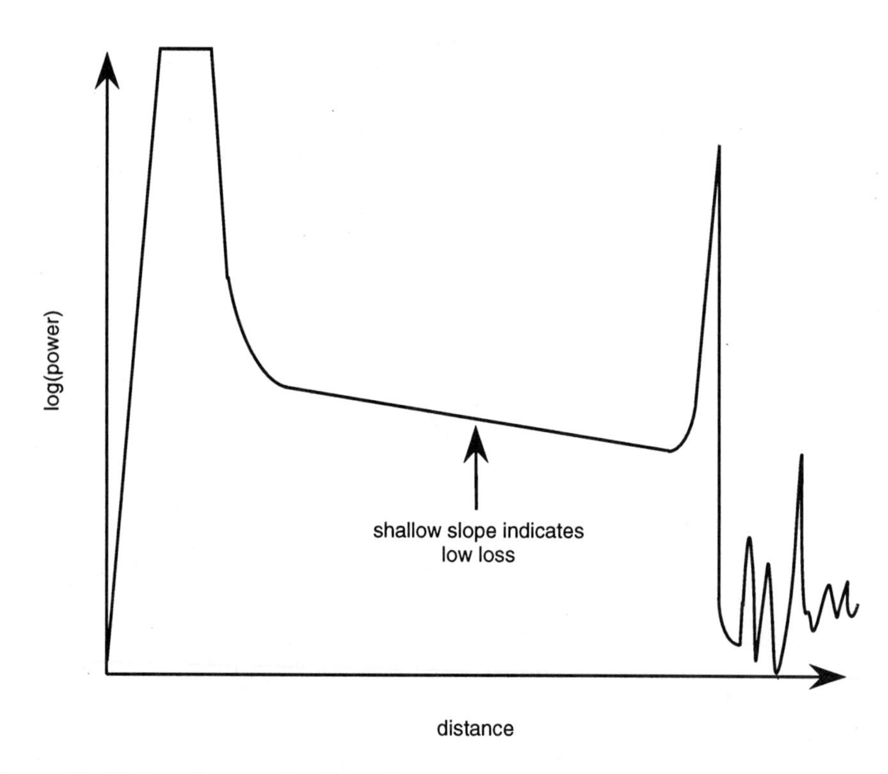

Figure 5–19 Low Attenuation Rate Trace

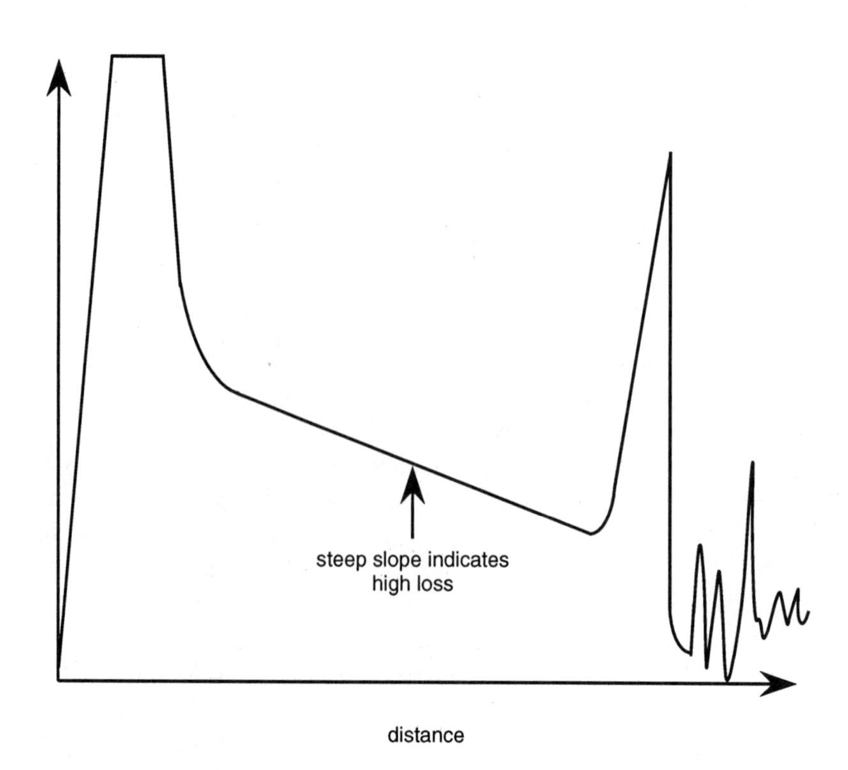

Figure 5–20 High Attenuation Rate Trace

This trace can result from a poorly made fusion splice. If the fusion splice has an air gap or a gas bubble, a Fresnel reflection will result.

This trace can result from a broken tight tube cable (Figure 1–28). In this situation, the tight buffer tube has forced the ends to remain in contact, but that contact will not be perfect.

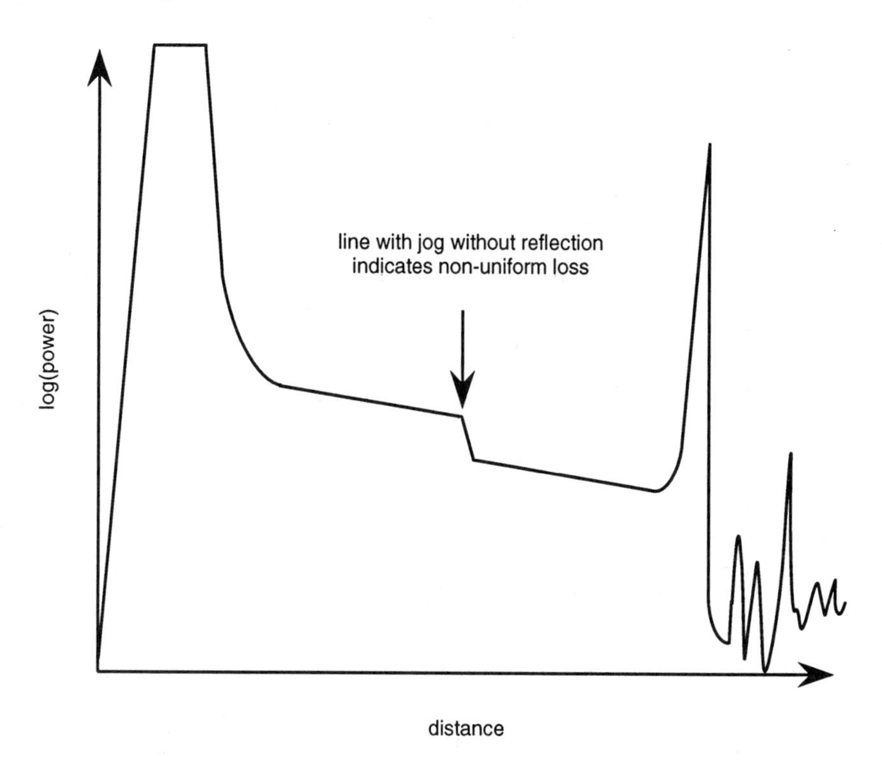

Figure 5–21 Non-Uniform Loss Trace, without a Reflection

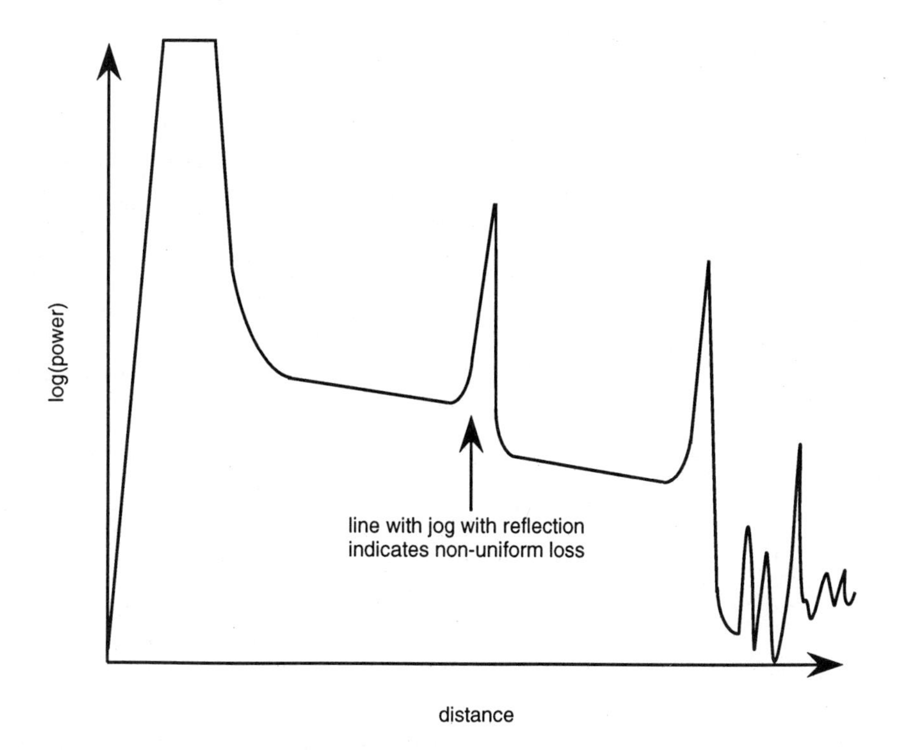

Figure 5–22 Non-Uniform Loss Trace, with a Reflection

The fifth possible interpretation is that this trace has multiple reflections (echoes) from a single length of cable. Echoes are created as follows: The pulse of light from the OTDR travels to the end of the cable. Most of the light energy exits, but some experiences a Fresnel reflection back to the OTDR. When this light travels back to the connector pair at the OTDR, most of the light passes into the OTDR, creating the end peak. But some of this light experiences a Fresnel reflection at the OTDR, causing some light to travel back towards the fiber end (a second trip). When this light reaches the end of the fiber, some experiences another Fresnel reflection (the third for this pulse). This light energy travels back to the OTDR, creating a second peak on the trace.

By this process of multiple Fresnel reflections, the high energy pulse from an OTDR can be reflected several times from the end of a cable. This phenomenon occurs in both multimode and singlemode cables.

Such multiple reflections can usually be easily identified by two characteristics: the lengths of each segment of cable are exactly equal (± 2m on a high quality OTDR such as the Tektronix Fiber Master™) and the loss at the center non-uniformity is high relative to the loss of a properly installed connector or splice.

Echoes can occur in cable segments of unequal length. In this case, the multiple reflection will not appear in the center of the segment.

How to Set Up an OTDR

The procedure for setting up an OTDR depends on the OTDR being used. However, some of the steps are common to all OTDRs. After turning on the OTDR, clean the connectors on both ends of the lead in cable. Select and input the values of the wavelength of testing, the index of refraction, the length measurement (feet or kilometers), and the pulse width.

The wavelength of testing is the wavelength of the transmitting optoelectronics. The index of refraction is the index of refraction of the fiber in the cable.

Since the horizontal axis of an OTDR trace is, in reality, time, the OTDR must be calibrated to the fiber under test. This calibration is provided by two factors: the index of refraction of the fiber under test and the difference between fiber length and cable length.[3] Should you know the manufacturer of the fiber in the cable under test, you can use the values in Table 1–1. Should you not know the index of refraction or the manufacturer of the fiber, you can use the values in Table 5–5, which are mid-points of the indices of refraction of four manufacturers (AT&T, Corning, Inc., Spectran Corporation, and Phillips N.V.). Use of these values should result in a small error in the fiber length.

The pulse width determines the power launched into the cable. The larger this width, the longer will be the length of cable that the OTDR can test accurately (Figure 5–17). However, the larger this width, the longer will be the optical dead zone and the event zone (Figure 5–13).

Fiber Type	Wavelength (nm)	Index of Refraction
50/125	850	1.48535
62.5/125	850	1.4982
50/125	1300	1.48145
62.5/125	1300	1.4938

Table 5–5 Indices of Refraction When Actual Values Unknown

How to Make Measurements with an OTDR

After setting up the OTDR according to the manufacturer's instructions, push the start/stop or test button. After the OTDR has completed its testing process, a trace will appear on the screen. From this trace, you will be able to measure lengths and distances, attenuation of fibers, loss of connectors, loss of splices, loss of microbends and macrobends, and reflectance.

Length and Distance. OTDRs have cursors, which you position to determine the locations of features. To determine the length of a single section of cable attached directly (without a lead in cable) to an OTDR, position the cursor at the lowest point of the straight line trace just before the end reflection (Figure 5–23). On most OTDRs, the OTDR will display the cursor position in feet or kilometers.

To determine the fiber length of a segment in a multi segment cable plant that contains mechanical splices or connectors (or a cable attached to a lead in cable), position one cursor at the lowest point of the straight line trace just before the first end reflection. Position a second cursor at the lowest point of the straight line trace just before the second end reflection (Figure 5–24). The segment length is the difference between the positions of these two cursors. On some OTDRs, such as the Tektronix FiberMaster™, the segment length will be displayed automatically.

Loss Measurements. The OTDR trace will enable you to make loss measurements of cable attenuation and attenuation rate, connector loss, microbend loss, and splice loss. Loss is determined by measuring the difference in signal strength between the locations of two cursors.

Cable loss is determined by placing one cursor at the lowest point of the straight line trace just before the end reflection and the second cursor at the highest part of the straight line trace, just after the initial peak (Figure 5–25).

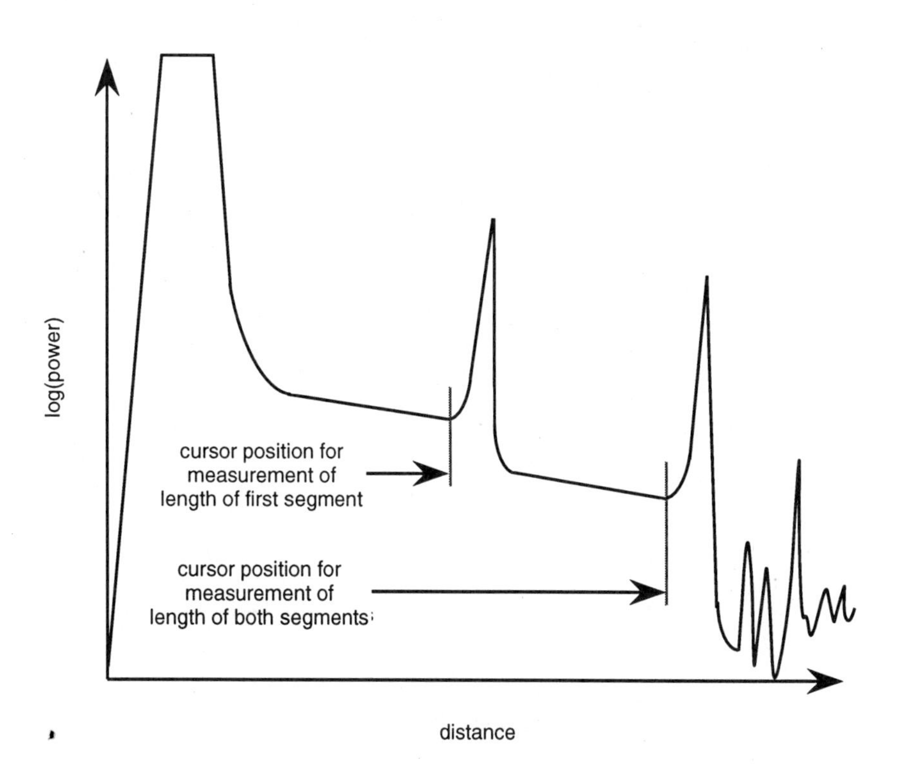

Figure 5–23 Cursor Positions for Measurement of Length

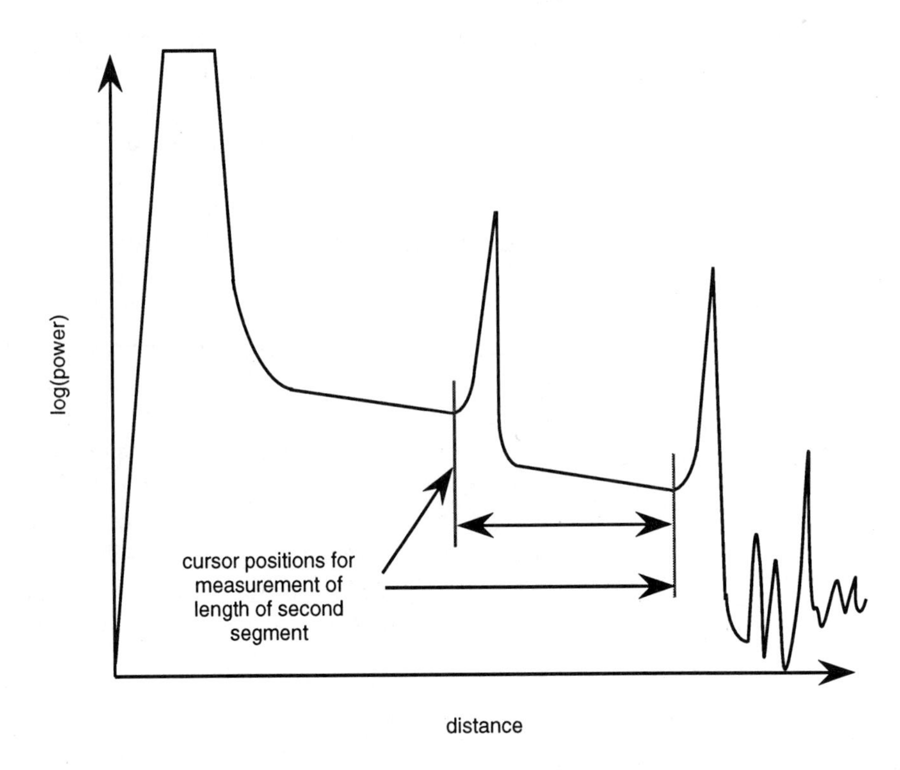

Figure 5-24 Cursor Positions for Measurement of Segment Length

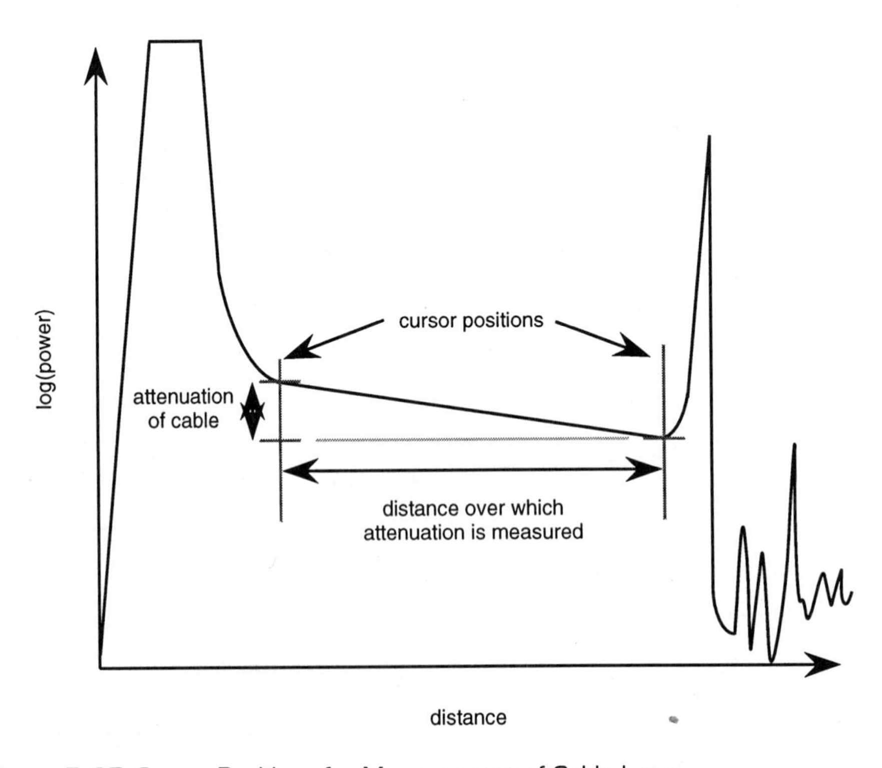

Figure 5-25 Cursor Positions for Measurement of Cable Loss

The loss of the segment is the difference in height at the points the two cursors cross the trace. You calculate the attenuation rate of the fiber by dividing the loss indicated on the OTDR by the distance between the cursors. This rate is automatically calculated by some OTDRs, such as those from Tektronix and Laser Precision Corporation.

Note that this measurement is the loss of the fiber between the cursors. This loss in not the total loss of the cable, since this loss does not include the loss of the cable in the optical dead zone.

There are two methods for determining connector loss. In the first method, one cursor is placed at the highest part of the straight line trace just after the connector. The second cursor is placed at the lowest point of the straight line trace just before the connector (Figure 5–26). This method includes the loss of cable between the two cursors.

A more accurate method for determining connector or splice loss is the splice loss method. This method does not include the loss of the cable between the two cursors. In this method, the OTDR makes a best-fit linear approximation of the linear traces on both sides of the connector or splice. The OTDR calculates the difference between the signal strength levels at the location of the connector or splice (Figure 5–27).

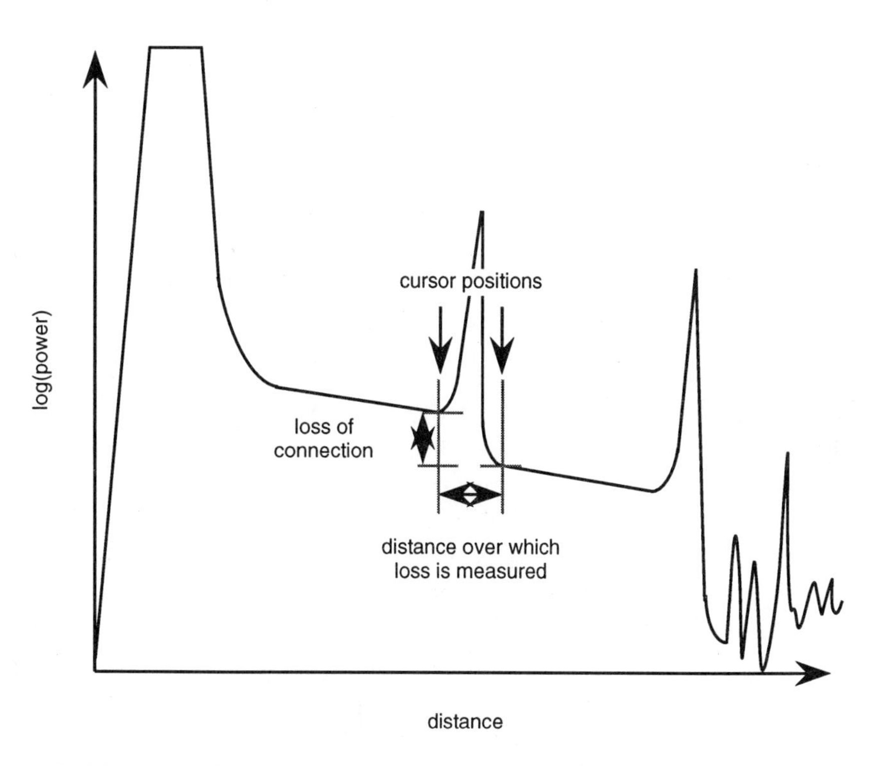

Figure 5–26 Cursor Positions for Measurement of Two-Point Connection Loss

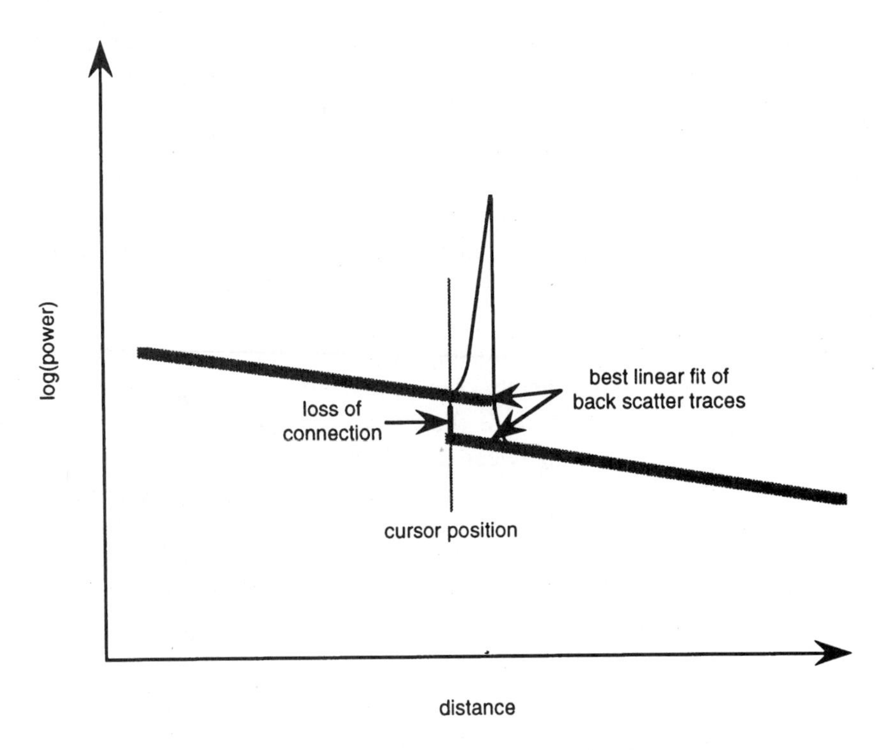

Figure 5–27 Cursor Position for Measurement of Splice Loss

Comparison of OTDR and Insertion Loss Measurements

OTDR testing and insertion loss testing are both forms of power loss measurements. As such, one might be tempted to compare the losses measured by these two types of test techniques. Such comparisons are difficult. This difficulty is due to four differences between these test techniques:

- Differences in the wavelengths
- Differences in the spectral widths
- Differences in the modal distribution of the light launched into the fiber
- Differences in test set-ups

Because of these differences, comparisons will rarely result in complete agreement.

Other OTDR Test Equipment

OTDRs are powerful diagnostic tools, both for certification of cable plants after initial installation and for maintenance and troubleshooting. These capabilities come at a relatively high cost, with OTDR mainframes ranging from $20,000 to $50,000. While OTDRs can be rented, renting is not always a convenient alternative. In recognition of the cost issue, OTDR manufacturers have designed three alternatives to the full, mainframe OTDRs. These alternatives, the mini-OTDR, the feature finder and the fault finder, are all forms of an OTDR.

Mini-OTDR. A mini-OTDR (Figures 5–28 and 5–29) is portable and battery-powered, with many of the features of a mainframe OTDR, but with slightly reduced capabilities and significantly reduced prices ($9,000 to $15,000). Mini-OTDRs can test at one or two wavelengths. Most will test either multimode or singlemode fibers, but not both with the same package (the Tektronix TFS3031 TekRanger is an exception). The distance measurement capability (dynamic range) of a mini-OTDR is lower than that of a mainframe OTDR. It takes longer

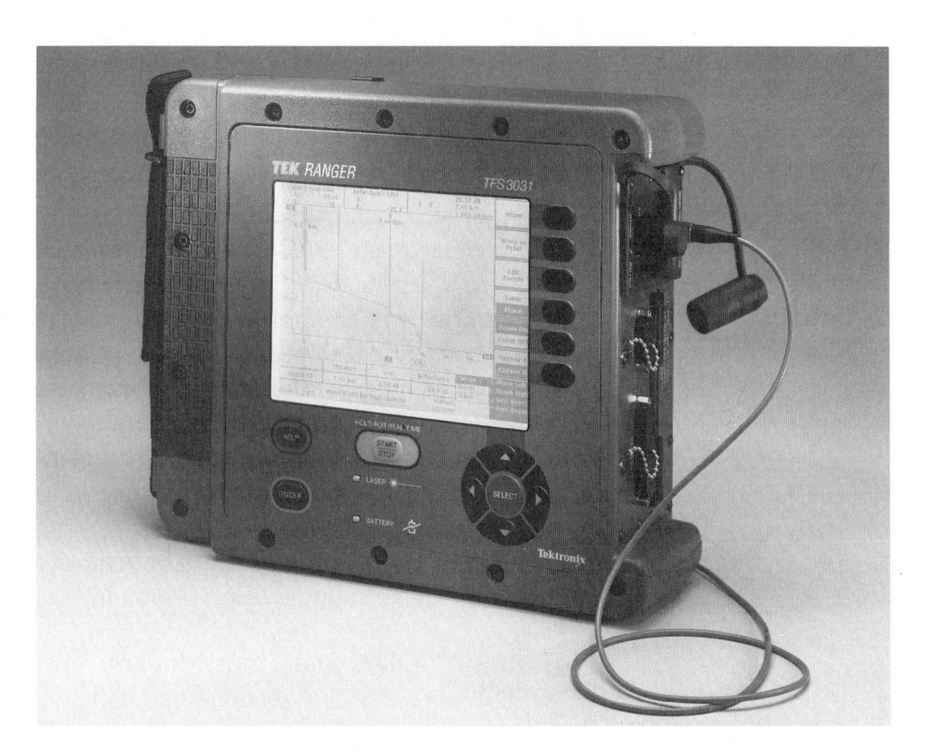

Figure 5–28 Mini-OTDR (Courtesy of Tektronix Inc.)

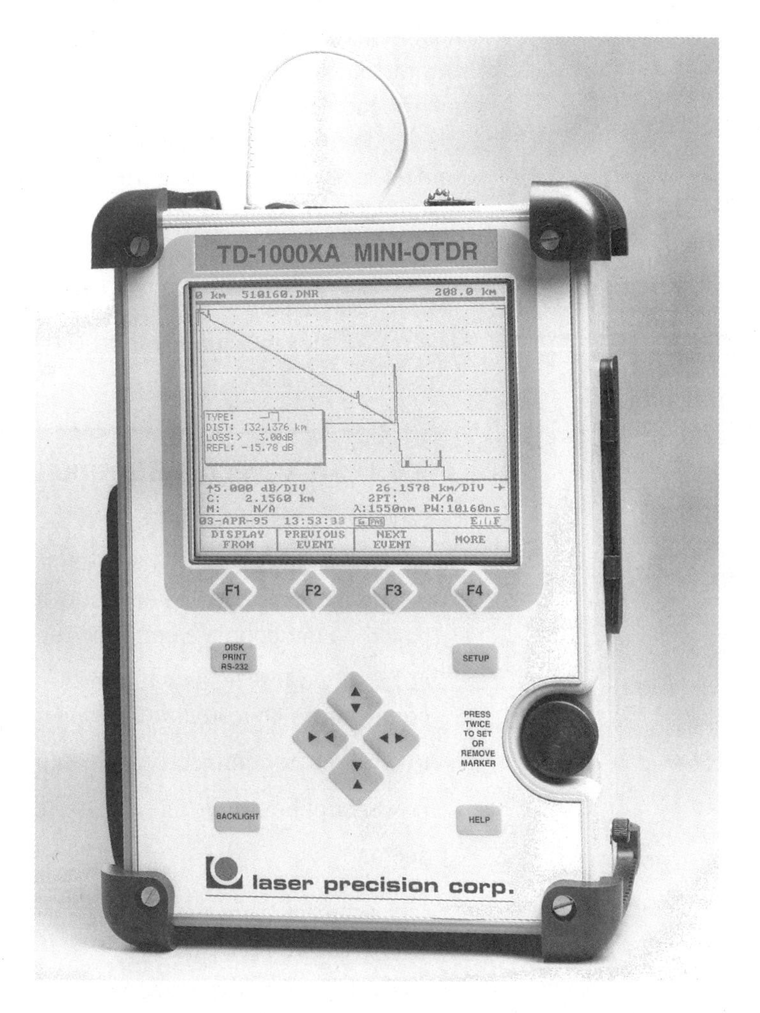

Figure 5–29 Mini-OTDR (Courtesy of Laser Precision Corp.)

to obtain a trace with a mini-OTDR than with a mainframe. Because of these advantages, mini-OTDRs are expected to experience a significant increase in use, with a corresponding decrease in use of mainframe OTDRs.

When we compared mainframe OTDRs to mini-OTDRs, we found the mini-OTDRs to be well suited to many, if not most, applications. This comparison revealed the mainframes to be the instrument of choice for the following four situations: when the time required to make measurements is important, the mainframe OTDR is preferable, because mainframes have faster processors, which draw more current. When long cables (up to 200 km) need to measured, the mainframe is required, because the lasers and detectors are cooled to produce increased dynamic range. When the OTDR needs to be controlled by a GPIB bus, the mainframe is required. When the connector attached to the OTDR needs to be tested, a mainframe is required (or lead in cable longer than the dead zone), since mini-OTDRs do not, at this time, contain an internal fiber. Other advantages of the mainframe OTDR are reduced sizes of the dead and event zones, easy to use color displays, high resolution displays, and internal printers.

Feature Finders and Fault Finders. Feature finders and fault finders use OTDR technology to obtain some of the information from a trace. These two products locate reflective events (Figure 5–24), measure distances, and, in some cases, provide measurements of the loss of an event. Because of their lower cost, they may not provide measurements of the loss of fiber in a cable and may be unable to find a non-reflective event with a loss lower than the threshold. With this limitation on threshold, feature finders and fault locators may not "see" microbend losses or losses from fusion splices. Because of their reduced capabilities and the ongoing reduction in costs of mini-OTDRs, feature finders and fault finders are expected to experience a significant reduction in sales and use. Some manufacturers have already stopped producing them.

THE REFLECTANCE MEASUREMENT PROCEDURE

The reflectance measurement procedure is defined in EIA/TIA-FOTP-107.

Tools and Supplies Required

reflectance test set, with FC/APC receptacle

at least two reflectance reference leads (rated better than the ones you intend to test) with an FC/APC connector on one end and the style of connector to be tested on the opposite end (one for spare)

at least two test leads with an FC/APC connector on one end and the style of connector to be tested on the opposite end

two barrels for connector style under test (one for spare)

power supplies and/or new batteries for light source and power meter

acetone

lens grade tissues

lens grade compressed air

cable assemblies to be tested

PROCEDURE 3: MEASURE RETURN LOSS

1. Obtain all equipment and supplies.
2. Turn on the return loss test set. Allow to stabilize for at least 15 minutes.
3. Clean both connectors on one of the reference leads and one of the test leads.
4. Clean the barrel with compressed air.
5. Connect FC/APC of the test lead to the return loss test set.
6. Connect the reference lead to the test lead with the clean barrel.
7. The "open" end of the reference lead is the end of the cable opposite to the end attached to the barrel and test lead. Wrap the open end of the reference lead around a pencil or a ¼-inch diameter dowel ten times.
8. Record this reference power level, in dBm, or push the reference button on the test set. Unwrap the reference lead. Detach the reference lead from the test lead. This value should be within 5 dB of the value at which it was purchased.
9. Clean the connector to be tested with acetone and tissues. For additional information on the effect of cleaning on reflectance measurements, see Luis M. Fernandez and Donald R. Cropper, "Evaluating Fiber Optic Connector Cleanliness for Accurate Measurements," *Fiber Optic Product News* (June 1995): 46–46.
10. Inspect the connector with a microscope. If the connector is clean, proceed to Step 11. If not, repeat Step 9 until the connector is clean.
11. Attach the connector to be tested to the barrel.
12. Wrap the open end of the cable under test around a pencil or a ¼-inch diameter dowel ten times.
13. Record this power level, in dBm. This is the return loss. Unwrap the cable under test.
14. Repeat Steps 9–13 on the opposite connector.

How to Troubleshoot the Reflectance Procedure

High Reflectance Reference Lead. A high reflectance reference lead can have three causes:

- A damaged or dirty test lead
- A damaged or dirty reference connector
- A dirty barrel

Microscopic inspection and cleaning of the connectors will reveal and solve this problem. If cleaning of the connectors does not eliminate high reflectance, the connector surfaces may need polishing. Repeat Steps 7–10 of Procedure 6 in Chapter 4. If two or three repetitions of Steps 7–10 of Procedure 6 in Chapter 4 do not eliminate high reflectance, the reference lead may need to be repolished through the entire Procedure 6 in Chapter 4. If repolishing through the entire Procedure 6 does not succeed, replace the reference lead.

A dirty barrel can be cleaned will compressed air or with a lint-free pipe cleaner moistened with isopropyl alcohol.

High Reflectance Measurements. A high reflectance measurement has four causes. The first three are the same as the causes of a high reflectance reference lead. The fourth is an undercut fiber.

An undercut fiber can be salvaged by repolishing through the entire polishing procedure. While repolishing can result in low reflectance, excessive repolishing can result in a short ferrule. A short ferrule will leave an air gap between mated connectors. This gap will result in high reflectance. A connector with a short ferrule must be replaced.

REVIEW QUESTIONS

1. Does attenuation of installed cables need to be checked after installation? If so, why? If not, why not?
2. Does pulse spreading need to be checked after installation? If so, why? If not, why not?
3. Why is insertion loss testing performed?
4. You are going to purchase a light source for measuring loss of installed cables and connectors. What are the requirements that this light source must meet?
5. You are going to purchase a light power meter for measuring loss of installed cables and connectors. What are the requirements that this light power meter must meet?
6. Besides the source and meter, what other equipment will you need to perform insertion loss tests of installed cables and connectors?
7. If you perform insertion loss testing of a number of short cable assemblies, all of which have good looking connectors on both ends, what loss value should you expect to see?
8. Should you expect the loss to be the same in both directions? If yes, why? If no, why not?
9. You have been testing a large number of short cable assemblies. A large percentage of these assemblies have positive losses? What should you do? What caused these positive losses?
10. You have a cable system with three segments, all connected with unkeyed connectors. Each connector pair exhibits an average range of 0.75 dB/pair. The difference between the initial insertion loss of this length and the current length is 1.5 dB. What troubleshooting action(s) should you take?
11. You have a cable system with three segments, all connected with keyed connectors. Each connector pair exhibits an average range of 0.35 dB/pair. The difference between the initial insertion loss of this length and the current length is 1.5 dB. What troubleshooting action(s) should you take?
12. Why, or when, will you need an OTDR? What information will the OTDR provide that the insertion loss test set will not provide?
13. What are the five parts of a typical OTDR trace? What causes each part?
14. What fiber characteristic do you need in order to measure fiber length accurately?
15. Why is the fiber length often not equal to the cable length?
16. What key feature in an OTDR trace differentiates the two groups of possible features? Group A is a violation of a cable performance characteristic or a properly made fusion splice; Group B, a connector pair, a mechanical splice, an improperly made fusion splice, or a break in a tight tube cable.
17. Describe the proper procedure for making a length measurement of a single length of cable connected to an OTDR.
18. Describe the proper procedure for making a length measurement of a segment of cable system connected to an OTDR when multiple lengths are attached to an OTDR.
19. Describe the proper procedure for making a measurement of the attenuation of a cable length.
20. Is this measurement a measurement of the complete loss of the fiber in the cable?

21. Describe two procedures for making a measurement of the loss of a splice or a connector.
22. What is an echo? How is an echo created?
23. What is the advantage of a splice loss measurement procedure for measurement of connector or splice loss over the two-cursor procedure?
24. You are performing an OTDR test of a seven segment link. The fourth and fifth segments have a total length of 15 m. The event zone is 20 m. Draw the OTDR trace you expect to see.
25. You needed to connect two long lengths of cable together. The cable connected directly to the OTDR has a 50 μm core. The second length has a 62.5 μm core. How much loss will you expect to see at the connectors connecting these two cables?
26. You need to connect two long lengths of cable together. The cable connected directly to the OTDR has a 62.5 μm core. The second length has a 50 μm core. How much loss will you expect to see at the connectors connecting these two cables?
27. You have an OTDR trace that looks like Figure 5–24. The connection in the center is a fusion splice. What do you know about that splice?

CHAPTER

6

How To Install Splices Properly

CHAPTER OBJECTIVES

From this chapter, you will be able to:

1. Set up and create low loss, high strength fusion splices.
2. Create fusion splices with unbuffered fiber and with pigtails.
3. Determine the optimum fusion splicing parameters.
4. Troubleshoot the splicing process.
5. Set up and create low loss mechanical splices.
6. Properly install splices in splice trays and enclosures.

INTRODUCTION

Splicing is the preferred method for making permanent connections between two fibers. Splices can be mid-span splices or pigtail splices. Mid-span splicing is performed during initial installation of two long lengths of fiber and during repair of broken cables.

Pigtail splicing is the splicing of a short length of cable (often 1 to 2 meters long) with a connector on one end to a longer length of installed cable. Pigtail splicing is performed for performance and/or cost reasons. The splicing of a factory-installed singlemode pigtail is performed to achieve low reflectance, since achieving low reflectance from field installed singlemode connectors can be difficult.

Splicing of pigtails is performed to reduce installation costs, since the cost of field-installed connectors can be two to three times the cost of factory-installed connectors. Splicing of pigtails is performed when the total cost of a pigtail and the splicing is less than the cost of a field-installed connector.

The two types of splicing are fusion and mechanical. Both types require five major steps:

- Preparation of the ends of the cable
- Preparation of the ends of the fiber (stripping and cleaving)
- Splicing
- Testing of the loss of the splice with an OTDR
- Installation of the splice in an appropriate splice tray and splice enclosure

Both types of splicing require *cleaving*. Cleaving is the process by which you create a fiber end which is perpendicular to the axis of the fiber and smooth (a “good” cleave). For both fusion and mechanical splicing, low loss is achieved with a good cleave. The difference between cleaving for fusion splicing and cleaving for mechanical splicing is in the length of bare fiber from the end of the buffer coating or buffer tube to the cleaved end.

Both types of splices require installation of the splice in a protective structure. This structure includes a splice tray (Figure 6–1) and a splice enclosure. The splice tray will be designed for the specific type of splice to be installed. The splice enclosure can be an indoor enclosure (Figure 6–2) or an outdoor enclosure (Figure 6–3).

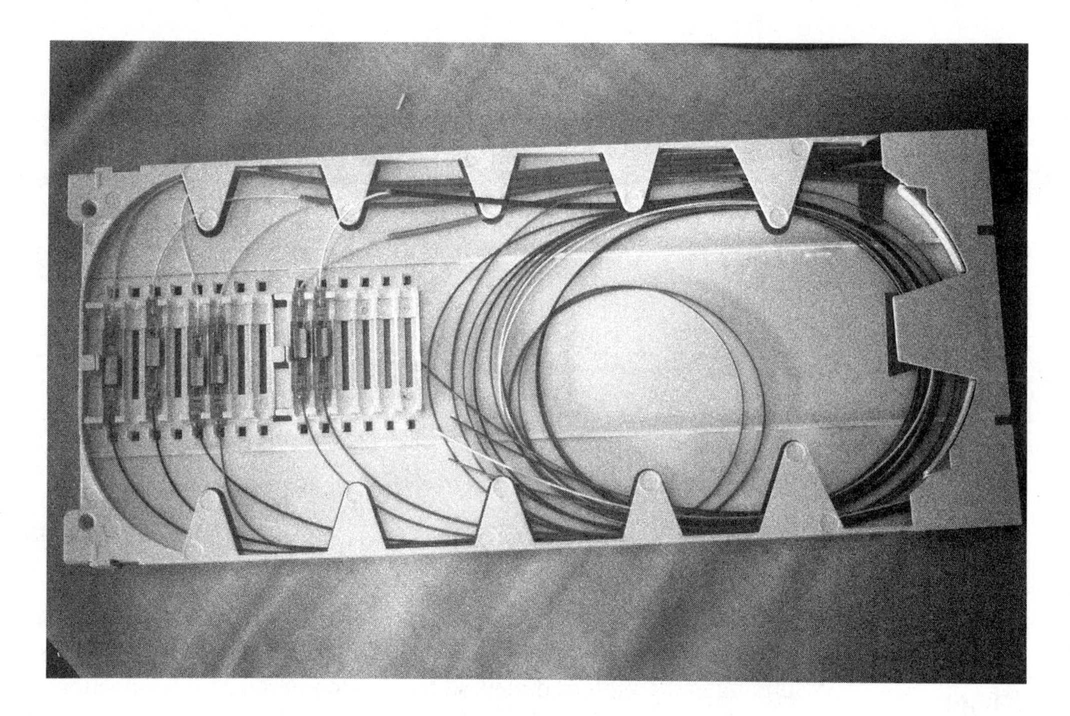

Figure 6–1 Mechanical Splices in Splice Tray (Courtesy of Pearson Technologies Inc.)

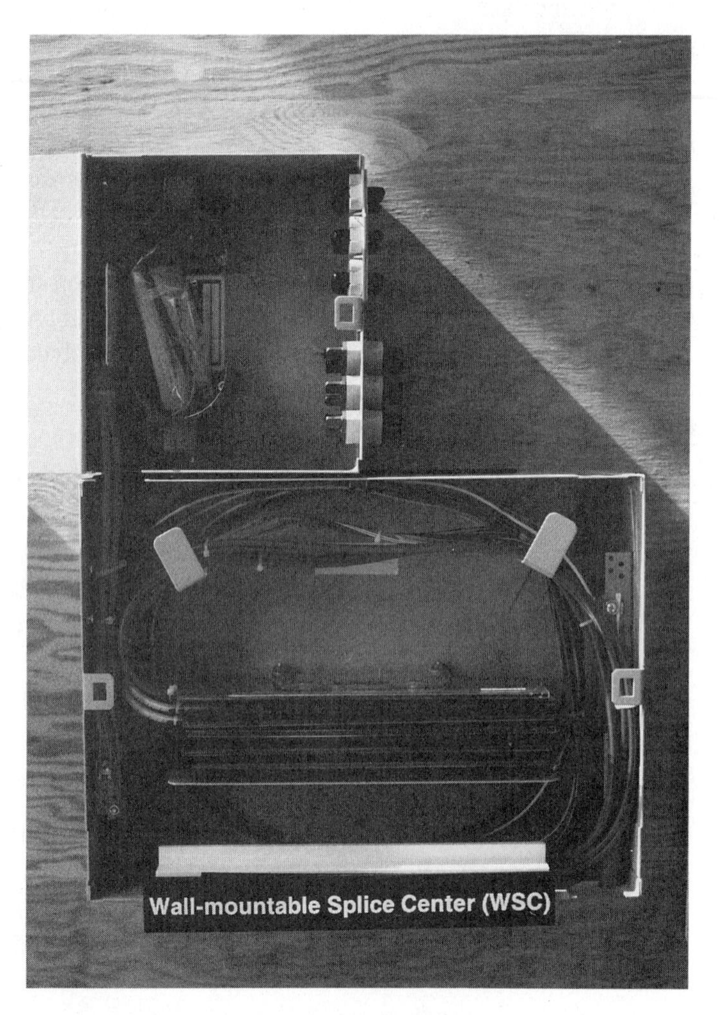

Figure 6–2 Indoor Splice Enclosure (Courtesy of Pearson Technologies Inc.)

Figure 6–3 Internal Structure of Outdoor Splice Enclosure (Courtesy of Pearson Technologies Inc.)

FUSION SPLICING

In this section, we present two procedures for fusion splicing. The first procedure is a practice procedure. A splice made by this procedure will perform as a splice of a cable pigtail near the transmitter end of a multimode cable system.

The second procedure is for splicing at a long distance from the input, or transmitter, end of a multimode fiber. A splice made by this procedure will perform as a mid-span splice or a pigtail splice made at the receiver end of a fiber.

We have developed these procedures for use with the Fiberlign Micro Fusion Splicer from Preformed Line Products (Figure 6–4) and for the Fitel cleaver (Figure 6–5), which are both relatively inexpensive, but perform excellently. To use these procedures with other fusion splicers, you will need to change the values of parameters of arc current, arc time, and overrun of the splicer. The correct values of these parameters will be in the instruction manuals of other splicers. While we have developed the procedures in this chapter for multimode fibers, the same procedures can be used for singlemode fibers. (For singlemode fiber splicing, use a singlemode fiber with a wavelength and light source calibrated at 1310 nm.) If you wish to practice a splicing technique for measurement at 1300 nm, use a 1300 nm light source and a power meter calibrated at 1300 nm.

Tools and Supplies Required

for Procedure 1, at least 15 feet of all-glass, multimode optical fiber

for Procedure 2, approximately 1,000 m of all glass, multimode optical fiber

two bare fiber adapters with same connector style as the power meter and light source

power meter with calibration at 850 nm

850 nm stabilized light source

Scotch Magic™ tape
fusion splicer (Preformed Line Products MicroFusion Splicer, Model MS-1008F)
buffer coating stripper (Clauss part number NN200 or equivalent)
cleaver (Fitel part number VL-310 or equivalent)
lens grade compressed air
91–95 percent isopropyl alcohol
heat shrink splice covers (ACP International Part Number DH1)
heat gun

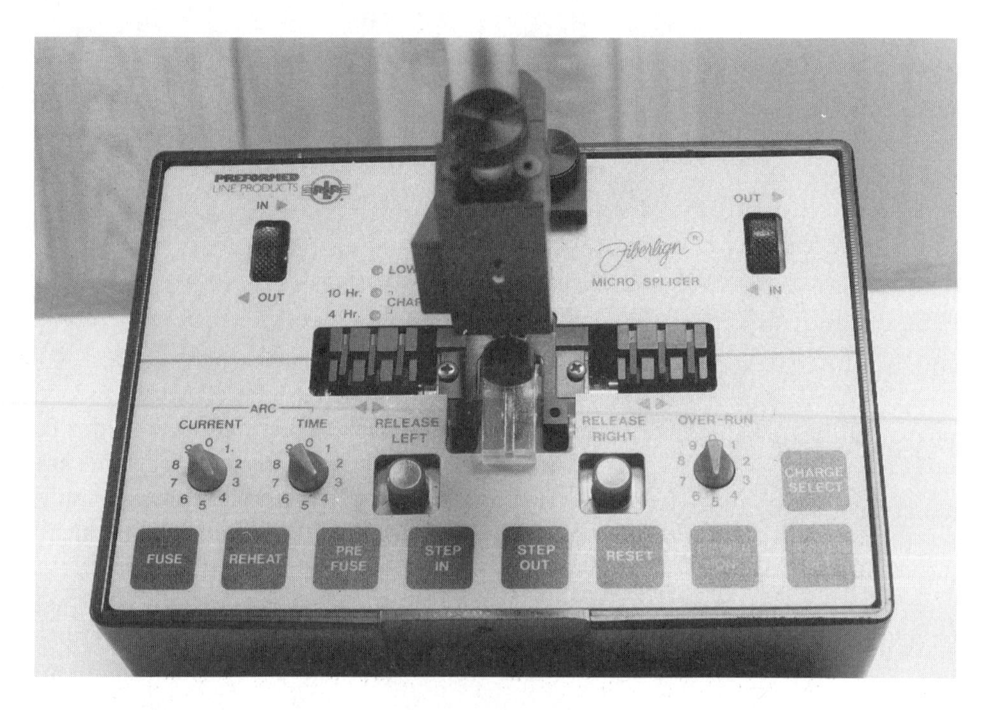

Figure 6–4 Fiberlign® Micro Fusion Splicer (Courtesy of Pearson Technologies Inc.)

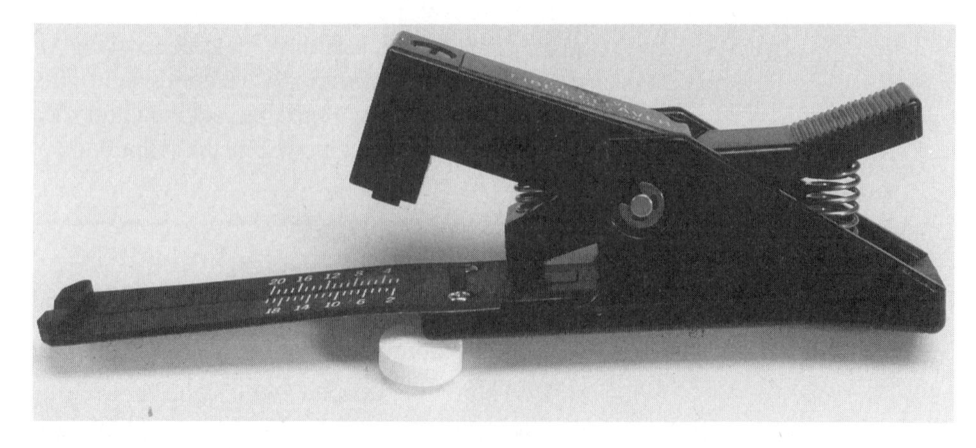

Figure 6–5 The Fitel Cleaver with Cleaving "Tail", Scribing Arm, Retaining Finger (Courtesy of Pearson Technologies Inc.)

Procedure 1: Splicing Short Lengths of Fiber

1. Obtain all tools and supplies.
2. Using the buffer stripper as in Steps 5–9 of Procedure 1, (Chapter 3), strip the buffer coating from one end of the fiber until 1.5 inches is bare.
3. Install the fiber in the cleaver: press down on the right end of the retaining finger (Figure 6–5) and slide the bare end of the fiber under the retaining finger until the buffer coating is at the 4 mm mark (Figure 6–6). Release the retaining finger onto the fiber.
4. With your thumb, press the fiber into the groove on the flexible tail of cleaver. Hold the fiber so that it is against one wall of the groove. Do not remove your thumb. Straighten the flexible tail of cleaver. The fiber will usually exhibit a slight bend. Press down on the right end of the retaining finger to allow the fiber to straighten out. Release the retaining finger onto the fiber. Gently push the scribing arm down until the scribing blade touches the fiber. Do not press the scribing arm down hard.
5. Press open the back end of the bare fiber adapter. Insert the bare fiber into the bare fiber adapter until it is flush with tip of the ferrule (Figure 6–7). Close the back end of the bare fiber adapter.

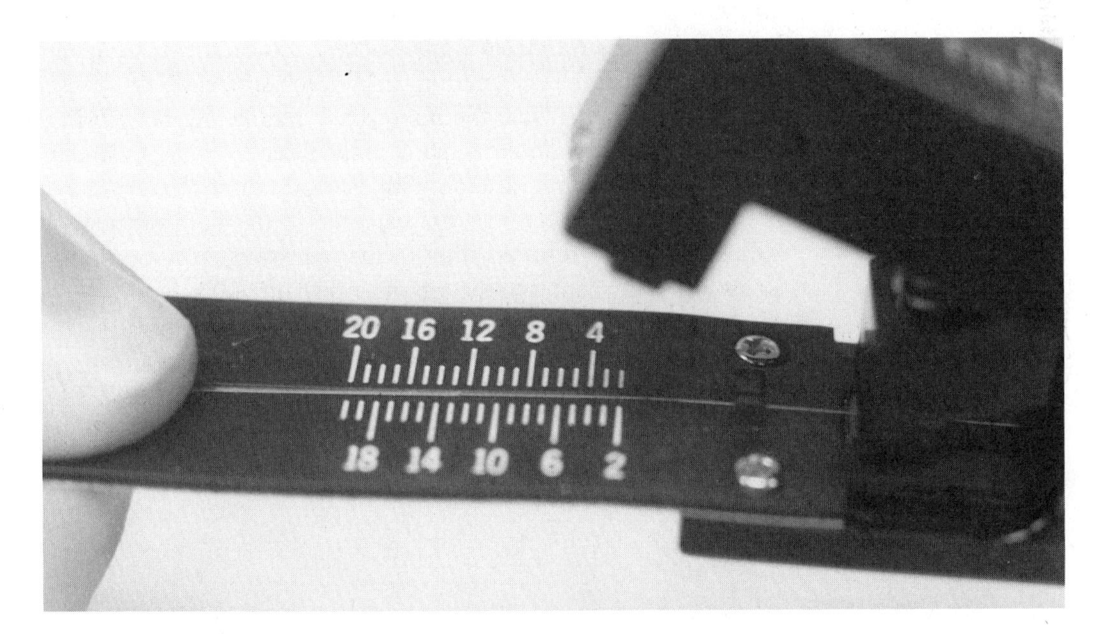

Figure 6–6 Fiber Properly Installed in Cleaver (Courtesy of Pearson Technologies Inc.)

Figure 6–7 Fiber in Bare Fiber Adapter (Courtesy of Pearson Technologies Inc.)

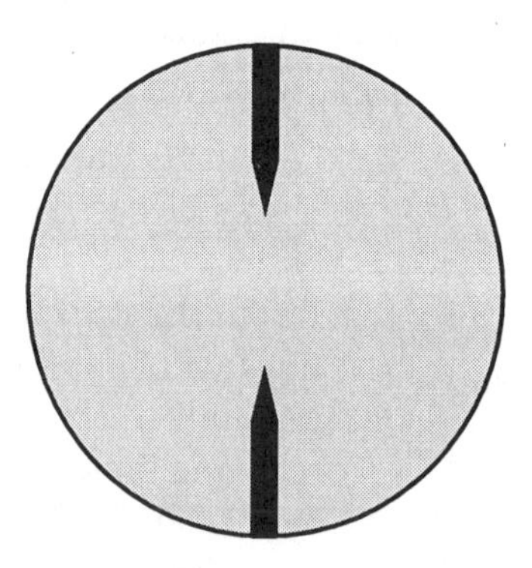

Figure 6–8 View when Microscope Properly Positioned

6. Insert the bare fiber adapter into the receptacle of the light source.
7. Tape down the fiber near the light source so that the fiber cannot be pulled from the bare fiber adapter accidently.
8. Repeat Steps 1–4 on the second end of the fiber.
9. Insert the bare fiber adapter into the power meter.
10. Tape down the fiber near the power meter. Check the fiber to ensure that there are no bend radii smaller than 2 inches in the fiber.
11. Set and record a reference signal (Step 7, Procedure 1, Chapter 5). Turn on the splicer. Push the reset button before each splice.
12. Set up the splicer. Install the microscope by placing the pin in the pin hole and by partially tightening the bolt in the threaded hole. Push the "power on" button. Focus the microscope with the top thumbscrew. Adjust the microscope so that the electrodes are centered in the field of view (Figure 6–8). Adjust the left to right position of the microscope by loosening the thumbscrew at the top center of the top plate (Figure 6–9).

    Adjust the coarse positioning thumb wheels (Figure 6–4) so that both 3-jaw clamp positioners are centered in the cut out area of the top plate (Figure 6–4). Set the recessed singlemode/multimode switch to "MM." Break the fiber near the middle. If you plan to use a heat shrinkable splice cover, install this cover on one end of the fiber.

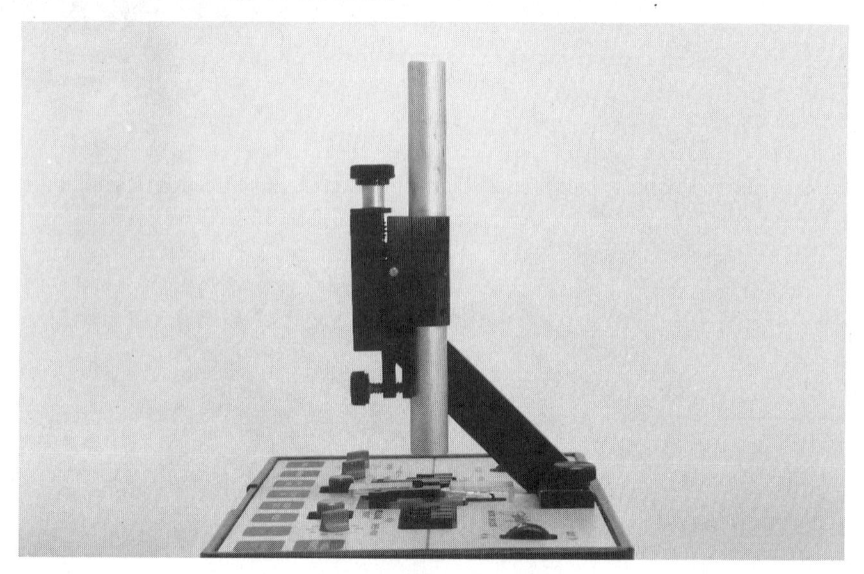

Figure 6–9 Microscope Position Adjustment Screws (Courtesy of Pearson Technologies Inc.)

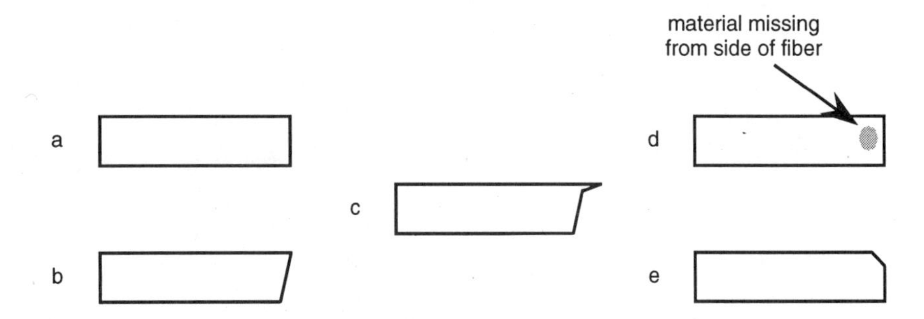

Figure 6–10 Good and Bad Cleaves

While pushing down both release buttons, blow off the "V-grooves" and the areas under the 3-jaw clamps with compressed air.

13. Using the buffer stripper, strip 1.5 inches of buffer coating from one end of the broken fiber.
14. Slide the fiber under the retaining finger of the cleaver until the buffer coating is at the 13 mm mark. Repeat Step 3 on first end of the broken fiber.
15. Push down the left release button of the splicer to lift the left 3-jaw clamp. Place the cleaved end into the central section of splicer (the splice area) so that the end roughly aligns with electrodes (Figure 6–8). Place the fiber under the 3-jaw clamp. Release the button for the 3-jaw clamp. The fiber should be centered between the electrodes (Figure 6–11).
16. With the microscope, examine the cleaved end of the fiber. The end should be flat and smooth (Figure 6–10 a), with no angle (Figures 6–8 and 6–10 b), excess material (Figure 6–10 c), or missing material (Figures 6–10 d and e). Rotate the fiber roughly 90° and examine the fiber again. If the end is not flat and smooth, repeat Steps 12–15. The end should be flat and smooth.
17. Repeat Steps 12 and 13 on the second end of the broken fiber.
18. Push down the right release button of the splicer to lift the right 3-jaw clamp. Place the cleaved end into the central section of the splicer so that the end roughly aligns with electrodes. Place the fiber under 3-jaw clamp. Release the button for the 3-jaw clamp.
19. Adjust the left positioner until the right image of the end of the fiber is aligned with the electrodes (Figure 6–11).
20. Adjust the right positioner until the left image is within one fiber diameter from the right image.
21. Repeat Step 16 on the left image.

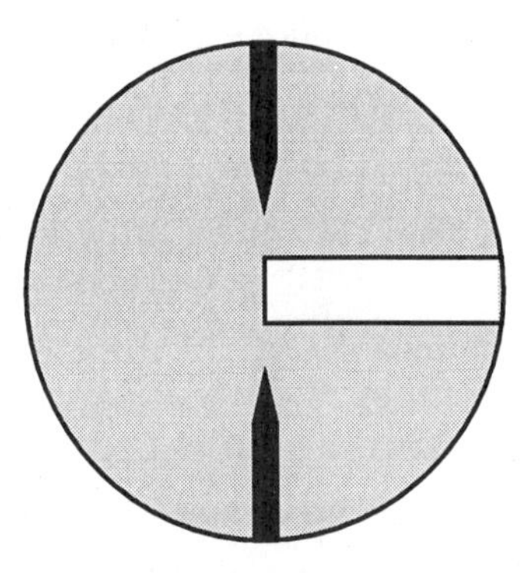

Figure 6–11 Alignment of First Fiber to Electrodes

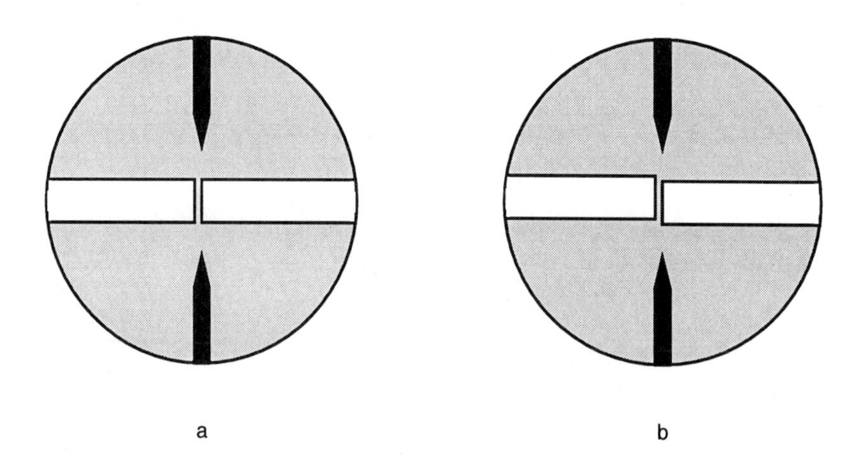

Figure 6–12 Acceptable and Unacceptable Alignment of Fiber Ends to One Another

22. Inspect the up down alignment of both fibers. The fibers should appear aligned with very little or no offset (Figure 6–12). If an offset appears, remove the fibers and wipe the stripped sections with lens grade tissue moistened with 91–95 percent isopropyl alcohol. With compressed air, blow off the V-grooves and the areas under both 3-jaw clamps. Reinstall both fibers according to Steps 14–22.
23. Set the controls according to Table 6–1. Enter the appearances of the cleaves in Table 6–2.
24. Press the reset button. Check the fibers under the microscope for cleanliness.
25. Press the PREFUSE button.

Control	Singlemode	Multimode
arc current	5	2
arc time	4	5
overrun	4	4
SM/MM switch	SM	MM

Table 6–1 Initial Settings for Fusion Splicer

	Cleave Appearance		Losses			
Splice #	Left	Right	Before	Initial	Reheat 1	Reheat 2
____	____	____	____	____	____	____
____	____	____	____	____	____	____
____	____	____	____	____	____	____
____	____	____	____	____	____	____
____	____	____	____	____	____	____
____	____	____	____	____	____	____
____	____	____	____	____	____	____
____	____	____	____	____	____	____

Table 6–2 Data Form for Splicing

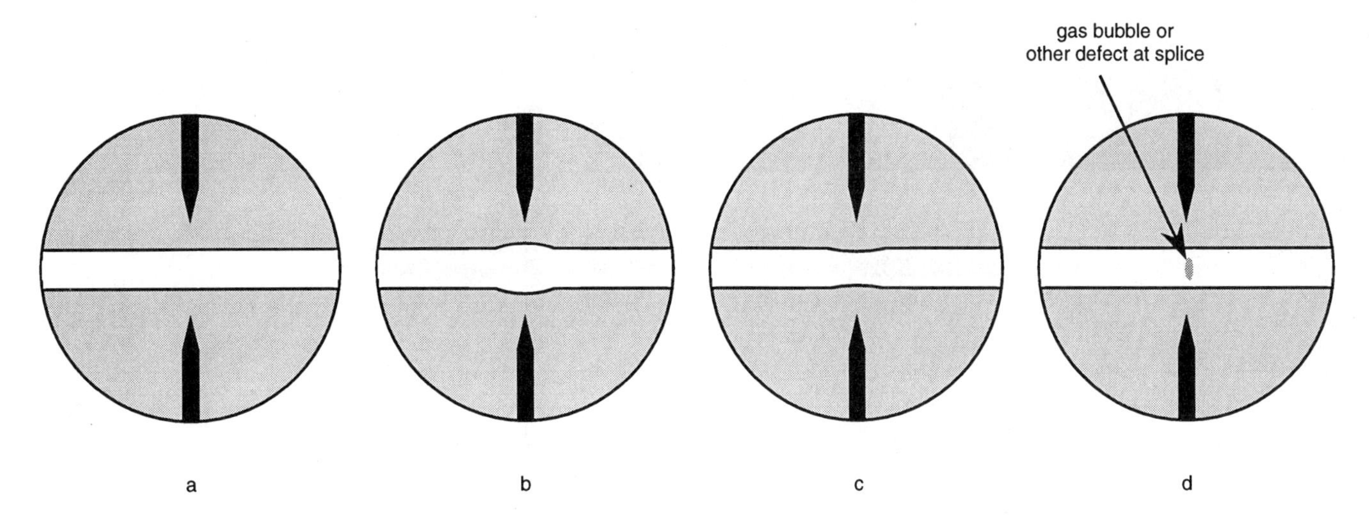

Figure 6–13 Proper and Improper Appearances of Fusion Splice

26. Push the STEP IN button repeatedly until the fiber ends just touch. If you push the STEP IN button too many times, push the STEP OUT button until you see an air gap between the fibers. Push the STEP IN button repeatedly until the fiber ends just touch.
27. Record the power meter reading in the column labeled "Before" (Table 6–2). Press the FUSE button to start the fusing process.
28. Inspect the fused fiber under the microscope. The fiber should be of uniform diameter (Figure 6–13 a), with no bulging (Figure 6–13 b), necking (Figure 6–13 c), or bubbling (Figure 6–13 d) in the area of the splice.
29. Measure and record the value on the power meter in the column labeled "Initial Loss"' (Table 6–2). If necessary, calculate the difference between the initial reading and the current reading to determine the splice loss.
30. If the splice is low loss (≤ 0.3 dB) and if the splice looks good (Step 28), remove the fiber by pressing both release buttons. Check the splice strength. Bend the bare fiber until it breaks. Compare the lengths of both of the broken ends. If the lengths of both ends are unequal, the splice is stronger than the virgin fiber. If the lengths of both ends are equal, the splice is weaker than the virgin fiber. If the ends are equal, increase the arc current by one setting and redo the splice (Steps 12–30).
31. If the splice loss is higher than the 0.3 dB, push the REHEAT button once. Measure and record the value on the power meter in the column labeled "Reheat 1" (Table 6–2). If the loss after reheating is still greater than 0.3 dB, repeat Steps 12–30. If the loss after reheating is unacceptable, increase the arc current setting by one digit.

32a. If you are practicing cleaving and splicing, repeat Steps 11–30 as many times as desired. However, allow at least 30 seconds between successive splices.

32b. If you are practicing the entire splicing procedure, install the splice cover. Slide the heat shrinkable splice cover until it is centered on the bare glass and covers all of the bare glass. Heat the splice cover with a heat gun until it has completely shrunk (Figure 6–14).

32c. Alternative procedure: remove the paper backing from an adhesive splice cover. Place the cover so that bare glass is centered in splice. Close the splice cover completely.

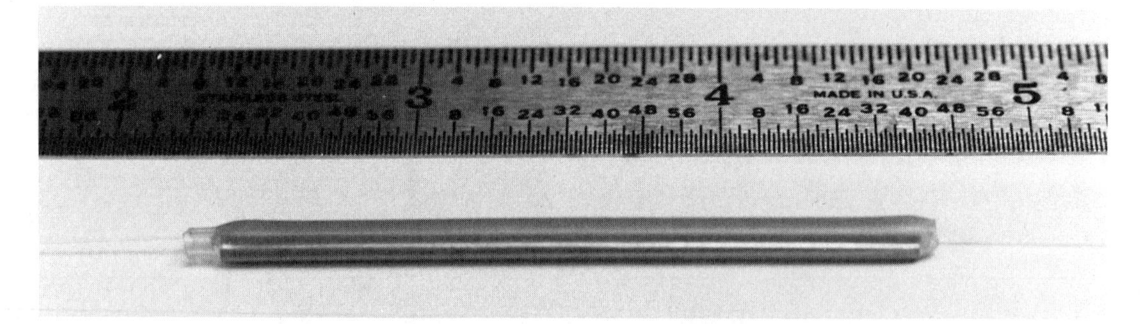

Figure 6–14 Fiber with Heat Shrink Splice Cover (Courtesy of Pearson Technologies Inc.)

Troubleshooting Cleaving

Occasional Bad Cleaves. Occasional bad cleaves are to be expected. Recleave.

Frequent Bad Cleaves. Frequent bad cleaves (when more than 5 percent of cleaves are visibly bad) indicate a problem with the cleaving technique or with the cleaver. To solve this problem, review three aspects of cleaving: how the fiber is held in the cleaver, straightness of the fiber before scribing, and how the fiber is bent during cleaving. The fiber must be held against one wall of the groove in the tail. The fiber must be straight before scribing. The fiber must be quickly bent, or snapped, so that the fiber is broken quickly.

If the problem persists after review of the procedure, the scribing blade may be worn or damaged. Replace this blade and recleave.

Troubleshooting Splicing

Offset Visible Prior to Splicing. Offset visible prior to splicing is usually due to contamination in the precision V-groove in the center of the splicer. Clean this groove with compressed air.

Infrequently, this offset may be due to buffer coating that was not removed from the fiber during stripping. Examine the bare fiber and remove any materials that remain on the fiber. Remove with an alcohol soaked lens-grade tissues.

Occasional High Loss. Occasional high loss is to be expected.

Consistent High Loss. Consistent high loss indicates a major deviation from optimum splicing conditions. Possible causes of this problem are a worn or damaged scribing blade in the cleaver or improper settings for the fiber being spliced (Table 6–1).

A worn or damaged scribing blade can create a cleave that looks good, but is actually not perpendicular to the axis of the fiber. Replace the scribing blade and attempt splicing. The lifetime of scribing blades depends on the cleaver and the care with which it is used. (During training, we have found the lifetime to be 1,000 cleaves.)

High loss from improper settings usually results in a splice with an improper visual appearance. This improper appearance is bulging, necking, or a "cold" or weak splice.

Gas Bubbles in Splice. Gas bubbles have two causes. The first cause is an end that is not perpendicular to the axis of the fiber. The second cause is gasses dissolved in the fiber. These gasses occasionally evolve during splicing.

Bulging in Splice. Bulging in the splice indicates an high overrun setting. Reduce the overrun control by one setting and resplice. If bulging persists, reduce the overrun control an additional setting until overrun disappears.

Necking in Splice. Necking in the splice indicates low overrun setting. Increase the over-run control by one setting and resplice. If necking persists, increase the overrun control setting until necking disappears.

If necking persists, check the procedure you are following. If you do not push the RESET button before each splice (Step 24 of Procedure 1), necking can occur. If you do not position the fibers so that they touch (Step 26 of Procedure 1), necking can occur.

"Cold" or Weak Splice. A cold or weak splice is usually, but not always, visible under the microscope. This splice will have a clearly definable line across the diameter of the fiber.

A weak splice indicates low arc current or low splice time. To test for low arc current, place one fiber in the splicer. Align the fiber so that its end is in line with an imaginary line connecting the electrode tips. Push REHEAT. The end should be slightly beaded, with a diameter slightly larger than that of the fiber. Slightly larger means an increase of less than 10 percent in diameter.

If the fiber will not bead slightly with several increases in arc current, the electrodes may need cleaning or replacement. The battery may have lost its charge.

Well Defined Splice. A well defined splice has a clearly defined interface between the two fibers. A well defined splice is a weak splice. The best made splices are not distinguishable from unspliced fiber. See Weak Splice.

Fiber Ends Not Centered between Electrodes. Fiber ends not centered between electrodes indicates the need for alignment of the electrodes. See instruction manual for the alignment procedure.

PROCEDURE 2: SPLICING LONG LENGTHS OF FIBER

1. Unreel approximately 20 m of fiber from the outer end of a long length of fiber on a reel.
2. Unspool the inner end of the fiber from the reel. Attach this end to the light source according to Steps 1–7 of Procedure 1.
3. Prepare the outer end of the fiber reel and attach this end to the power meter according to Steps 8–10 of Procedure 1.
4. Repeat Step 11 of Procedure 1.
5. Break the fiber approximately 10 m from the power meter.
6. Repeat Steps 12–31 of Procedure 1.

Troubleshooting

See Troubleshooting of Procedure 1.

Question

Were the losses from Procedure 2 greater than or less than those from Procedure 1? In general, the losses from Procedure 2 will be less than those from Procedure 1. The reason for this reduction is differential modal attenuation.

Splicing Fiber to Tight Tube Cables

Tight buffer tube cables or tight tube pigtails can be spliced to fibers from loose buffer tube cables. These tight tube cables will be either a single fiber cable or a single fiber from a premise cable design. When splicing loose tube fibers to tight tube cables, the length from the end of the buffer tube to the cleave must be controlled to avoid misalignment. Misalignment can result from the large difference between the diameter of the buffer coating of the fiber (250 μm) and the diameter of the buffer tube of the cable (900 μm). The specific length of cleaved fiber will depend on the splicer. The 13 mm dimension can be used with 900 μm cables and the Preformed Line Products Fiberlign® Micro Splicer.

MECHANICAL SPLICING

In this section, we present the procedure for mechanical splicing. This procedure is designed around the Finger Splice from AMP Inc. With minor modifications, this procedure can be used with other mechanical splices. One of the modifications will be the length of the cleaved fiber. If you wish to practice singlemode fiber splicing, use a singlemode fiber with a light source and wavelength calibrated at 1310 nm. If you wish to practice a splicing technique for measurement at 1300 nm, use a 1300 nm light source and a power meter calibrated at 1300 nm.

TOOLS AND SUPPLIES REQUIRED

- at least 15 feet of all-glass, 250 μm buffer coated multimode optical fiber
- two bare fiber adapters with same connector style as the power meter and light source
- power meter with calibration at 850 nm
- 850 nm stabilized light source
- buffer coating stripper (Clauss part number NN200 or equivalent)
- Scotch™ tape
- cleaver (Fitel part number VL-310 or equivalent)
- index matching gel (Fiber Instrument Sales part number F1-0001)
- lens grade compressed air
- 91–95 percent isopropyl alcohol
- AMP Finger Splice (part number 608119-1)

PROCEDURE 3: MECHANICAL SPLICING

1. Obtain all tools and supplies
2. Follow Steps 1–11 of Procedure 1 to obtain a reference signal.
3. Break the fiber in the middle.
4. Follow Steps 12 and 13 of Procedure 1, but place the end of the buffer coating at 6–7 mm, instead of 13 mm.
5. Clean the cleaved end by placing the end of the fiber against a clean piece of Scotch Magic™ tape.
6. Slide both clamping rings so that they are near the ends of the splice (Figure 6–15). If this is not the first use of this splice, insert this fiber end into index matching gel.
7. With your thumb, press down on one end of the clear plastic section in the center of the splice (the "nose," Figure 6–16).
8. While maintaining pressure on the nose, insert the fiber into the end of the splice opposite to your thumb.
9. While sliding the fiber into the splice, rotate the splice back and forth. Insert the fiber until it meets the limit stop inside the splice.

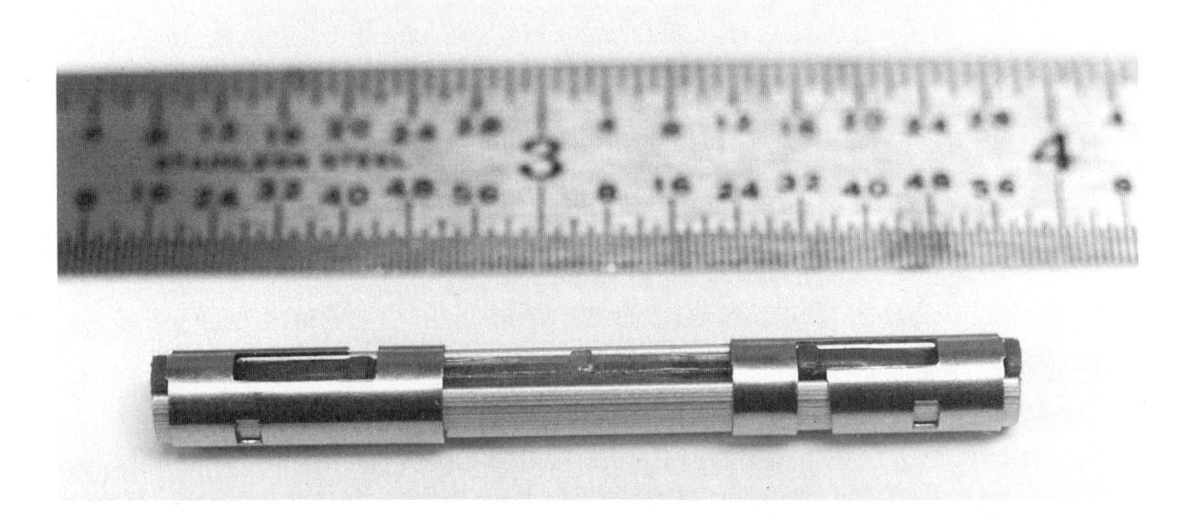

Figure 6–15 Mechanical Splice with Clamps at Ends (Open Position) (Courtesy of Pearson Technologies Inc.)

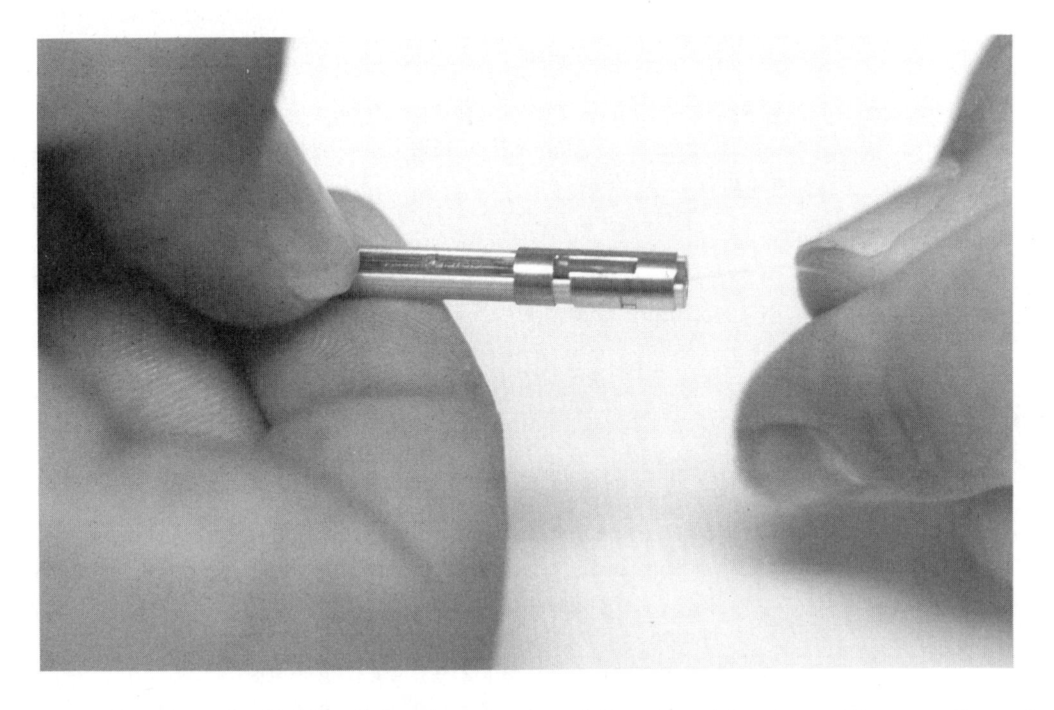

Figure 6–16 Proper Technique for Holding Splice for Insertion of First Fiber (Courtesy of Pearson Technologies Inc.)

10. Withdraw the fiber approximately 0.5 mm (0.020 inch).
11. Move the clamping ring toward the center of the splice until it will move no further. The fiber is fixed in the splice.
12. Repeat Steps 4 and 5 on the second end of the fiber.
13. Insert the second fiber end into the splice as in Step 8.
14. Place pressure on the fiber to keep it in the splice. Slide the second clamping ring towards the center of the splice until it will move no further.
15. Record the power meter reading and calculate the loss.
16. If the loss is less than 0.5 dB, remove both fibers and repeat the entire process. When repeating the process with this splice, dip the first fiber end in index matching gel. Dip just the end. If you put excessive gel on the sides of the fiber, the gel will prevent the two fiber ends from contact.

Troubleshooting Mechanical Splicing

If the loss is greater than 0.5 dB, slide one of the clamping rings to the end position. Rotate the fiber until you achieve a lowest loss reading on the power meter. Slide this clamping ring to the center position.

If the loss is still greater than 0.5 dB, slide the second clamping ring to the end position. Rotate the fiber until you achieve the lowest loss reading on the power meter. Slide this clamping ring to the center position.

If the loss is still greater than 0.5 dB, slide one clamping ring to the end position, remove the fiber, and repeat Steps 12–16.

If the loss is still greater than 0.5 dB, slide the second clamping ring to the end position, remove the fiber, and repeat Steps 12–16.

INSTALLING SPLICES IN TRAYS AND ENCLOSURES

Installing splices in trays requires actions before and after splicing. Before splicing, the buffer tube (usually a loose buffer tube from a stranded, multiple fiber per tube cable) is placed in and fastened to the tray, with a service loop of fiber. The service loop is approximately 3 feet long, but can be longer to allow removal of the fiber for splicing. After splicing, the tray will appear as in Figure 3–11.

Installing Splices in Trays

When placing the fiber in the tray, do not twist the fiber. In addition, do not bend the fiber to less than a 2.5-inch diameter. Twisting and bending will result in excess splice loss.

Installing Splice Trays in Enclosures

Installing splice trays in enclosures is straightforward. Most enclosures have a threaded stud. The splice tray will fit on the stud (Figures 6–2 and 6–3).

REVIEW QUESTIONS

1. For fusion splicing, did you find a correlation between the loss before splicing and the loss after splicing?
2. What correlation did you find between appearance of the cleaves and loss after fusion splicing? How do you explain the correlation you found?
3. Did you find a relationship between splices made at the transmitter end of a multimode cable and splices made at the receiver end of a multimode cable? How do you explain the correlation you found?
4. Describe the appearance of a good (low loss) cleave.
5. Describe the appearance of a bad (high loss) cleave.
6. Describe the appearance of a fusion splice made with excessive overrun.
7. Describe the appearance of a fusion splice made with insufficient overrun.
8. How can you test a splice to determine that the splice was a cold splice?
9. What is the function of the prefuse step?
10. What is the function of a splice cover on a fusion splice?

CHAPTER

7

How To Certify and Troubleshoot Fiber Systems

CHAPTER OBJECTIVES

From this chapter, you will be able to:

1. Identify the questions to ask before beginning troubleshooting.
2. Establish a logical sequence of troubleshooting steps.
3. Interpret the results of these steps.
4. Establish and verify your conclusions.
5. Recognize the most commonly occurring problems.

This chapter provides students with a guide to the processes of certifying and troubleshooting fiber optic cable and connector systems. It also provides a series of examples, from which students will reinforce their understanding of the methods of operation and failure of such systems. Finally, this section provides students with realistic problems, from which they can recognize the most-commonly occurring problems.

INTRODUCTION AND OVERVIEW OF THE SEVEN QUESTIONS

Certifying and troubleshooting might involve checking both optical power loss and pulse spreading in installed fiber optic cable systems. However, as demonstrated in Chapter 1, installation errors can both reduce and improve system performance. Installation errors can reduce system performance by reducing the optical power delivered to the receiver. We call this reduction excess loss. Installation errors can improve system performance by reducing the rate of pulse spreading. Since pulse spreading performance cannot be reduced, it will be ignored. Thus, certifying and troubleshooting is focused on interpretation of power loss measurements.

These measurements are insertion loss measurements and optical time domain reflectometer (OTDR) measurements. The insertion loss measurement simulates the power loss that the transmitter receiver pair will experience. The OTDR measurements reveal proper and improper installation techniques by measuring the loss of each element in the cable, splice, and connector system.

We presented the procedures for making these measurements in Chapter 5. In this chapter, we present the interpretation of these measurements.

Certifying and troubleshooting are similar, yet different. The procedures for certifying and troubleshooting fiber optic cable systems are the same: both require making and interpreting measurements. The difference between certifying and troubleshooting is the time at which such measurements and interpretations are done: certification is performed after the initial installation of the system; troubleshooting is performed during or after the initial installation.

The basic questions answered by certifying and troubleshooting are different. In certifying, the basic questions are: Is the system acceptable? Has the system been properly installed? In troubleshooting, the basic question is: Where is/are the location(s) of unacceptable (high) loss conditions?

DETAILED REVIEW OF THE SEVEN QUESTIONS

In order to certify and troubleshoot a specific fiber optic link, you must answer seven questions. These seven questions need to be answered in approximately the order presented. The first, and most important question is: What are the link characteristics. These characteristics are listed in Table 7–1. Knowledge of these characteristics will enable you to set up the test equipment properly, to calculate the maximum loss, to calculate the expected loss, and to determine whether the optoelectronics will function at the specified accuracy.

The second question is: Is the link continuous? If the link is not continuous, the power delivered to the receiver will, obviously, be insufficient.

The third question is: If the link is continuous, is the loss sufficiently low? "Sufficiently low" can have two meanings: is the loss sufficiently low to allow proper operation of the optoelectronics; and, is the loss sufficiently low to indicate proper installation of the cables and connections? The difference between these two meanings indicates the relative advantages and disadvantages of insertion loss and OTDR measurements. An improperly installed cable system can have an insertion loss less than both the maximum loss and the optical power budget. Therefore, the optoelectronics will function. However, an improperly installed cable system will usually have a loss greater than the expected loss. In this case, the system will function properly, but may fail at some time in the future, due, usually, to long-term stress placed on the fiber.

The fourth question is: If the loss is high (not sufficiently low), where is it high? OTDR testing (testing by OTDR or mini-OTDR) is required to reveal such locations of unnecessary, excess loss. Insertion loss testing will not reveal such locations.

The fifth question is: If the link is not continuous, where is the discontinuity? While some discontinuities can be located without an OTDR, OTDR testing is usually required.

core diameter (μm)
cladding diameter (μm)
NA
wavelength of operation of the optoelectronics (nm or μm)
maximum attenuation rate of the fiber at the wavelength of operation of the optoelectronics (dB/km)
maximum loss/pair of the connectors (dB/pair)
maximum loss/splice (dB/splice)
typical attenuation rate of the fiber (dB/km)
typical loss/pair of the connectors (dB/pair)
typical loss/splice (dB/splice)
repeatability of the connectors (dB/pair)
the index of refraction of the fiber at the wavelength of the optoelectronics
distance between the transmitter and receiver
number of splices
number of connectors
the difference between cable length and fiber length
and the optical power budget (dB)

Table 7–1 Link Characteristics

The sixth and seventh questions to be answered apply to troubleshooting. These questions are: What were the as-built measurements? And, what are the symptoms?

Occasionally, due to a design error, you will need to answer an eighth question: Is the loss too low? If the loss is too low, the receiving optoelectronics will have excessive optical power. Such overloaded optoelectronics will not function properly.

In order to answer the fourth through seventh questions, you need to perform testing. In this section, we will review the tests you can, and should, conduct, in order to identify the source of the problem.

Question 1: What Are the Link Characteristics?

The core diameter and NA of the cables to be tested must be the same as those of the lead in and lead out cables used in the insertion loss test. If the core diameter and NA are not proper, the loss measurements will be higher than anticipated and not representative of the loss that the optoelectronics will experience.

Question 2: Is the Link Continuous?

A continuity test is the first test to make on a link that does not function. A continuity test is performed before a power loss measurement for two reasons: first, a continuity test is the simplest to make; and second, an extremely high loss indicative of lack of continuity will require such a test.

There are three types of continuity tests: white light, red light, and power meter.[1] White light testing is done with a high intensity light at one end of a fiber. If the fiber is continuous, the white light will appear at the output end of the fiber. While we do not know the maximum length to which white light testing can be done, we have performed this testing to the following distances: up to 1,500 feet with a 50 μm fiber; up to 4,000 feet with a 62.5 μm fiber; and up to 3,100 feet with a 100 μm fiber.

These distances can be increased by using a microscope. It is easier to use a microscope for white light testing after the connectors are installed.

It is also possible to to perform continuity tests with a high intensity red light source. However, some of these sources are lasers. Microscopes cannot be used with laser sources.

For very long cables, for which white light or red light cannot be used for continuity testing, a power meter and light source of the appropriate wavelength can be connected to the cable under test. After connection, the light source is turned on and off. If the cable is continuous, the light meter reading will increase when the source is on.

Question 3: Is the Loss Sufficiently Low?

In order to answer this question, you will perform two calculations and compare their results to the measured insertion loss. The first calculation is of the maximum loss that the link can have. Perform this calculation as shown (Equation 7–1).

(Equation 7–1)

$$\text{maximum loss} = \text{maximum cable loss} + \text{maximum connector loss}$$
$$= (\text{length of cable, km}) \times (\text{maximum loss/km}) +$$
$$(\text{number of connector pairs}) \times (\text{maximum loss/pair, dB/pair})$$

You can obtain these values from the person who designed the system or from data sheets for the products installed. The maximum loss must be less than the optical power budget of the optoelectronics. Otherwise, the link will not function.

You can obtain the optical power budget from the person who designed the system or from data sheets for the products installed.

The second calculation is of the expected loss that the link should have. Perform this calculation as shown (Equation 7–2).

(Equation 7-2)

expected loss = average cable loss+ average connector loss
= (length of cable, km) × (average loss/km, dB/Km) +
(number of connector pairs) × (average loss/pair, dB/pair)

The measured loss should, in general, be closer to the expected loss than to the maximum loss. A measured loss close to the expected loss indicates proper installation and maintenance. A measured loss close to the maximum loss indicates improper installation and maintenance.

For example, consider the cable link in Figure 7–1. If this link contained ST connectors and 62.5/125 fiber specified at 850 nm, the maximum and average connector losses would be 1.0 dB/pair and 0.3 dB/pair. The maximum and average cable attenuation rates would be 3.75 dB/km and 3.2 dB/km. With these values, the maximum loss would be 9.5 dB; the average loss would be 6.8 dB. This range of loss values means that the cable could be improperly installed with an excess loss of up to 2.7 dB without exceeding the maximum loss.

Question 4: Where Is the Loss High?

In order to answer this question, you will need to make an OTDR trace. From that trace, you determine the losses of each cable segment and of each connector pair (and splice) in the link. You will compare the OTDR loss of each cable segment to the maximum loss and expected loss for that segment. You will compare the OTDR loss of each connector pair to the maximum and average connector loss. From these comparisons, you will be able to identify the location(s) at which the loss is higher than expected.

Question 5: Where Is the Discontinuity?

From an OTDR trace, you will identify the location at which an unexpected end reflection appears. With the OTDR calibrated to the index of refraction of the fiber under test, the OTDR will provide a length measurement. You will reduce the OTDR distance measurement with a correction factor. This correction factor is required because fiber length is greater than cable length.[2]

Question 6: What Were the As-Built Measurements?

During troubleshooting, you will compare current insertion loss and OTDR measurements to those made after the initial installation. If there is any significant difference between current and initial measurements, you will need to continue your

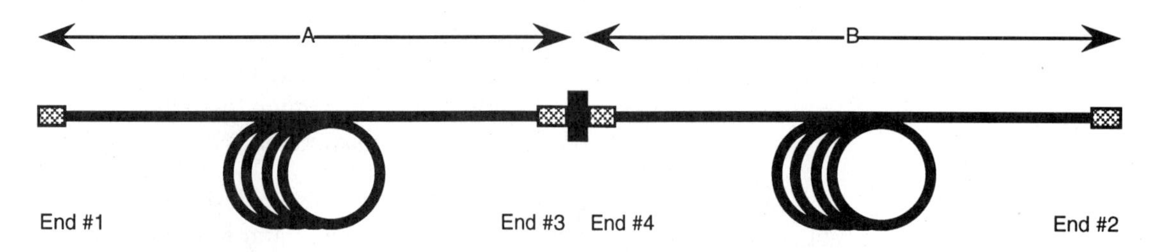

Figure 7–1 Cable Link for Loss Calculations

troubleshooting of the cable system. If there is no significant difference, you will troubleshoot the optoelectronics.

The key to answering this question is the term "significant difference." A significant difference is a difference larger than the value that the repeatability of the connectors would lead you to expect.

The significant difference is different for insertion loss measurements and for OTDR measurements. For insertion loss measurements, the significant difference is equal to the number of connector pairs in the link times the repeatability. For example, ST-compatible style connectors have a typical repeatability of 0.2 dB/pair. The range (the difference between the maximum and minimum loss measured during multiple measurements of the same cable assembly) is usually double the repeatability (0.4 dB/pair). Since the cable system in Figure 7–1 has two pairs, the range of loss measurements could be as large as 0.8 dB. Thus, an increase in insertion loss of more than 0.8 dB is interpreted as a problem requiring troubleshooting.

However, our work in training indicates ST-compatible style connectors have an average range of 0.31 dB/pair (Table 1–7). Using this value, any increase in loss of more than 0.62 dB can be interpreted as a problem requiring additional troubleshooting.

For OTDR measurements, the repeatability is applied to each connector pair. If an OTDR measurement of the center pair in Figure 7–1 exhibited an increase in loss of less than 0.31 dB, there would be no reason for additional troubleshooting.

In some cases, the as-built measurements will not be available. In such cases, you will compare the measured losses to those calculated with Equations 7–1 and 7–2.

Question 7: What Are the Symptoms?

Symptoms of a malfunctioning system can either assist you in identifying the cause of the problem or work against you. If the symptoms are accurately described, the symptoms will assist you. Since fiber optics is a relatively new technology, many people do not how to describe the malfunction accurately. Because of this newness, the symptoms may misdirect you and work against rapid identification of the problem.

IDENTIFY COMMON PROBLEMS IN ELEVEN CABLE SYSTEMS

In this section, we will present insertion loss measurements and OTDR traces for eleven cable systems. From these measurements and traces, we will be able to identify locations of high loss and locations of discontinuity.

Common System Characteristics

Unless otherwise indicated, the following characteristics apply to all troubleshooting examples. The systems consist of 62.5 μm fibers, with ST-compatible connectors at each end. All length tolerances are -0, +10 m. The link optoelectronics operate at a wavelength of 850 nm. The fiber NA is 0.275 nominal.

The ST-compatible connectors have a maximum loss of 1.0 dB/pair and an average loss of 0.3 dB/pair. The fiber has a maximum attenuation rate of 4.0 dB/km and a typical rate of 3.2 dB/km.

All insertion loss measurements were performed according to EIA/TIA-526-14. The test leads had 62.5 μm, 0.275 NA fibers, and low loss ST-compatible connectors.

All OTDR traces (for which we used a Tektronix FiberMaster TFP2) were made with a 25 m lead in cable with the same core diameter and NA as in the segment(s). This 25 m lead in cable has low loss ST-compatible connectors.

Troubleshooting—Example 1

Common system characteristics apply to Example 1. The two 100 m segments are spliced together (Figure 7–2). The splice was made against a maximum loss of 0.3 dB, with an expected average loss of 0.2 dB.

From this information and Equations 7–1 and 7–2, we can calculate the maximum and expected losses. The maximum loss is:

cable loss = (0.200 + 0.02) km × 4 dB/km =	0.88 dB
connector loss = 1 pair × 1 dB/pair =	1.00 dB
splice loss =	0.30 dB
	2.18 dB

The expected loss is:

cable loss = (0.200 + 0.02) km × 3.2 dB/km =	0.70 dB
connector loss = 1 pair × 1 dB/pair =	0.30 dB
splice loss =	0.20 dB
	1.20 dB

This link passes a white light test. It is continuous.

Since the typical optical power budget for the 62.5 μm fiber at a wavelength of 850 nm is 16 dB (Table 1–9), this link should function. However, we do not know whether this link has been installed correctly.

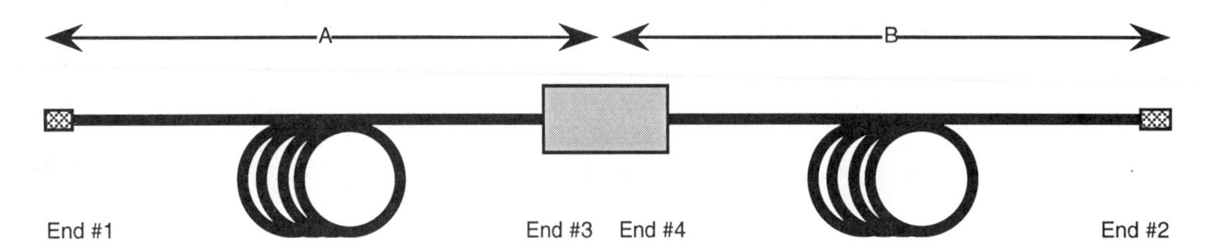

Figure 7–2 Map of Troubleshooting Examples 1–4

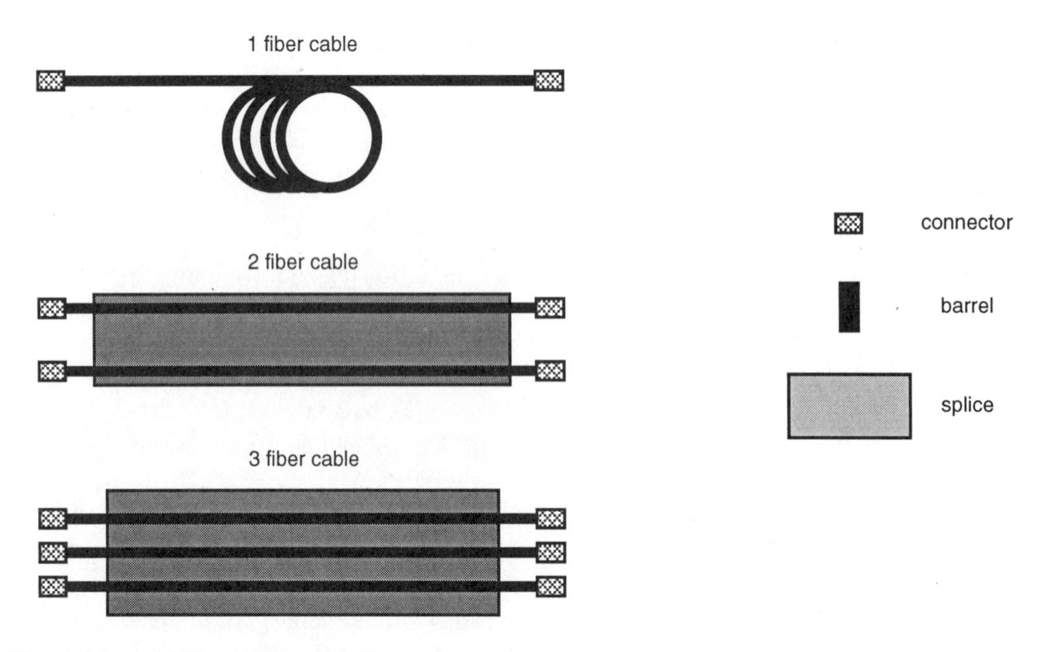

Figure 7–3 Key to Maps of Troubleshooting Examples

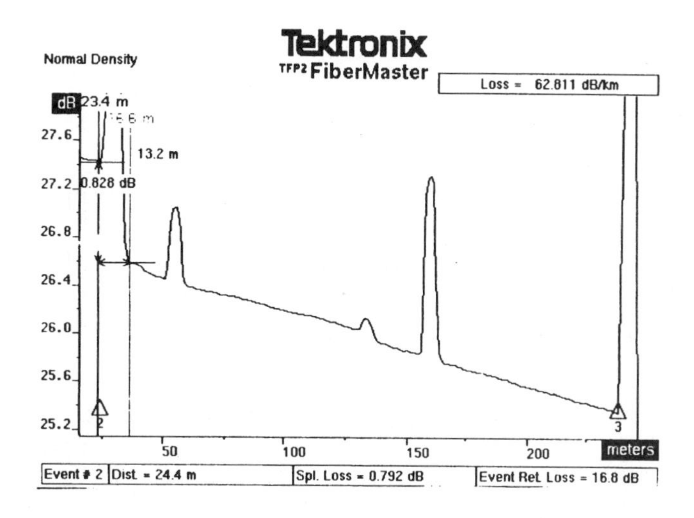

Figure 7–4 OTDR Trace, Example 1, End 1, Connector Loss

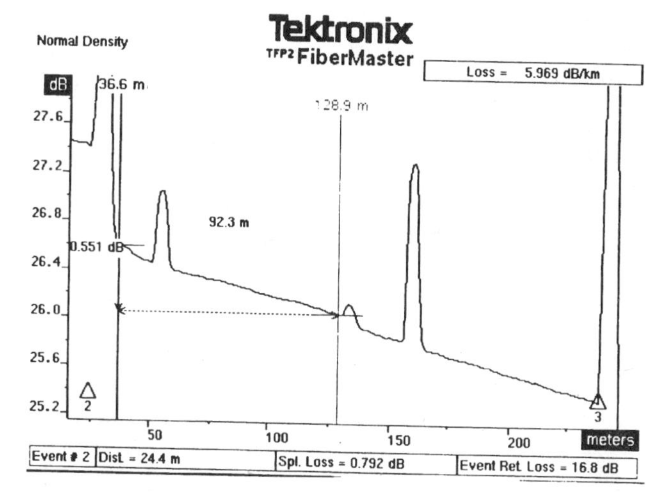

Figure 7–5 OTDR Trace, Example 1, Segment A, Cable Attenuation Rate

A power meter insertion loss test provides losses of 3.8 dB with the light source at End 1 and 3.9 dB with the light source at End 2. These losses are higher than anticipated.

From End 1 (Figure 7–4), the link apparently has three peaks, or reflections. Reflections can result from mechanical splices, improperly made fusion splices, connector pairs, from a break in a tight tube cable, or from multiple reflections (echoes). We will present an evaluation of these peaks after we review the trace from End 2.

From End 1 (Figure 7–4), the loss of the connector pair at End 1 is 0.828 dB. This loss is higher than the average, but less than the maximum specified (1 dB/pair).

The attenuation rate of Segment A (Figure 7–5) is 5.969 dB/km. This rate is higher than specified.

Note that the cursor at 36.6 m is not correctly placed: it is in the decay of the peak, not in the linear section. However, this slight misplacement will not result in a significant increase in the attenuation rate.

From End 2 (Figure 7–6), we see no reflections. From this trace, we can conclude that the peaks evident from End 1 (Figure 7–4) are echoes. If these three reflections were not echoes, reflections would appear in both directions.

The loss of the connector at End 2 (Figure 7–6) is 0.663 dB. This loss is higher than the average, but less than the maximum.

The attenuation rate of Segment B (Figure 7–7) is 4.926 dB/km. This rate is higher than specified.

Figure 7–8 indicates a non-reflective, non-uniform loss at a distance of 127.9 m from the end. This distance means that the splice is 102.9 m from End 2 (127.9 - 25). This distance is in agreement with the map (100 m, -0 m, +10 m).

Since the splice exhibits no reflection, the splice must be a fusion splice. The two-point loss measurement of this splice (Figure 7–8) is 0.145 dB. This splice loss is less than both the maximum and average values. As such it is acceptable. This splice loss is approximately the same in both directions, since the loss from End 1 is 0.12 dB (Figure 7–9).

Summary, Example 1. In conclusion, the connectors at Ends 1 and 2 are acceptable. The splice in the center is also acceptable. However, the cable in Segments A and B have unacceptably high attenuation rates. Since the traces are linear without any non-uniform losses, the cable is probably installed correctly.

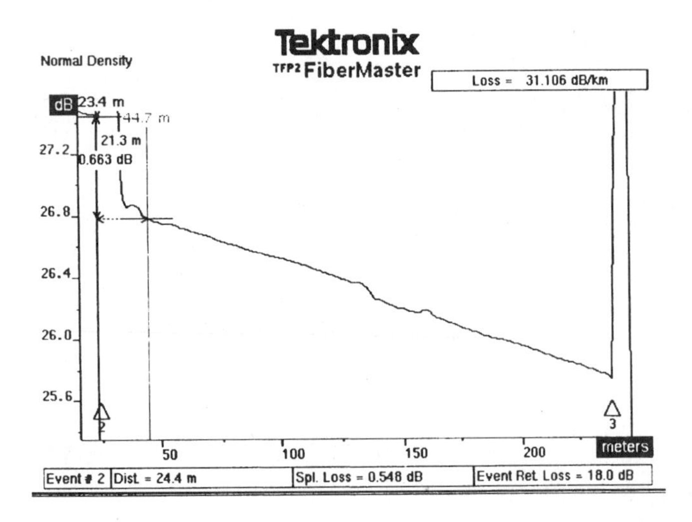

Figure 7-6 OTDR Trace, Example 1, End 2, Connector Loss

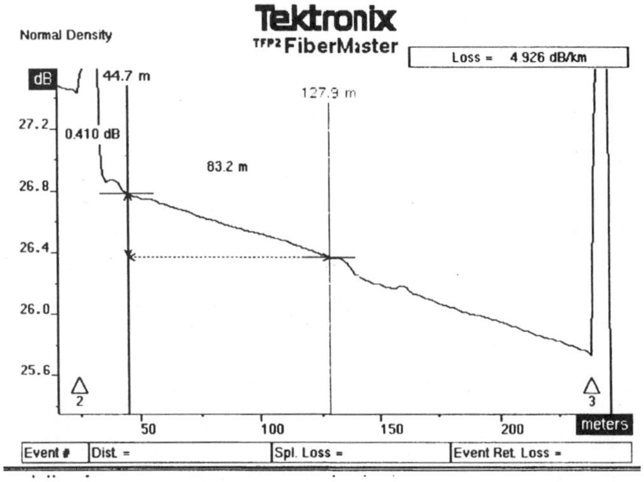

Figure 7-7 OTDR Trace, Example 1, End 2, Segment B, Cable Attenuation Rate

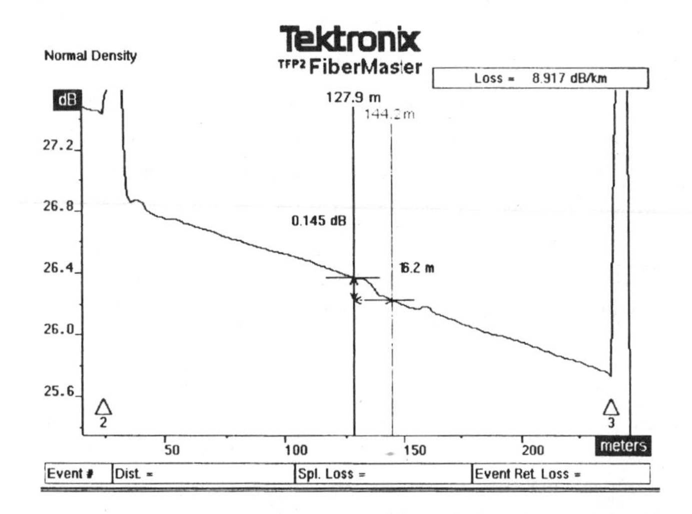

Figure 7-8 OTDR Trace, Example 1, End 2, Splice Loss

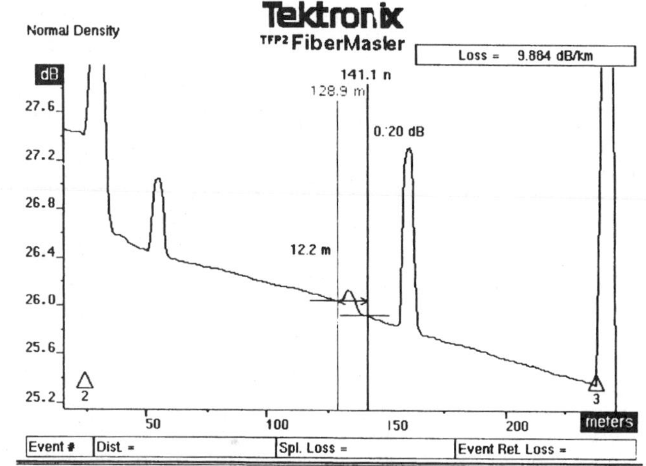

Figure 7-9 OTDR Trace, Example 1, End 1, Splice Loss

Therefore, the cable was probably high loss before installation. (The cable could also be under uniform stress along its entire length, but this is very unlikely.)

If we use the OTDR loss measurements to create a total, we obtain the following:

cable loss = (0.1039) km × 5.569 dB/km =	0.58 dB
cable loss = (0.1029) km × 4.926 dB/km =	0.51 dB
connector loss, End 1 =	0.83 dB
connector loss, End 2 =	0.66 dB
splice loss =	0.15 dB
	2.73 dB

This total double counts the connector at the end of the 25 m cable attached to the OTDR. This double counting results in a total higher than reality. It is not possible to remove this double counting without assuming the loss of the OTDR connector. We have no way of knowing whether or not this assumed loss is correct.

However, this total is less than the insertion loss of 3.8 and 3.9 dB. Note that we cannot completely correlate the insertion loss and OTDR measurements.

Some of this difference could be due to differences between the connector on the end of the 25 m OTDR cable and the connectors on the insertion loss lead in and lead out cables. Such differences will be small, since all connectors were 3M connectors with ceramic ferrules. Such connectors have low losses, with an average of 0.3 dB/pair.

We suspect that most of this difference is due to differential modal attenuation. The reference signal for insertion loss was made with lead in cable of approximately 1 m, while the OTDR connector loss was made with a total of 125 m. There is an internal fiber of 100 m and an external cable of 25 m. This difference in lengths results in a focusing of the light closer to the core in the OTDR lead in cable than in the insertion loss lead in cable. This focusing can result in a lower connector loss from the OTDR lead than from the insertion loss lead.

There will be an additional focusing effect of the connectors at the OTDR and at the end of the 25 m cable. These connectors strip the light from the outer regions of the core. Such stripping will result in reduced loss measurements from connectors after these connectors.

Since we cannot account for most of the excess loss from the OTDR measurements, we must perform a microscopic inspection of the connectors at Ends 1 and 2. This inspection reveals the connectors to be damaged. From this inspection, we can conclude that the high loss of this cable system is due to high loss from damaged connectors and high attenuation rate cable segments.

Troubleshooting—Example 2

Common system characteristics apply to Example 2. The segments are spliced together (Figure 7–2). The splice was made against a maximum loss of 0.3 dB, with an expected average loss of 0.2 dB.

From this information and Equations 7–1 and 7–2, we can calculate the maximum and expected losses. As shown for Example 1, the maximum loss is 2.18 dB and the expected loss is 1.20 dB.

This link passes a white light test. It is continuous.

An insertion loss test provides losses of 3.0 dB with the light source at End 1 and 2.8 dB with the light source at End 2. These losses are higher than expected.

Since the typical optical power budget for the 62.5 μm fiber at a wavelength of 850 nm is 16 dB (Table 1–9), this link should function. However, we do not know whether this link has been installed correctly.

From End 1 (Figure 7–10), the link exhibits two segments of uniform attenuation with a non-uniform loss in the middle. No peaks appear between the beginning and end.

From this end (Figure 7–10), the loss of the connector pair at End 1 is 0.787 dB. This loss is higher than the average, but less than the maximum specified (1 dB/pair).

The attenuation rate of Segment A is 3.417 dB/km (Figure 7–11). This value is slightly higher than the average, but less than the maximum. As such, this rate is acceptable.

The two-point loss measurement of the splice (Figure 7–12) is 0.335 dB. Since this value is higher than the maximum allowable, this splice appears to be unacceptable.

The loss of Segment B (Figure 7–13) is 5.102 dB/km. Since this value is higher than the maximum allowable, this segment is unacceptable.

From End 2 (Figure 7–14), the splice loss measured by the splice loss technique is 0.256 dB. The splice loss technique determines the loss by creating a linear approximation of the cable on both sides of the splice. The loss is determined

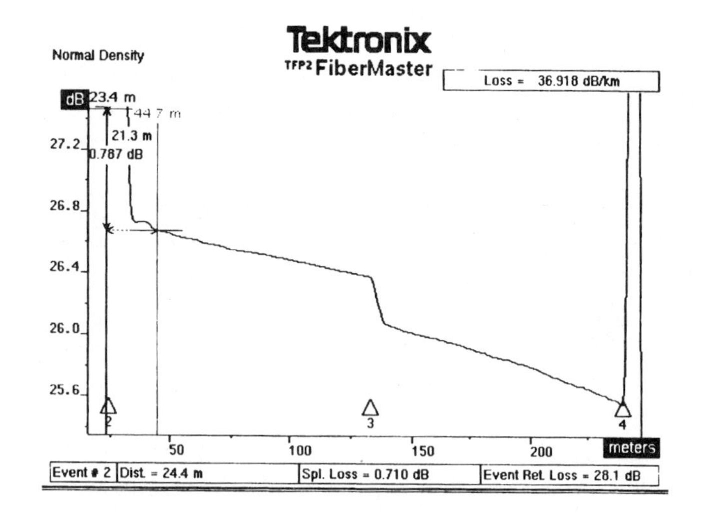

Figure 7–10 OTDR Trace, Example 2, End 1, Connector Loss

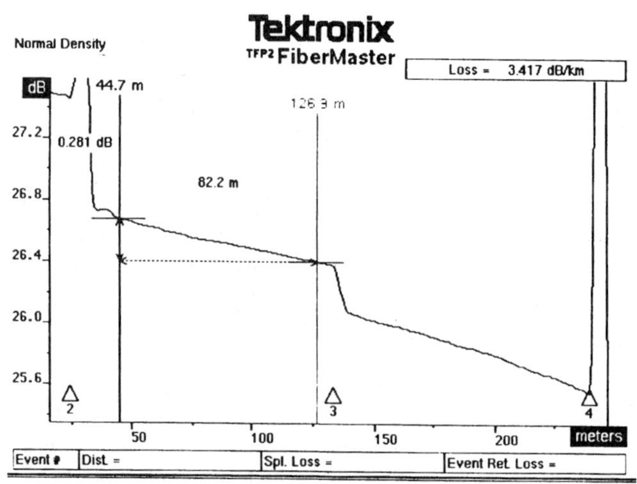

Figure 7–11 OTDR Trace, Example 2, End 1, Segment A Cable Attenuation Rate

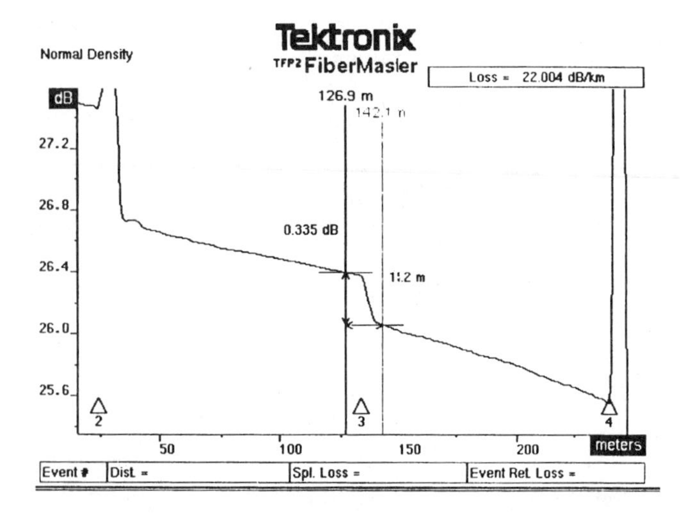

Figure 7–12 OTDR Trace, Example 2, End 1, Two-Point Splice Loss

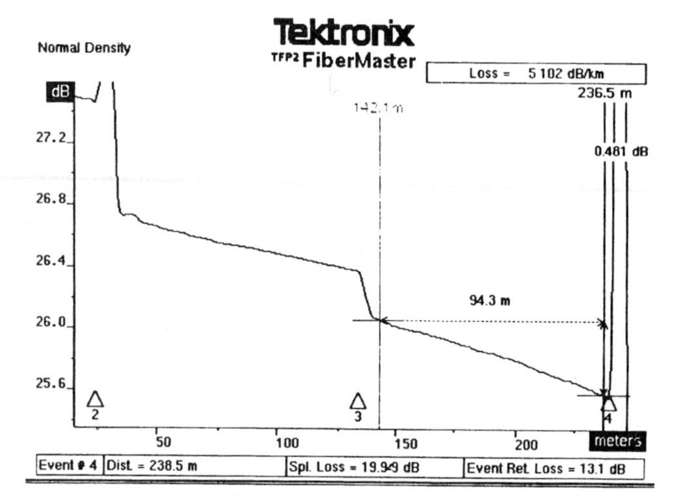

Figure 7–13 OTDR Trace, Example 2, End 1, Segment B Cable Attenuation Rate

from the difference in the linear approximations at the location of the splice. Note that the two-point (Figure 7–12) and splice loss (Figure 7–14) techniques result in different answers. Part of this difference is due to the inclusion of cable loss between the cursors. This loss is included in the two-point technique, but not the splice loss technique. Part of this difference is due to the difference in direction. Splices and connectors usually have different losses in different directions.

This example reveals the importance of specifying the technique, either two-point or splice loss, by which splice loss is to be measured. Common practice is measurement of the splice loss in both directions. If the loss in one direction exceeds the limit, the splice can be accepted on the basis of the average of the losses in both directions. On short length cables with some OTDRs, the splice loss technique cannot be used.

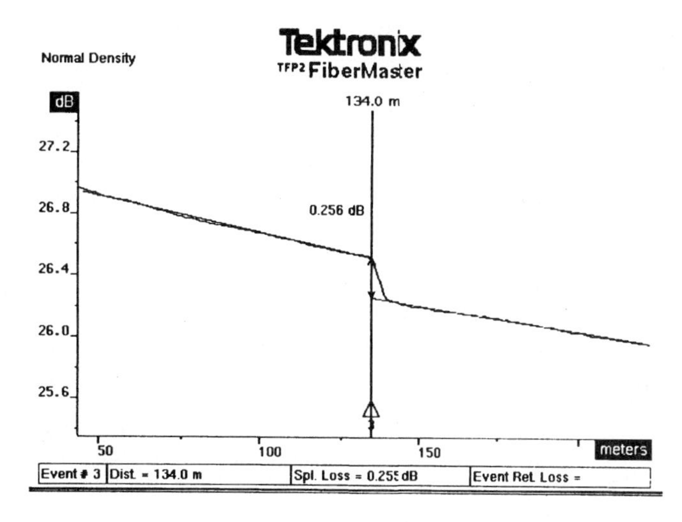

Figure 7–14 OTDR Trace, Example 2, End 2, Splice Loss

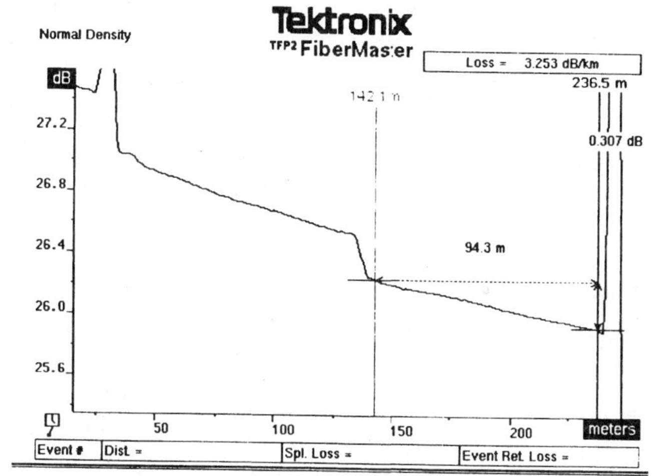

Figure 7–15 OTDR Trace, Example 2, End 2, Segment B Cable Attenuation Rate

Figure 7–14 clearly indicates a non-reflective, non-uniform loss at a distance of 134.0 m from the end. This position places the splice at a distance of 108.0 m from the end (134.0 m - 25 m). This distance is in agreement with the map (100 m -0 m, +10 m). Since the splice exhibits no reflection, it must be a fusion splice.

OTDR measurements exhibit directional differences, as do insertion loss measurements. Compare the loss of Segment A from End 1 (3.417 dB/km, Figure 7–11) to that from End 2 (3.253 dB/km, Figure 7–15).

From End 2 (Figure 7–16), the loss of the connector pair at End 2 is 0.401 dB. This loss is higher than the average, but less than the maximum specified (1 dB/pair).

As in Example 1, we cannot completely explain the high insertion loss from the OTDR measurements. The slightly high OTDR losses of the connectors at Ends 1 and 2 are suspect, as in Example 1. Microscopic examination of these connectors reveals damaged cores.

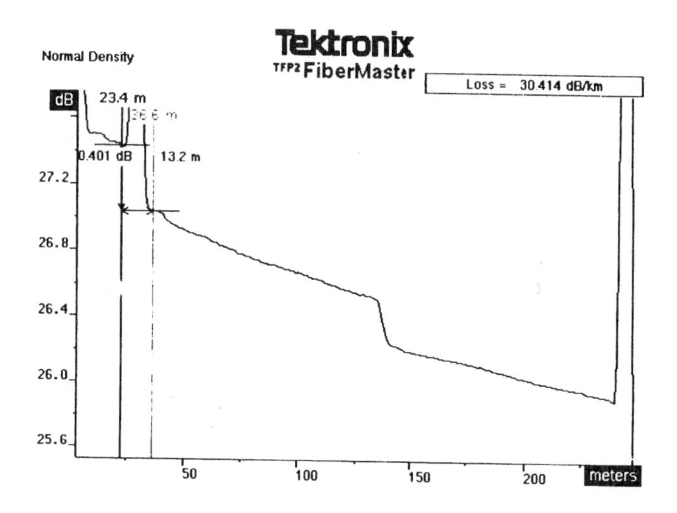

Figure 7–16 OTDR Trace, Example 2, End 2, Connector Loss

Summary, Example 2. In summary, the connectors at Ends 1 and 2 are unacceptable. The splice in the center may or may not be acceptable, depending on how the splice is measured and on whether or not averaging is allowed. The cable in Segment B has an unacceptably high attenuation rate. Since the traces are linear with no non-uniform losses, the cable is probably installed correctly. Therefore, Segment B was high loss before installation.

Troubleshooting—Example 3

Common system characteristics apply to Example 3. The segments are spliced together (Figure 7–2). The splice was made against a maximum loss of 0.3 dB, with an expected average loss of 0.2 dB.

From this information and Equations 7–1 and 7–2, we can calculate the maximum and expected losses. As shown for Example 1, the maximum loss is 2.16 dB and the expected loss is 1.20 dB.

This link passes a white light test. It is continuous.

An insertion loss test provides losses of 2.8 dB with the light source at End 1 and 2.8 dB with the light source at End 2. These losses are unacceptably high.

Since the typical optical power budget for the 62.5 μm fiber at a wavelength of 850 nm is 16 dB (Table 1–9), this link should function. However, we do not know whether this link has been installed correctly.

From End 1 (Figure 7–17), the link apparently has one peak, or reflection. This reflection indicates a mechanical splice or a poorly made fusion splice. The mechanical splice is the correct identification.

From this end (Figure 7–17), the loss of the connector pair at End 1 is 0.666 dB. This loss is higher than the average, but less than the maximum specified (1 dB/pair).

The attenuation rate of Segment A (Figure 7–18) is 3.646 dB/km. This rate is acceptable, since it is lower than that specified.

Figure 7–18 indicates a reflective, non-uniform loss at a distance of 128.9 m from the end. This distance means that the splice is 103.9 m from the end (128.9 m - 25 m). This distance is in agreement with the map (100 m -0 m, +10 m). The two point loss is 0.337 dB (Figure 7–19), which is greater than the specified value. This splice appears to be unacceptable.

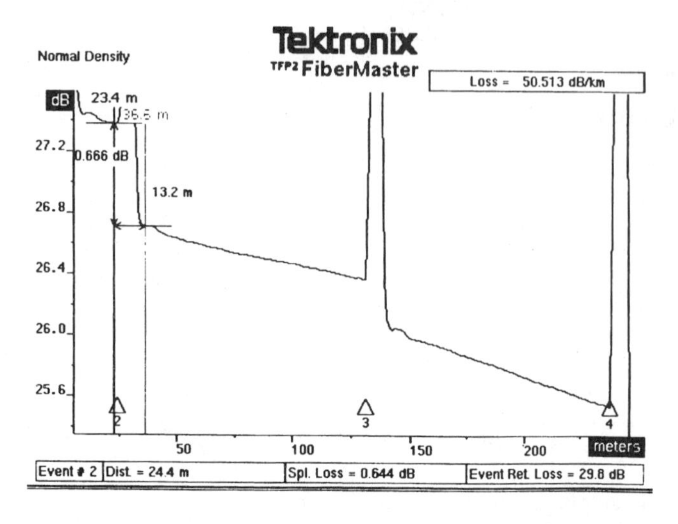

Figure 7–17 OTDR Trace, Example 3, End 1, Connector Loss

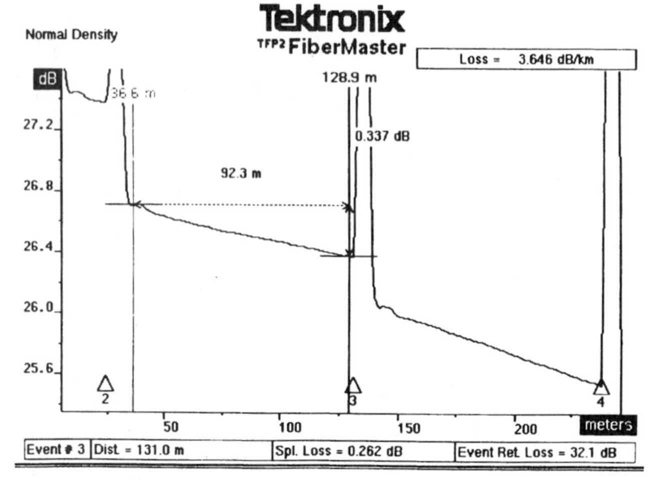

Figure 7–18 OTDR Trace, Example 3, End 1, Segment A Cable Attenuation Rate

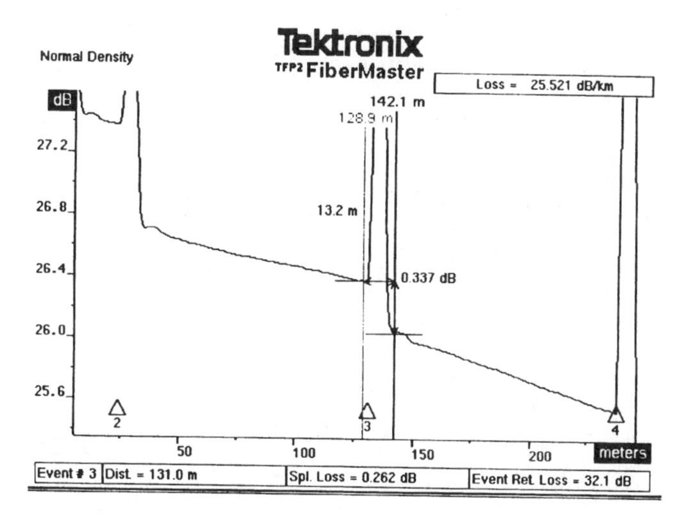

Figure 7–19 OTDR Trace, Example 3, End 1, Two-Point Splice Loss

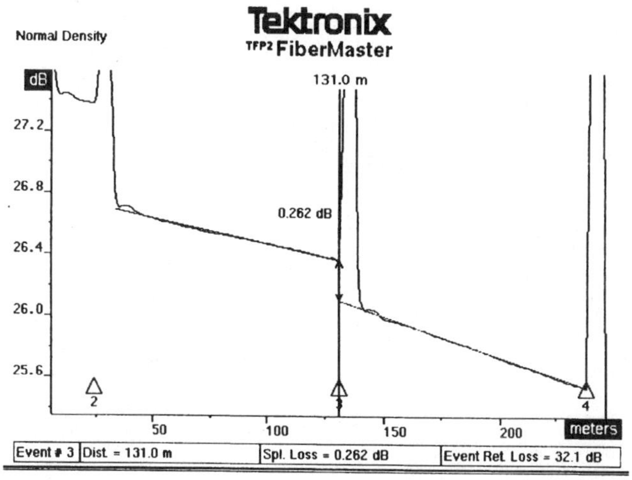

Figure 7–20 OTDR Trace, Example 3, End 1, Splice Loss

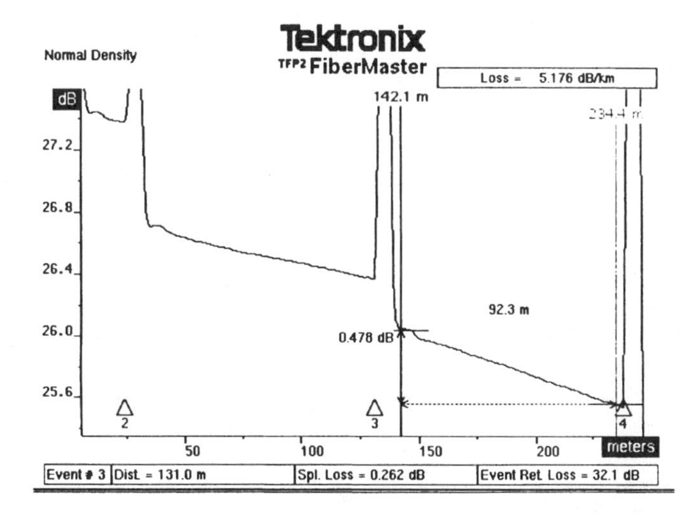

Figure 7–21 OTDR Trace, Example 3, End 1, Segment B Cable Attenuation Rate

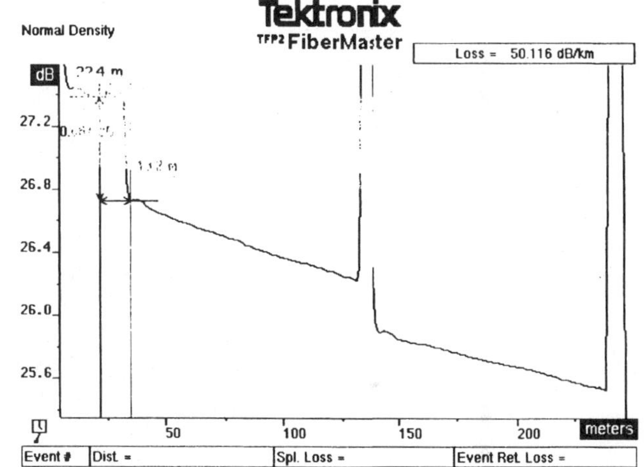

Figure 7–22 OTDR Trace, Example 3, End 2, Connector Loss

However, a splice loss measurement (Figure 7–20) indicates a lower loss, 0.262 dB. As in Example 2, acceptance or rejection of the splice depends on the measurement technique specified.

Finally, the attenuation rate of Segment B (Figure 7–21) is unacceptably high, 5.176 dB/km.

From End 2 (Figure 7–22), we see only the reflection in the center. The loss of the connector at End 2 (Figure 7–22) is 0.661 dB. This loss is higher than the average, but less than the maximum.

Summary, Example 3. In summary, the connectors at Ends 1 and 2 have losses higher than average, but acceptable. The splice in the center may be unacceptable, depending on the measurement method specified. The cable in Segment B has an unacceptably high attenuation rate. Since the traces are linear without any non-uniform losses, the cable is probably installed correctly. Therefore, the cable Segment B had high loss before installation.

From Figures 7–17 to 7–22, we can investigate directional effects. The attenuation rate of Segment A from End 2 (Figure 7–23) is 3.665 dB/km. This rate is approximately the same as that from End 1, 3.646 dB/km (Figure 7–18).

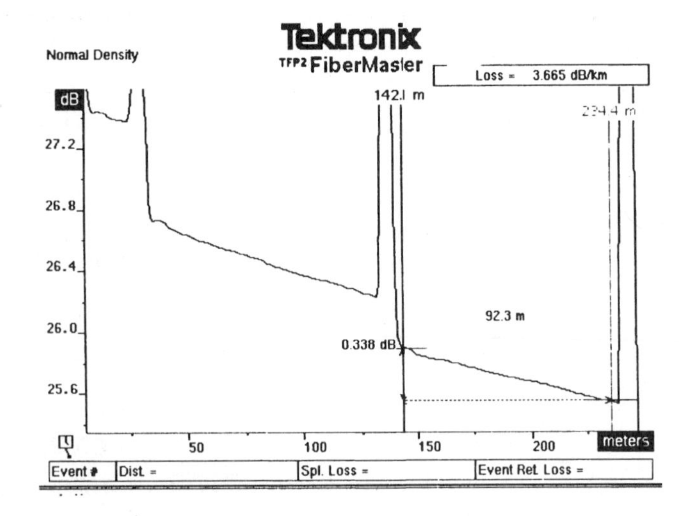

Figure 7–23 OTDR Trace, Example 3, End 2, Segment A Cable Attenuation Rate

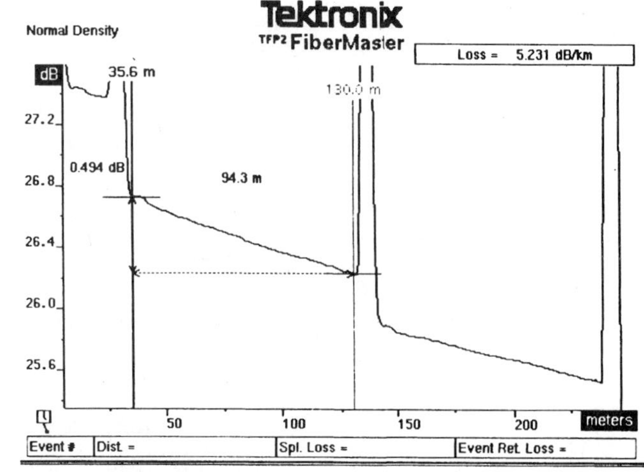

Figure 7–24 OTDR Trace, Example 3, End 2, Segment B Cable Attenuation Rate

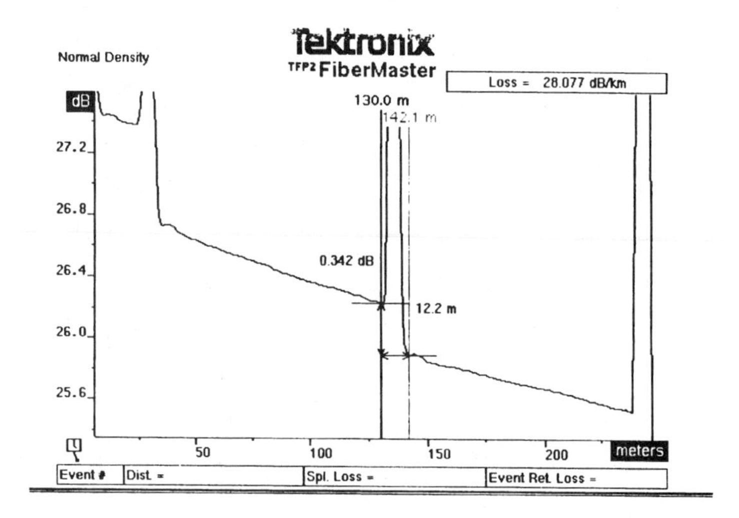

Figure 7–25 OTDR Trace, Example 3, End 2, Two-Point Splice Loss

The attenuation rate of Segment B from End 2 (Figure 7–24) is 5.231 dB/km. This rate is approximately the same as from End 1, 5.176 dB/km (Figure 7–21).

The two-point loss measurement of the splice (Figure 7–25) is 0.342 dB. This loss is approximately the same as from End 1, 0.337 dB (Figure 7–19). From these three comparisons, we can see that the directional effects, though present, are small.

Troubleshooting—Example 4

Common system characteristics apply to Example 4. The segments are spliced together (Figure 7–2). The splice was made against a maximum loss of 0.3 dB, with an expected average loss of 0.2 dB.

From this information and Equations 7–1 and 7–2, we can calculate the maximum and expected losses. As shown for Example 1, the maximum loss is 2.16 dB and the expected loss is 1.20 dB.

This link fails a white light test. It is not continuous.

An insertion loss test provides losses of 14.0 dB with the light source at End 1 and 14.0 dB with the light source at End 2. These losses are very high. (Note that failure of a white light test does not always indicate a discontinuous link.)

Since the typical optical power budget for the 62.5 μm fiber at a wavelength of 850 nm is 16 dB (Table 1–9), this link should function. However, we do know that this link has been installed incorrectly.

From End 1 (Figure 7–26), the link apparently has one peak, or reflection. Reflections can result from mechanical splices, poorly made fusion splices, connector pairs, from a break in a tight tube cable or from multiple reflections (echoes). This trace has two linear sections, which indicate uniform loss in the two segments.

From this end (Figure 7–26), the loss of the connector pair at End 1 is 0.535 dB. This loss is higher than the average, but less than the maximum specified (1 dB/pair).

The attenuation rate of Segment A (Figure 7–27) is 3.569 dB/km. Because the attenuation is linear and less than the maximum, this segment is acceptable.

The splice loss in the center is 2.658 dB (Figure 7–28), unacceptably high.

The attenuation rate of Segment B (Figure 7–29) is 4.175 dB/km. Because the attenuation is greater than the maximum, this segment is unacceptable.

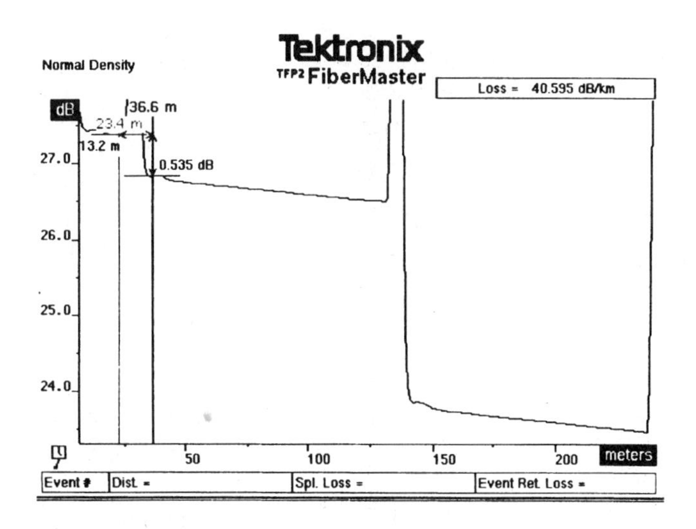

Figure 7–26 OTDR Trace, Example 4, End 1, Connector Loss

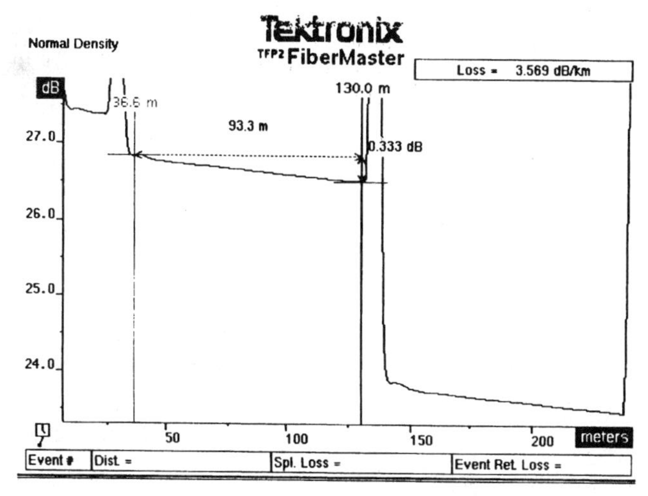

Figure 7–27 OTDR Trace, Example 4, End 1, Segment A Cable Attenuation Rate

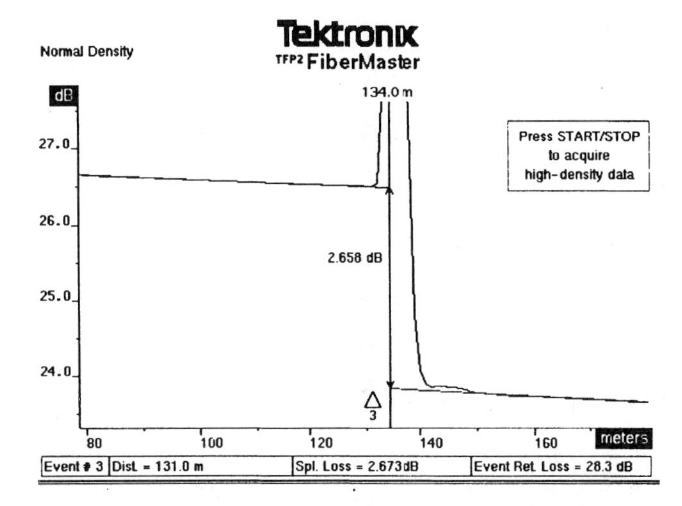

Figure 7–28 OTDR Trace, Example 4, End 1, Splice Loss

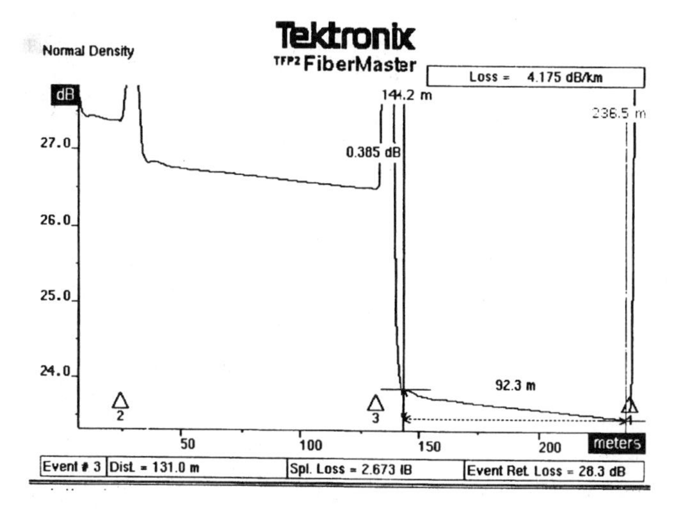

Figure 7–29 OTDR Trace, Example 4, End 1, Segment B Cable Attenuation Rate

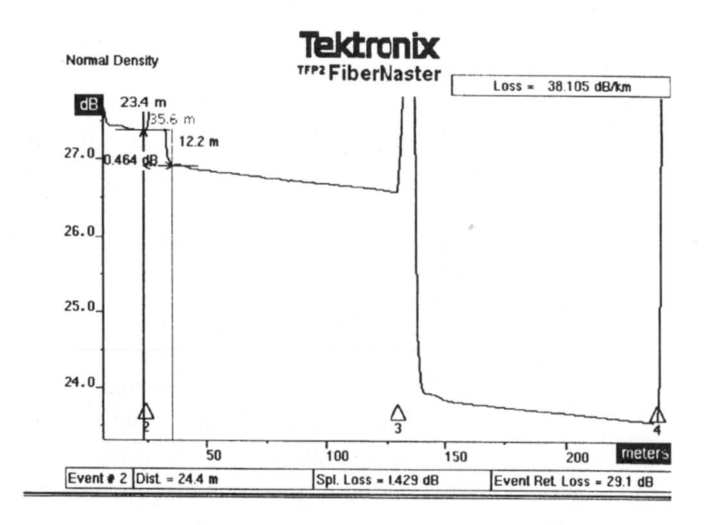

Figure 7–30 OTDR Trace, Example 4, End 2, Connector Loss

From End 2, the connector loss is 0.464 dB (Figure 7–30). This loss is higher than the average, but less than the maximum specified (1 dB/pair).

Summary, Example 4. In summary, the connectors at Ends 1 and 2 are acceptable. The splice in the center is unacceptable. The cable in Segment B has an unacceptably high attenuation rate. Since the traces are linear without any non-uniform losses, the cable Segment B is probably installed correctly. Therefore, the cable was high loss before installation.

Troubleshooting—Example 5

This system consists of one segment of tight tube cable with a 960/1000 μm plastic optical fiber (POF). The segment is 165 m long (Figure 7–31), with ST-compatible style connectors at each end. The link optoelectronics operate at 660 nm. The fiber NA is 0.51 nominal.

This link did not operate when connected to the RS-232 optoelectronics. The link passed a white light test. It is continuous. The electronics worked when attached by a 1 m length of cable from the same reel from which the 165 m length was cut.

Since the optoelectronics operate at 19.2 kbps, pulse spreading is not the problem. However, POF is limited to short distances (<100 m) due to high attenuation rates (125–250 dB/km). Since this link is more than 100 m in length, the likely cause of failure is excessive loss.

The loss calculations from Equations 7–1 and 7–2 support this conclusion. The maximum loss is:

cable loss = (0.165) km × 250 dB/km =	41.25 dB
connector loss = 1 pair × 2 dB/pair =	2.00 dB
	43.25 dB

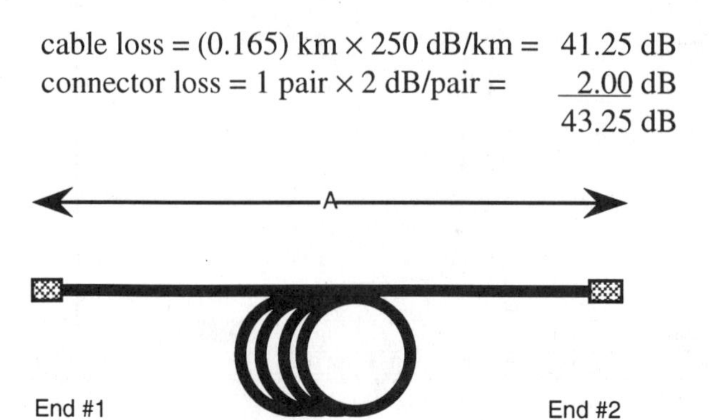

Figure 7–31 Map of Troubleshooting Example 5

The minimum loss is:

cable loss = (0.165) km × 125 dB/km =	20.63 dB
connector loss = 1 pair × 1 dB/pair =	1.00 dB
	21.63 dB

Troubleshooting—Example 6

This system consists of two segments of tight tube cable with a 62.5 μm fiber (Figure 7–32). The segments are connected together. Each of Segments A and B is 100 m, -0 m, +10 m long, with ST-compatible style connectors at Ends 1 and 2. Ends 3 and 4 have 906 SMA connectors. The link optoelectronics operate at 850 nm. The fiber NA is 0.275 nominal.

The ST-compatible style connectors have a maximum loss of 1.0 dB/pair and an average loss of 0.3 dB/pair. The 906 SMA connectors have a maximum loss of 2.0 dB/pair and an average loss of 1.0 dB/pair. The fiber has a maximum attenuation rate of 4.0 dB/km and a typical rate of 3.2 dB/pair.

Both of these cable assemblies had low loss when tested prior to installation. They were shipped to a customer, who connected them together. When connected together, the loss was much higher than expected. What is wrong?

From this information and Equations 7–1 and 7–2, we can calculate the maximum and expected losses. The maximum loss is:

cable loss = (0.200 + 0.02) km × 4 dB/km =	0.88 dB
ST-compatible connector loss = 1 pr × 1 dB/pr =	1.00 dB
SMA connector loss = 1 pair × 2 dB/pair =	2.00 dB
	3.88 dB

The expected loss is:

cable loss = (0.200 + 0.02) km × 3.2 dB/km =	0.70 dB
ST-compatible connector loss = 1 pr × 1 dB/pr =	0.30 dB
SMA connector loss = 1 pair × 1 dB/pair =	1.00 dB
	2.00 dB

This link fails a white light test. It is, apparently, not continuous.

An insertion loss test provides losses of 12.2 dB with the light source at End 1 and 12.1 dB with the light source at End 2. These losses are very high. Note that the insertion loss test result is inconsistent with the failure of the white light test.

From End 1 (Figure 7–33), the link apparently has one peak at the center. Note that this trace is consistent with the map.

From this end (Figure 7–33), the loss of the connector pair at End 1 is 0.91 dB. This loss is higher than specified, but less than the maximum specified (1 dB/pair).

The attenuation rate of Segment A (Figure 7–34) is 4.279 dB/km. This rate is higher than specified.

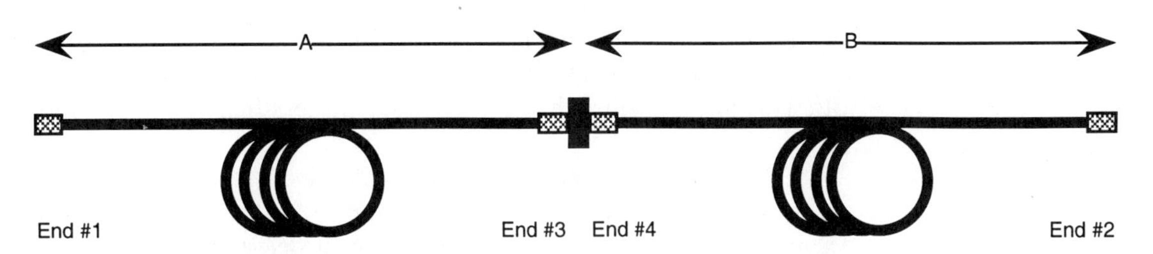

Figure 7–32 Map of Troubleshooting Example 6

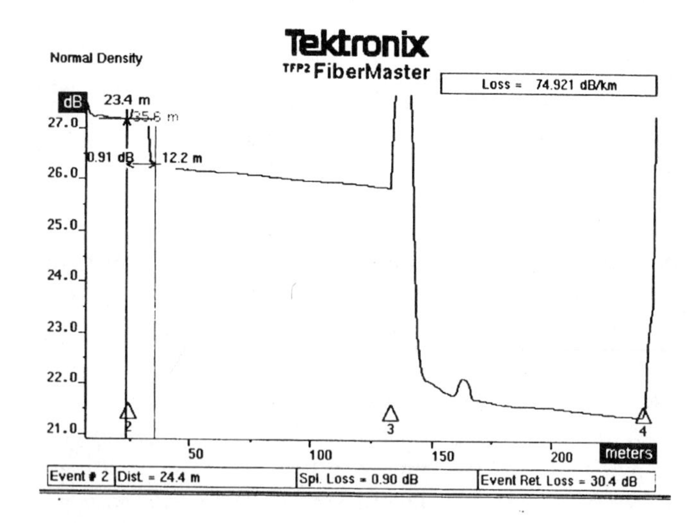

Figure 7–33 OTDR Trace, Example 6, End 1, Connector Loss

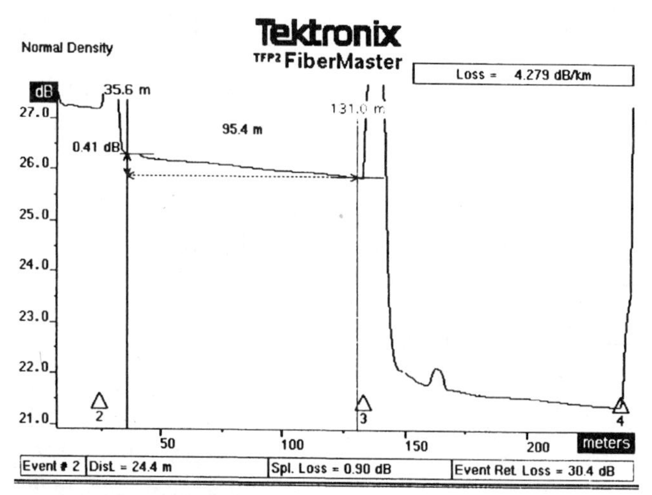

Figure 7–34 OTDR Trace, Example 6, End 1, Segment A Cable Attenuation Rate

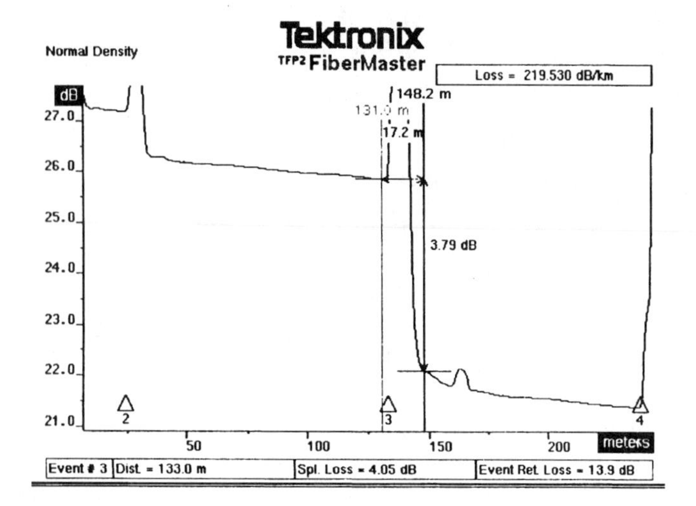

Figure 7–35 OTDR Trace, Example 6, End 1, Central Connector Loss

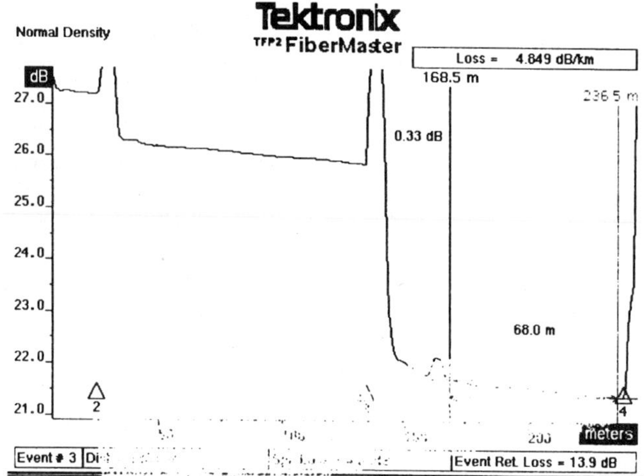

Figure 7–36 OTDR Trace, Example 6, End 1, Segment B Cable Attenuation Rate

The two-point loss measurement of the central connector (Figure 7–35) is 3.79 dB. This connector is well in excess of the acceptable loss.

The attenuation rate of Segment B (Figure 7–36) is 4.849 dB/km. This rate is higher than specified. Note that the cursor is placed after the small peak at approximately 165 m. This peak is an echo, which interferes with obtaining an accurate attenuation rate.

When Segment B is measured from End 2, the attenuation rate is 4.451 dB/km (Figure 7–37). This value is close to that obtained from Figure 7–36. Thus, the placement of the cursor in Figure 7–36 is justified.

From End 2 (Figure 7–38), the loss of the connector pair at End 2 is 0.64 dB. This loss is higher than the average, but less than the maximum specified (1 dB/pair).

Examination of the connector pair, Ends 3 and 4 reveals the cores to be round, clear, and featureless. Examination of the SMA barrel reveals the alignment sleeve to be missing. Insertion of the alignment sleeve will bring these connectors into specification.

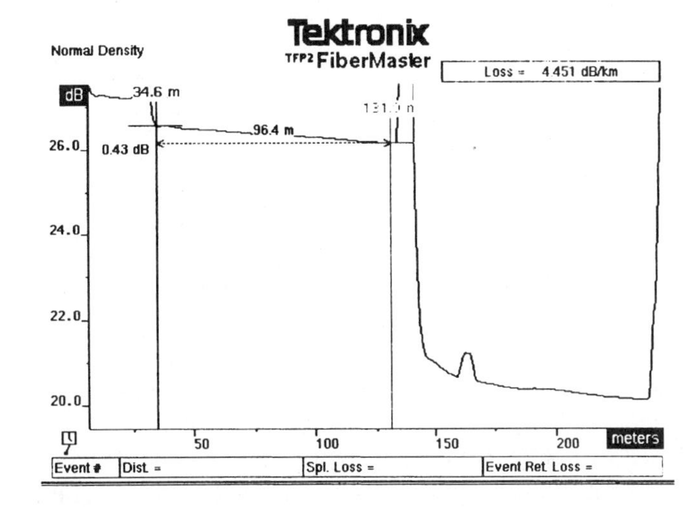

Figure 7–37 OTDR Trace, Example 6, End 2, Segment B Cable Attenuation Rate

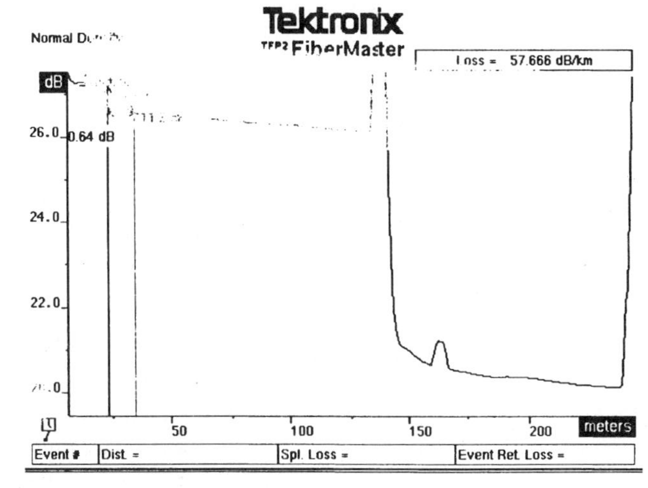

Figure 7–38 OTDR Trace, Example 6, End 2, Connector Loss

Summary, Example 6.
In summary, the connectors at Ends 1 and 2 are acceptable. The connector pair in the center is not acceptable. The cable in Segments A and B have attenuation rates higher than acceptable. Since the traces are linear without any non-uniform losses, the cable is probably installed correctly. Therefore, the cable was high loss before installation.

Troubleshooting—Example 7

This system consists of one segment of three-fiber, tight tube cable with a 62.5 μm fiber (Figure 7–39). The Segment A is 50.m -0 m, +10 m long, with ST-compatible style connectors at each end. The link optoelectronics operate at 850 nm. The fiber NA is 0.275 nominal.

The ST-compatible style connectors have a maximum loss of 1.0 dB/pair and an average loss of 0.3 dB/pair. The fiber has a maximum attenuation rate of 4.0 dB/km and a typical rate of 3.2 dB/pair.

All fibers transmitted light prior to the connectors being installed. However, there was no light transmitted after the connectors were installed.

From this information and Equations 7–1 and 7–2, we can calculate the maximum and expected losses. The maximum loss is:

$$\text{cable loss} = (0.05 + 0.01)\ \text{km} \times 4\ \text{dB/km} = 0.24\ \text{dB}$$
$$\text{connector loss} = 1\ \text{pair} \times 1\ \text{dB/pair} = \underline{1.00\ \text{dB}}$$
$$1.24\ \text{dB}$$

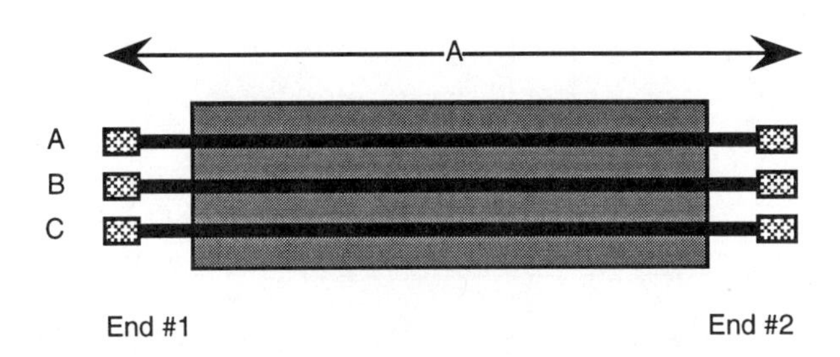

Figure 7–39 Map of Troubleshooting Example 7

The expected loss is:

cable loss = (0.05 + 0.01) km × 3.2 dB/km =	0.19 dB
connector loss = 1 pair × 1 dB/pair =	0.30 dB
	0.49 dB

All three fibers in this link fail the white light test. It is not continuous. An insertion loss test is not useful.

The OTDR traces were made with a 25 m lead in cable with the same core diameter and connector style as in the two segments.

From End 1 (Figures 7–40 to 7–42), the link apparently has an end reflection at 79.2 m. This reflection indicates a cable length of 54.2 m (79.2 m - 25 m). This length agrees with the map length.

From End 2 (Figures 7–43 to 7–45), the link apparently has an end reflection at 54.2 m. This reflection indicates a cable length of 55 m (79.2 - 25 m). This length agrees with the map length.

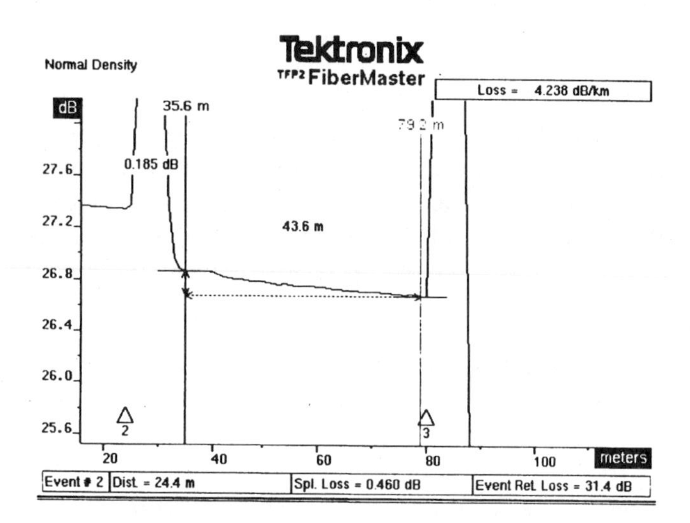

Figure 7–40 OTDR Trace, Example 7, End 1, Channel A Cable Attenuation Rate

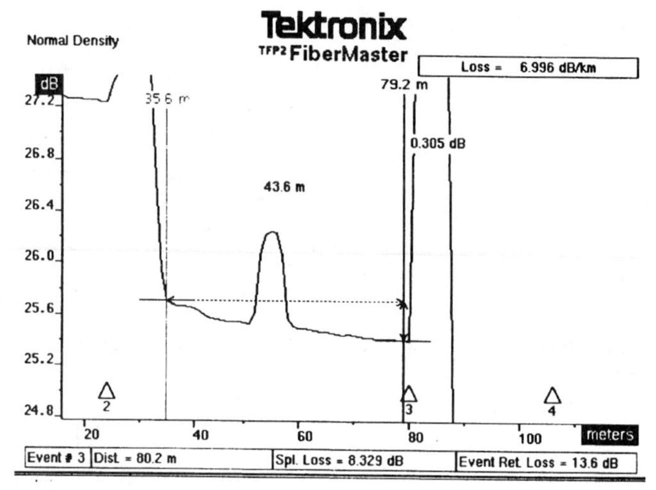

Figure 7–41 OTDR Trace, Example 7, End 1, Channel B Cable Attenuation Rate

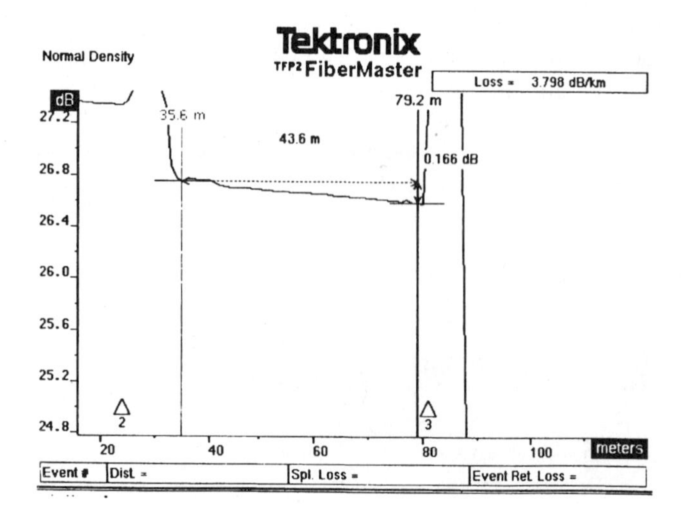

Figure 7–42 OTDR Trace, Example 7, End 1, Channel C Cable Attenuation Rate

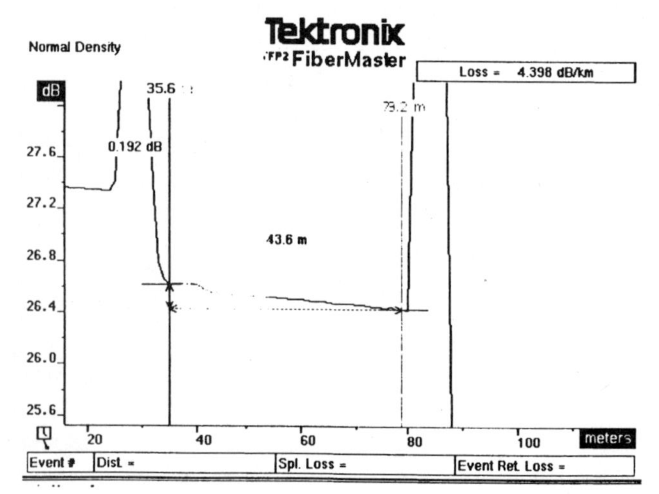

Figure 7–43 OTDR Trace, Example 7, End 2, Channel A Cable Attenuation Rate

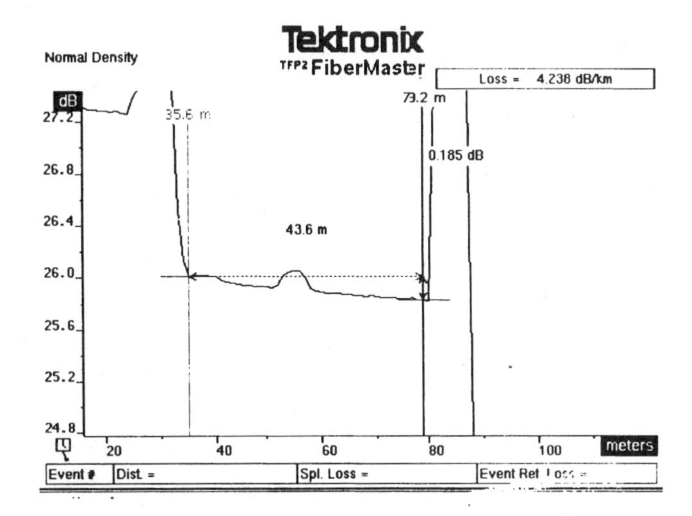

Figure 7–44 OTDR Trace, Example 7, End 2, Channel B Cable Attenuation Rate

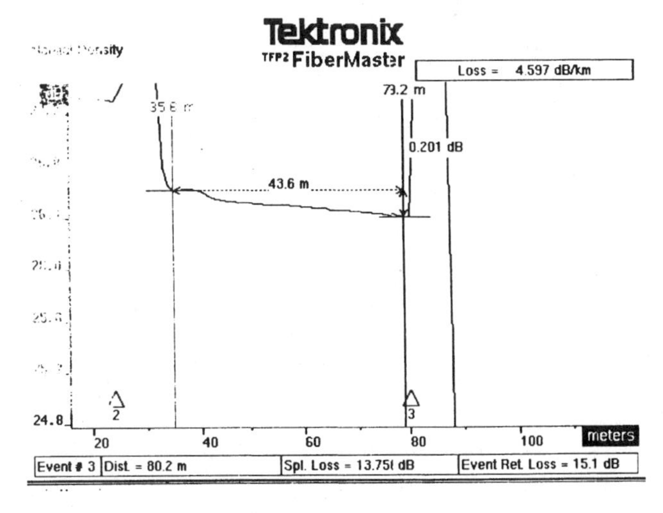

Figure 7–45 OTDR Trace, Example 7, End 2, Channel C Cable Attenuation Rate

There are two possible explanations for failure of the white light test. The first is breakage of all three fibers near or at the connectors. However, breakage of fiber would result in a bad launch. Therefore, breakage cannot be the cause of this problem.

The second possible explanation for failure of the white light test is mislabeling of the connectors at the patch panels at the ends of this cable. Cross checking proves this to be the explanation.

Comparison of Figures 7–41 and 7–42 reveals a difference: Figure 7–41 exhibits a peak or reflection in the middle of the link, while Figures 7–40 and 7–42 lack such a reflection. This reflection could be interpreted as a break in the tight tube cable. This interpretation would be strengthened by Figure 7–44, which is Figure 7–41 in the opposite direction. Figure 7–44 exhibits such a peak at approximately the same location.

However, this peak is an echo, which results from either or both of the fiber internal to the OTDR and the 25 m lead in cable. Identification of this reflection as an echo is straightforward: create a trace of the 25 m lead in cable. This peak will appear in the same location as it does in Figures 7–41 and 7–44 (Figure 7–46).

Often, echoes have the characteristic of very low, or no, apparent loss. If you measure the loss across the echo, it will be very low (Figure 7–47).

Note that echoes appear when the connector attached to the OTDR or to the lead in cable is relatively high loss. We can see this from Figures 7–48 to 7–53. In Figures 7–48, 7–50, 7–51, and 7–53, the loss of the connector at the OTDR end of the cable is low. No central peak occurs. However, in Figures 7–49 and 7–52, the loss of the connector at the OTDR end of the cable is relatively high, 1.259 dB and 1.531 dB, respectively. These high loss connectors reduce the level of the back scattered signal, allowing the lower level echo reflections to appear.

Summary, Example 8. In conclusion, the ends were mislabeled. Four of the six connectors are acceptable. The two connectors on the ends of Channel 2 are not. In addition, the attenuation rates of Channels 1 and 2 are unacceptably high (Figures 7–40 and 7–41).

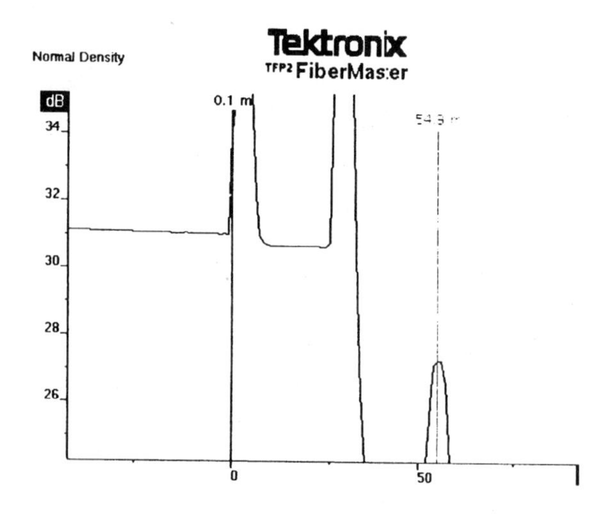

Figure 7–46 OTDR Trace of Lead In Cable

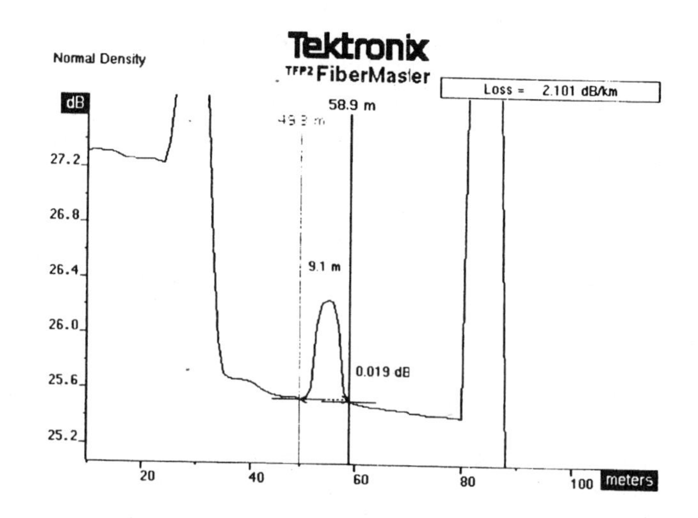

Figure 7–47 OTDR Trace Loss of Ghost Reflection

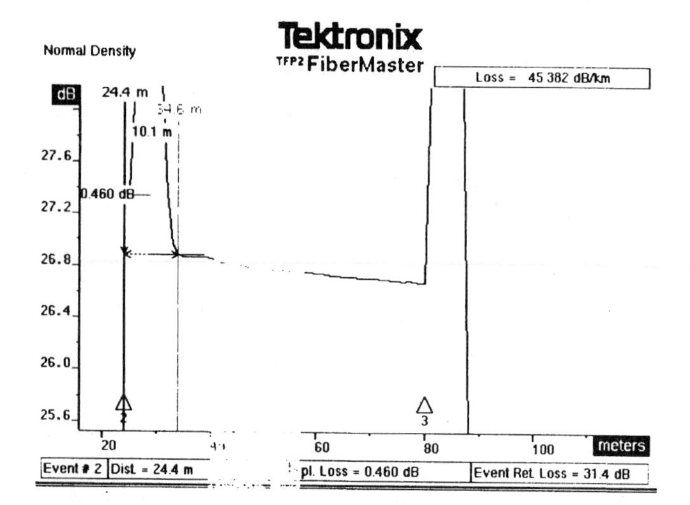

Figure 7–48 OTDR Trace, Example 7, End 1, Channel A Connector Loss

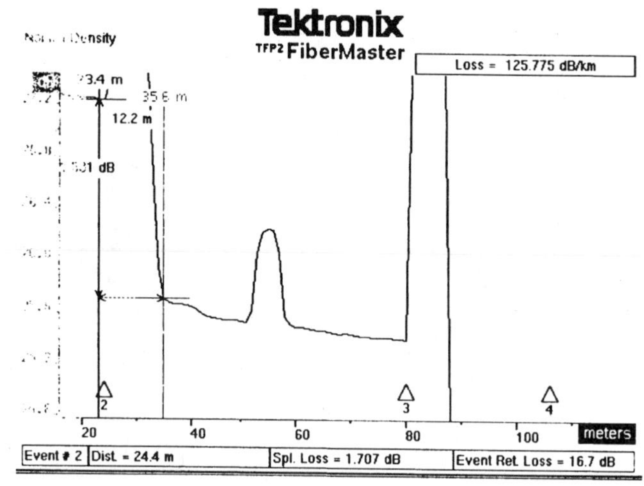

Figure 7–49 OTDR Trace, Example 7, End 1, Channel B Connector Loss

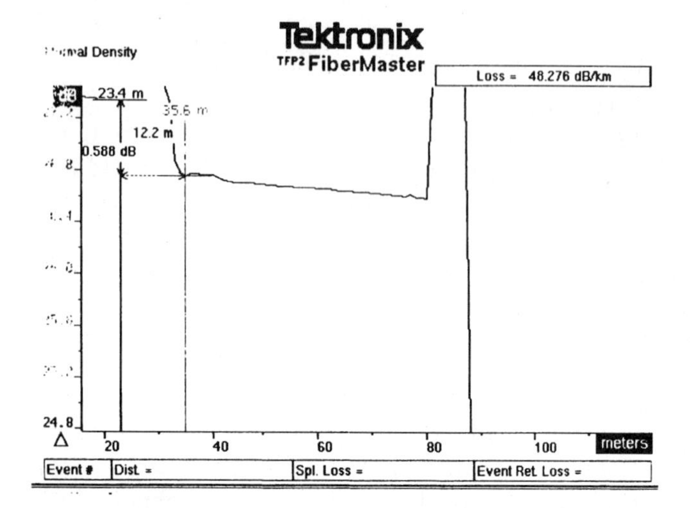

Figure 7–50 OTDR Trace, Example 7, End 1, Channel C Connector Loss

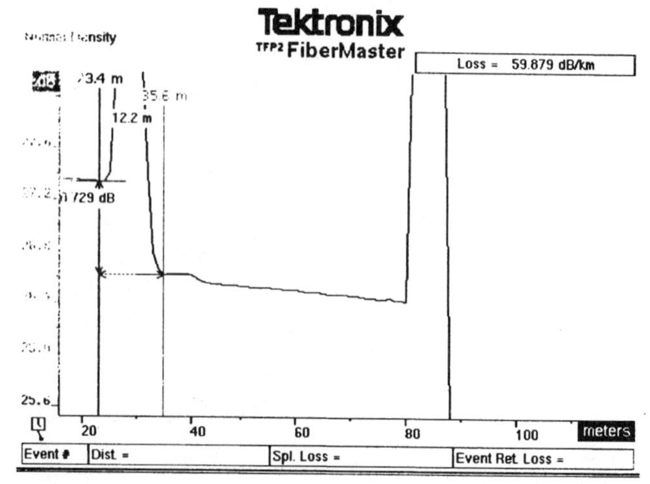

Figure 7–51 OTDR Trace, Example 7, End 2, Channel A Connector Loss

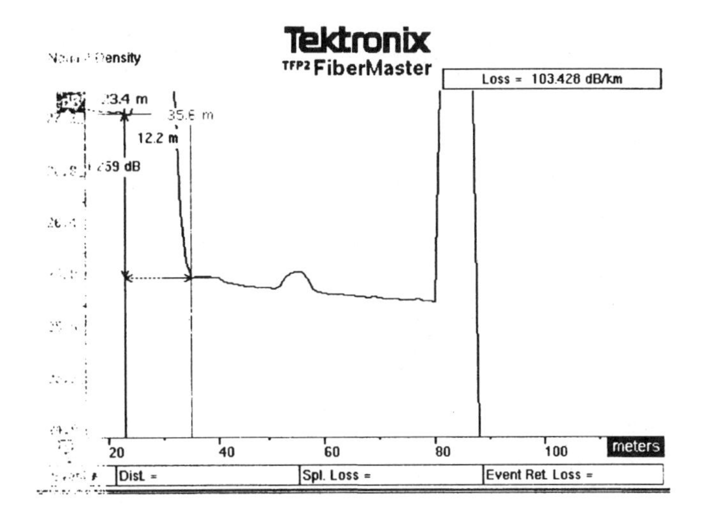

Figure 7–52 OTDR Trace, Example 7, End 2, Channel B Connector Loss

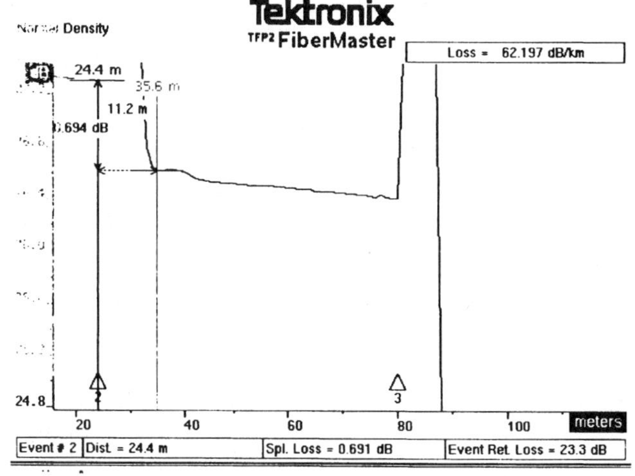

Figure 7–53 OTDR Trace, Example 7, End 2, Channel C Connector Loss

Troubleshooting—Example 8

This system consists of two segments of tight tube cable with a 62.5 μm fiber (Figure 7–54). The segments are connected together. Each of the Segments A and B is 100 m, -0 m, +10 m long, with ST-compatible style connectors at each end. The link optoelectronics operate at 850 nm. The fiber NA is 0.275 nominal. Standard characteristics apply.

Both of these cable assemblies had low loss when tested prior to installation. They were shipped to a customer, who installed and connected them together.

From this information and Equations 7–1 and 7–2, we can calculate the maximum and expected losses. The maximum loss is:

cable loss = (0.200 + 0.02) km × 4 dB/km =	0.88 dB
connector loss = 2 pairs × 1 dB/pair =	2.00 dB
	2.88 dB

The expected loss is:

cable loss = (0.200 + 0.02) km × 3.2 dB/km =	0.70 dB
connector loss = 2 pair × 0.3 dB/pair =	0.60 dB
	1.30 dB

This link fails a white light test. It is not continuous. An insertion loss test is unnecessary.

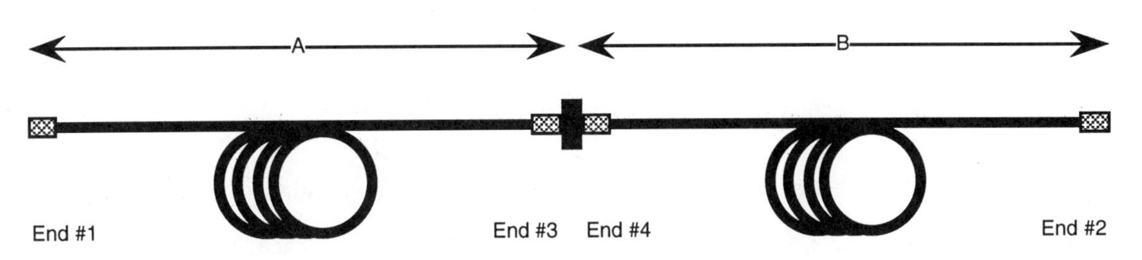

Figure 7–54 Map of Troubleshooting Example 8

From End 1 (Figure 7–55), the link apparently has two sections. The first section is 106 m (= 131.0 m - 25 m). The length of the first section agrees with the map (100 m, -0 m, +10 m). The second section is 51.7 m (181.7 - 131 m, Figure 7–56). Apparently, there is a break at 51.7 m from the central connector.

The attenuations of Section A and the first 51.7 m of Section B are acceptable at 3.237 dB/km and 3.967 dB/km (Figures 7–55 and 7–56).

The connector losses at Ends 1 (Figure 7–57), 3 and 4 (Figure 7–58) are acceptable at 0.902 dB and 0.209 dB.

From End 2 (Figure 7–59), we see a reflection at a distance of 58.3 m (83.3 m - 25 m). Apparently, there is a break at 58.3 m from End 2. Addition of the two sections of Segment B (58.3 m + 51.7 m) produces a total length of 110 m. This total length agrees with the map (100 m, -0 m, +10 m). Note that there is an echo in the center of the trace in Figure 7-59.

The attenuation of this section of Segment B is unacceptably high at 4.469 dB/km. The connector loss at End 2 is acceptable at 0.542 dB (Figure 7–60).

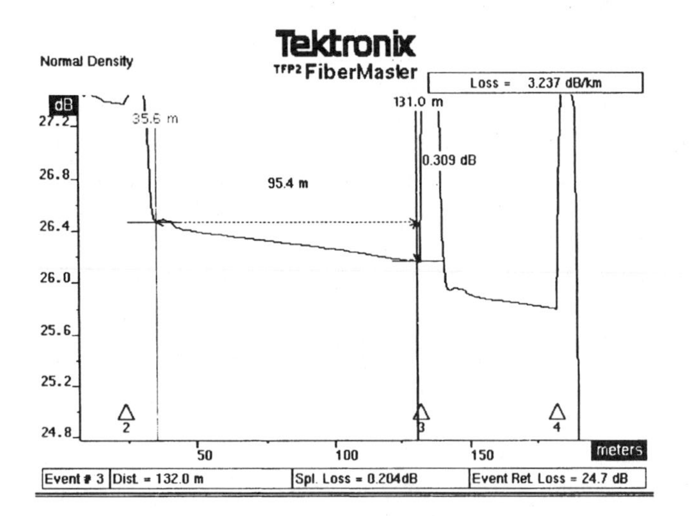

Figure 7–55 OTDR Trace, Example 8, End 1, Segment A Cable Attenuation Rate

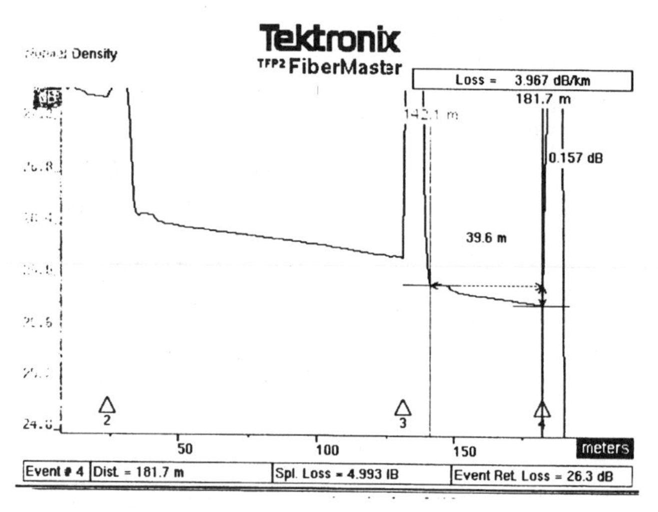

Figure 7–56 OTDR Trace, Example 8, End 1, Segment B Cable Attenuation Rate

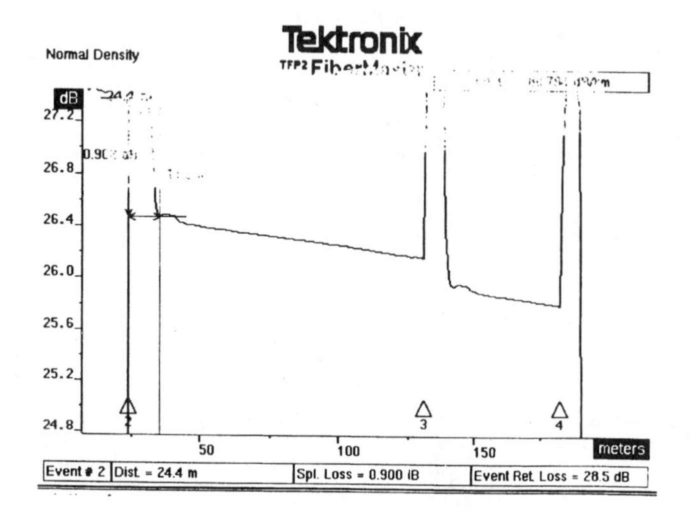

Figure 7–57 OTDR Trace, Example 8, End 1, Connector Loss

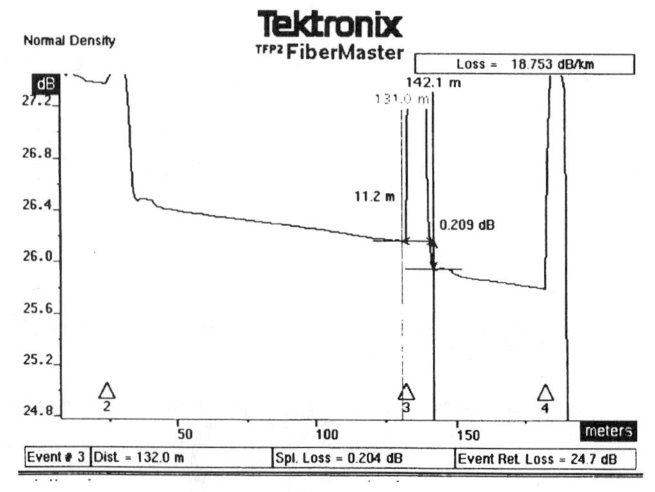

Figure 7–58 OTDR Trace, Example 8, Ends 3 and 4, Connector Loss

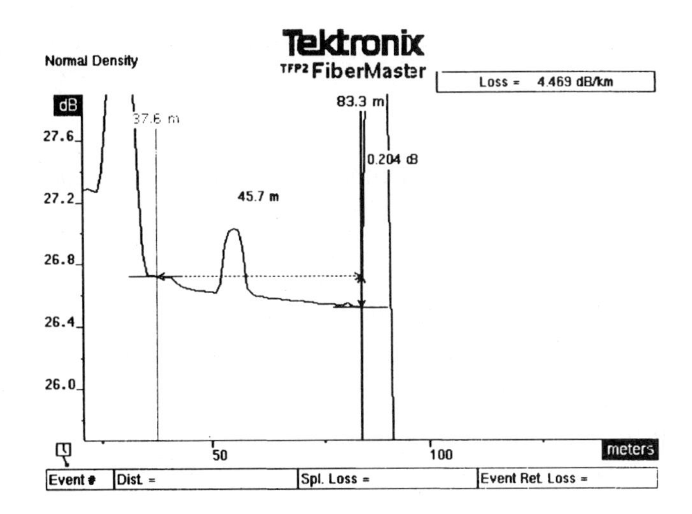

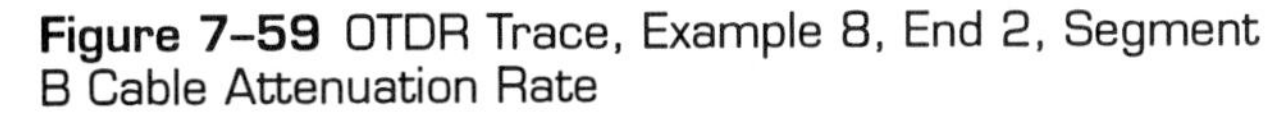

Figure 7–59 OTDR Trace, Example 8, End 2, Segment B Cable Attenuation Rate

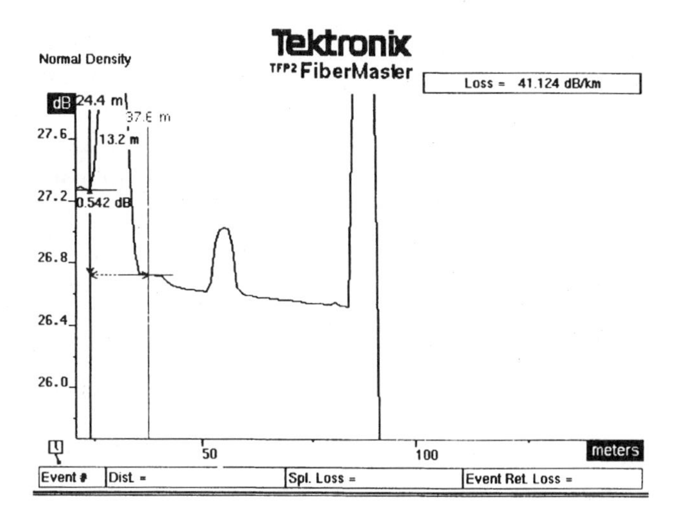

Figure 7–60 OTDR Trace, Example 8, End 2, Connector Loss

Summary, Example 8.

In conclusion, all connectors are acceptable. The cable in Segment A and the first part of Segment B have acceptable attenuation rates. However, the cable in Segment B was broken during installation. The section of Segment B near End 2 has unacceptably high attenuation rate.

Troubleshooting—Example 9

This system consists of two segments of tight tube cable with a 62.5 μm fiber (Figure 7–54). The segments are connected together. Each of the Segments A and B is 100 m, -0 m, +10 m long, with ST-compatible style connectors at each end. The fiber NA is 0.275 nominal. The optoelectronics have an operating wavelength of 660 nm.

This cable did not work when connected to the modems. However, these modems worked when connected to a 1-meter length of cable cut from the same reel as were these two segments.

From this information and Equations 7–1 and 7–2, we can calculate the maximum and expected losses. The maximum loss and expected loss are the same as those in Example 8, 2.88 dB and 1.30 dB, respectively.

This link passes a white light test. It is continuous.

An 850 nm insertion loss test provides losses of 4.5 dB with the light source at End 1 and 4.2 dB with the light source at End 2. These losses are unacceptably high.

However, the modems operate at a wavelength of 660 nm. This wavelength is characteristic of plastic optical fiber. POF often has a core diameter of 960 μm, which is much larger than the 62.5 μm core diameter of this cable link. In addition, the NA of POF is larger, 0.51 vs. 0.275 for the 62.5 μm core glass fiber. These two factors result in a significant reduction of the power coupled into the 62.5 μm core fiber from that into a 960 μm POF.

Summary, Example 9. In summary, there is a mismatch between the fiber and the modems. Both need to be chosen to operate at the same wavelength. The solution is to choose modems that operate at 850 nm, since the length of the link, 200 m, is greater than that at which most POF modems will function.

Troubleshooting—Example 10

This system consists of two segments of tight tube cable with a 62.5 μm fiber (Figure 7–61). The cables have two fibers. Each of the Segments A and B is 50 m, -0 m, +10 m long, with ST-compatible style connectors at each end. The link opto-electronics operate at 850 nm. The fiber NA is 0.275 nominal.

The ST-compatible style connectors have a maximum loss of 1.0 dB/pair and an average loss of 0.3 dB/pair. The fiber has a maximum attenuation rate of 4.0 dB/km and a typical rate of 3.2 dB/km.

This duplex cable worked when installed. When disconnected from and reconnected at the central patch panel, it did not work. What is wrong?

From this information and Equations 7–1 and 7–2, we can calculate the maximum and expected losses. The maximum loss is:

$$\begin{aligned} &\text{cable los s} = (0.100 + 0.02)\ \text{km} \times 4\ \text{dB/km} = && 0.48\ \text{dB} \\ &\text{connector loss} = 2\ \text{pair} \times 1\ \text{dB/pair} = && \underline{2.00\ \text{dB}} \\ & && 2.48\ \text{dB} \end{aligned}$$

The expected loss is:

$$\begin{aligned} &\text{cable loss} = (0.100 + 0.02)\ \text{km} \times 3.2\ \text{dB/km} = && 0.38\ \text{dB} \\ &\text{connector loss} = 1\ \text{pair} \times 0.3\ \text{dB/pair} = && \underline{0.60\ \text{dB}} \\ & && 0.98\ \text{dB} \end{aligned}$$

This link fails the white light test of both the A and B channels. Both channels are apparently not continuous.

From End 1, Connector A (Figure 7-62), the link has a length of 54.7 m (79.7 m - 25 m). This length agrees with that of the first segment of the map (Figure 7–61). The cable attenuation rate of the fiber in Channel A is acceptable at 3.929 dB/km (Figure 7–62).

From End 1, Connector B (Figure 7–63), the link has two segments. The first segment has a length of 54.7 m (79.7 m - 25 m). The second segment has approximately the same length. The lengths of these two segments agree with those in the map (Figure 7–61). These data (failure of the white light test and an OTDR trace that reveals two segments) suggest that Channel A has been cross-connected at the center connection. A white light check between End 1, Connector B, and End 2, Connector A confirms this suggestion. The cable attenuation rate of the fiber in Channel B is unacceptable at 4.52 dB/km (Figure 7–62).

From End 1, Connector B (Figure 7–64), the loss of the connector pair at the center is 0.528 dB. This loss is less than the maximum specified (1 dB/pair).

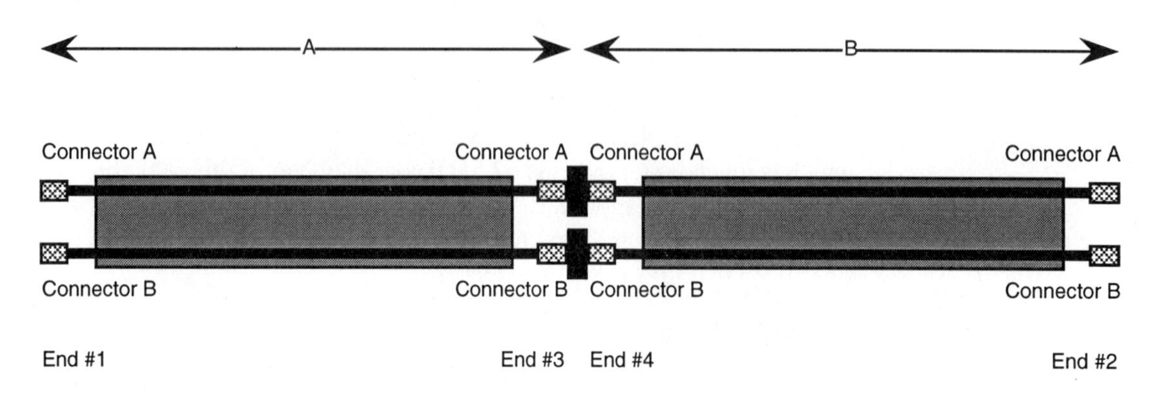

Figure 7–61 Map of Troubleshooting Example 10

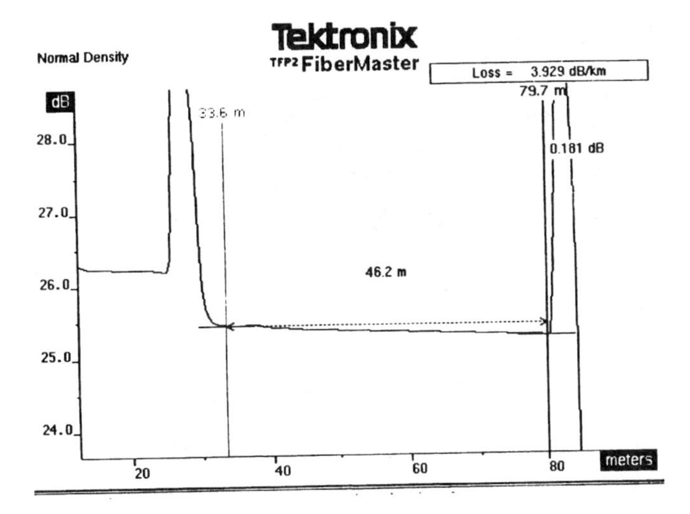

Figure 7–62 OTDR Trace, Example 10, End 1, Channel A Cable Attenuation Rate

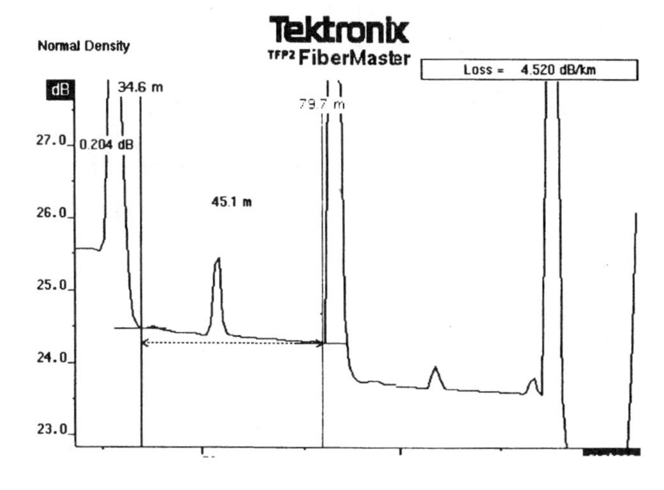

Figure 7–63 OTDR Trace, Example 10, End 1, Channel B Cable Attenuation Rate

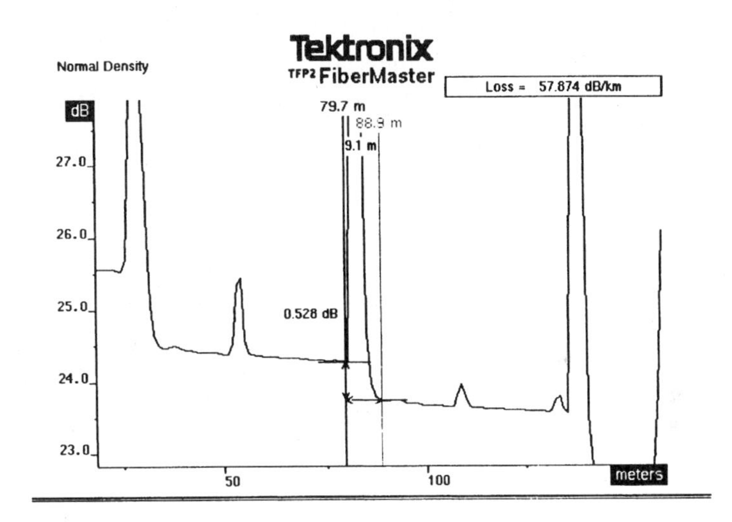

Figure 7–64 OTDR Trace, Example 10, End 1, Connector B, Central Connector Loss

Summary, Example 10. In summary, the connectors at Ends 3 and 4 have been cross connected. In addition, Connector A, End 3, or Connector B, End 4 is broken. White light testing of these two segments will reveal which is broken.

Troubleshooting—Example 11

This system consists of two segments of tight tube cable with a 62.5 μm fiber (Figure 7–65). Segment A is 50 m -0 m, +10 m long. Segment B is 100 m, -0 m, +10 m long. Standard characteristics apply. This link does not pass white light.

From this information and Equations 7–1 and 7–2, we can calculate the maximum and expected losses. The maximum loss is:

cable loss = (0.150 + 0.02) km × 4 dB/km =	0.68 dB
connector loss = 2 pair × 1 dB/pair =	2.00 dB
	2.68 dB

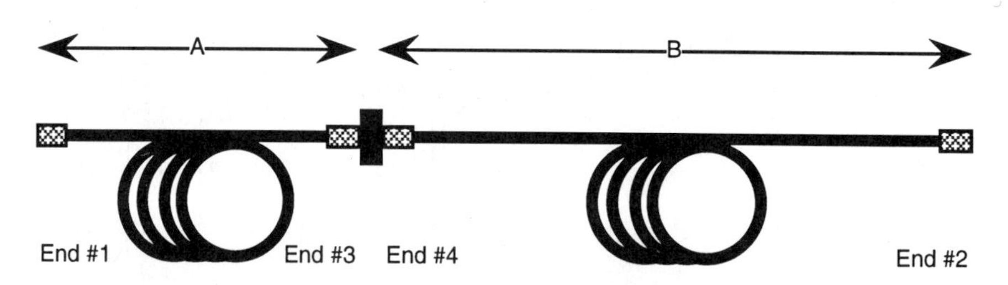

Figure 7–65 Map of Troubleshooting Example 11

The expected loss is:

cable loss = (0.150 + 0.02) km $\times$ 3.2 dB/km =	0.54 dB
connector loss = 2 pair $\times$ 0.3 dB/pair =	0.60 dB
	1.14 dB

The trace from End 1 (Figure 7–66) reveals two segments. The length of the first segment is 52.2 m (77.2 m - 25 m, Figure 7–67). The length of the second segment is 106.5 m (183.7 m - 77.2 m, Figure 7–68). These two lengths agree with those of the map, 50 m, -0 m, +10 m and 100 m, -0 m, +10 m. From these two figures, we could conclude that the lack of continuity is due to a broken connector at End 2.

From End 2, we obtain a trace that indicates a segment of approximately 50 m in length (Figure 7–69). This trace conflicts with our conclusion based on Figures 7–66 through 7–68. The only explanation of this conflict is an error in the map. This is the correct explanation.

In addition to the map error, a connector or cable is broken. End 1 to the second patch panel passes a white light test. End 2 to the second patch panel fails a white light test. Therefore, the connector at the end of Segment B.

The actual map of the link is Figure 7–70.

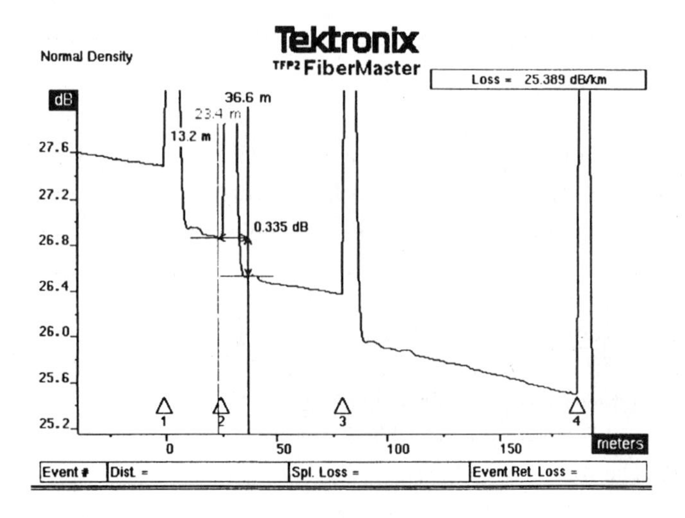

Figure 7–66 OTDR Trace, Example 11, End 1, Connector Loss

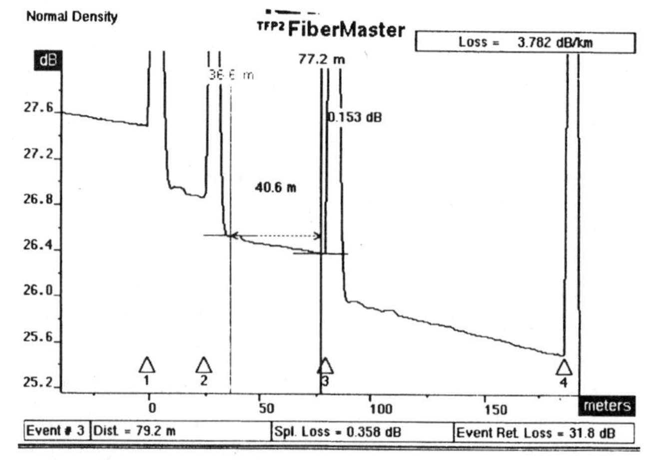

Figure 7–67 OTDR Trace, Example 11, End 1, Segment A Cable Attenuation Rate

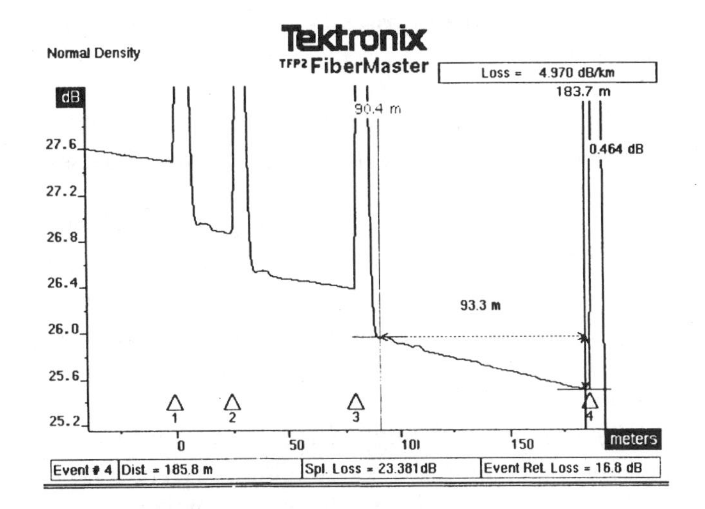

Figure 7–68 OTDR Trace, Example 11, End 1, Segment B Cable Attenuation Rate

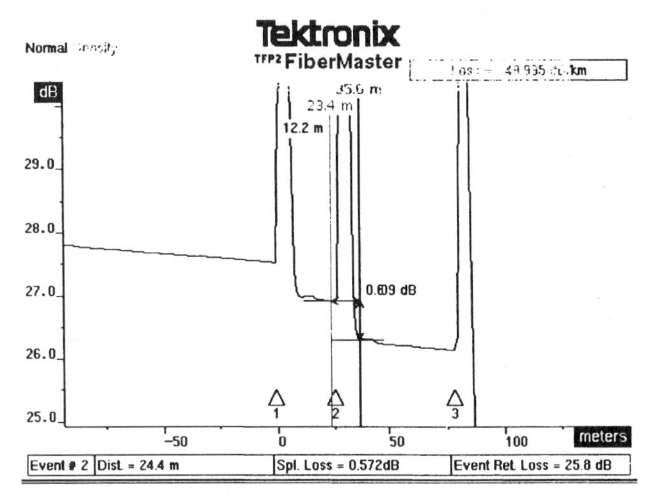

Figure 7–69 OTDR Trace, Example 11, End 2, Connector Loss

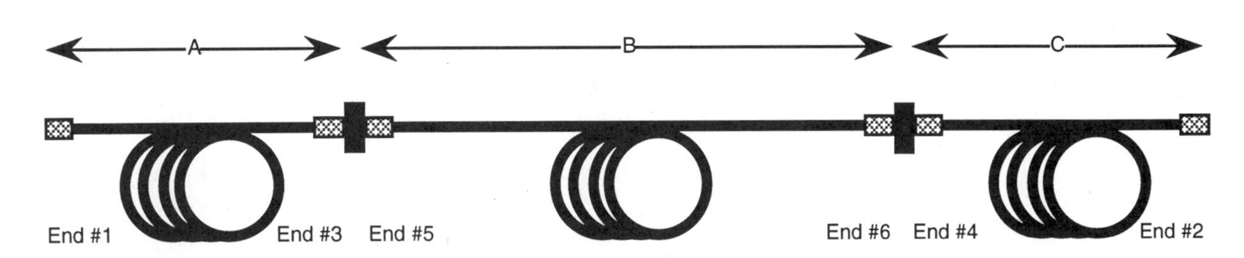

Figure 7–70 Map of Troubleshooting Example 11

REVIEW QUESTIONS

1. If the cable in Figure 7–1 had 50/125 fiber, with a maximum and average attenuation rates of 3.0 dB/km and 2.55 dB/km, respectively (at 850 nm), what would be the maximum and expected losses? Assume the connectors are STs, with maximum and average losses of 1 dB/pair and 0.3 dB/pair.
2. If the cable in Figure 7–1 had 50/125 fiber, with a maximum and average attenuation rates of 1.0 dB/km and .65 dB/km, respectively (at 1300 nm), what would be the maximum and expected losses? Assume the connectors are STs, with maximum and average losses of 1 dB/pair and 0.3 dB/pair.
3. Why is the two-point technique for measuring of connector loss conservative (result in higher than actual values)?
4. Why should a lead in cable used with an OTDR have fiber with the same NA and core diameter as those of the fiber being tested?
5. You must connect an installed cable directly to the OTDR. The OTDR has a dead zone and event zone of 30 m (32.8 feet). Your cable has a length of 25 m. What loss characteristics of the cable can you test with the OTDR?
6. You are connecting two 300 m segments with a jumper cable which has a length of 25 m. The OTDR has a dead zone and event zone of 30 m (32.8 feet). What loss characteristics of the jumper cable can you test with the OTDR? If

you can test any characteristics, will these losses be lower or higher than actual losses? Justify your answer.

7. Your cable system consists of a 350 m length of cable fusion spliced to a 400 m cable. The 400 m cable is connected to another 350 m cable. The OTDR has a dead zone and event zone of 30 m (32.8 feet). This cable system passes a white light test. Sketch the OTDR trace you expect to see.
8. The OTDR is connected to the cable system in the previous question. The connector attached to the OTDR is broken. Sketch the OTDR trace you expect to see.
9. The OTDR is connected to the cable system in the previous question. The connector on the end of the second 350 m cable segment is broken. Sketch the OTDR trace you expect to see.
10. How can you differentiate between a real reflective feature and an echo?
11. Refer to Troubleshooting Example 9. If the loss of power due to NA mismatch is described by Equation 7–3, what is the reduction in power coupled into the 62.5 μm fiber due to this mismatch?

(Equation 7–3)

$$\text{NA loss} = 10 \log [(NA_1/NA_2)^2], \text{ where } NA_1 < NA_2$$

12. Refer to Example 9. If the loss of power due to core diameter mismatch is described by Equation 7–4, what is the reduction in power coupled into the 62.5 μm fiber due to this mismatch.

(Equation 7–4)

$$\text{core diameter loss} = 10 \log [(\text{diameter}_1/\text{diameter}_2)^2], \text{ where } \text{diameter}_1 < \text{diameter}_2$$

13. Your 350 m cable link is that in Figure 7–71. The fiber has a 50 μm core and an NA of 0.20. The fiber has a maximum attenuation rate of 3.0 dB/km and an average attenuation rate of 2.75 dB/km. The connectors are ST-compatibles with maximum and average losses of 1.0 and 0.3 dB/pair. However, your insertion loss test leads have a core of 62.5 μm and an NA of 0.275. Use Equations 7–3 and 7–4 to calculate the maximum and average losses you would expect to measure with your test leads.
14. Your cable link is that in Figure 7–71. The singlemode fiber has an 8.5 μm core and an NA of 0.11. The fiber has a maximum attenuation rate of 3.0 dB/km and an average attenuation rate of 2.75 dB/km. The connectors are ST-compatibles with maximum and average losses of 1.0 and 0.3 dB/pair. However, your insertion loss test leads have a core of 62.5 μm and an NA of 0.275. Calculate the maximum and average insertion losses you would expect to measure with your test leads.

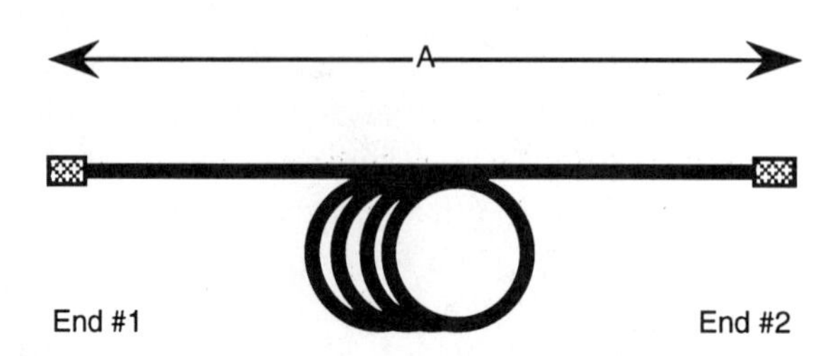

Figure 7–71 Map of Review Question 14

APPENDIX

1

Fiber Optic Terms

905 SMA: see SMA 905.

adapter: a device for mating two connectors.

APD: avalanche photodiode. This device converts an optical signal to an electrical signal. It operates to lower optical power levels and higher speeds than does its cousin, the photodiode.

armor: a layer of material, usually stainless steel, which is placed around a cable core to prevent damage from gnawing rodents. The armor is usually corrugated and covered with a layer of plastic. The plastic is usually heat sealed to itself.

attenuation, attenuation rate: the loss of light power or intensity as the light travels in a fiber. It is expressed in units of decibels. When used to describe fibers or cables, it is expressed as a rate in decibels/kilometer. It is specified at a specific wavelength under precisely defined test conditions. The usual method of measurement used by fiber manufacturers is called the "cut-back" test method. Attenuation or attenuation rate can be determined with an OTDR. In this book, attenuation refers to reduction in signal strength in a fiber.

average loss/pair: the average loss/pair you will experience when you install connectors correctly.

back reflection: an outdated term, which was used to mean return loss or reflectance. See reflectance and return loss.

backshell: the portion of a connector in back of the retaining nut.

bandwidth: a measure of the information transmission capacity of an analog transmission system.

bandwidth-distance product: the product of the length of a fiber and the bandwidth that fiber can transmit over that length. It is specified at a specific wavelength under precisely defined test conditions, and is expressed in units of MHz-km for multimode fibers. It is not used for singlemode fibers. Instead, the term "dispersion rate" is used.

bend radius, long-term: see bend radius, minimum unloaded.

bend radius, minimum loaded: the smallest radius to which a cable can be bent during installation at the maximum recommended installation load without any damage to either the fiber or the cable materials. Typically, this radius is 20 times the diameter.

bend radius, minimum recommended: the smallest radius to which a cable can be bent without any damage to either the fiber or the cable materials.

bend radius, minimum unloaded: the smallest radius to which a cable can be bent without any damage to either the fiber or the cable materials while the cable is unloaded. Typically, this radius is 10 times the diameter for the life of the cable.

bend radius, short-term: see bend radius, minimum loaded.

biconic: a style of connector.

binding tape: a cable component. This tape holds buffer tubes together during the jacket extrusion operation.

binding yarn: a cable component. This yarn holds buffer tubes together during the jacket extrusion operation.

bit error rate: a measure of the accuracy of a digital fiber optic system. The BER is the rate of errors produced by the optoelectronics.

bit rate: the data transmission rate of a digital transmission system.

boot: a plastic device that slides over the cable and the backshell of a connector. It limits the radius of curvature of the cable as it exits the backshell.

break out: a style of cable composed of subcables, each of which contains a single fiber.

buffer coating: a layer of plastic placed around the clad by the fiber manufacturer.

buffer coating diameter: the diameter of the layer of plastic placed around the fiber by the fiber manufacturer. Typical diameters are 250 and 500 µm.

buffer tube: a layer of plastic that surrounds a fiber or a group of fibers.

bulkhead: see adapter.

butt coupling: a method of transmitting light from one fiber to another by precise mechanical alignment of the two fiber ends without the use of lenses.

BWDP: bandwidth-distance product.

cable: the structure that protects an optical fiber or fibers during installation and use.

cable core: the structure of fibers, buffer tubes, fillers, and strength members that reside inside the innermost jacket of a cable.

cable end boxes: the enclosures placed on the end of a cable to protect the buffer tubes and fibers.

cap: a plastic structure that protects the end of a connector ferrule from dust and damage when the connector is not in use.

CBT: central buffer tube.

central buffer tube: a cable design in which all fibers reside in a single, centrally located buffer tube.

central strength member: a strength member that resides in the center of a cable.

chromatic dispersion: the spreading of pulses of light due to rays of different wavelengths traveling at different speeds through the core.

clad, cladding: the region of an optical fiber that confines the light to the core and provides additional strength to the fiber.

clad diameter: the outer diameter of the clad, measured in micrometers.

clad non-circularity: the degree to which the clad and the core deviate from perfect circularity.

cleaver: device to create a flat and perpendicular surface on the end of a fiber.

cleaving: the process of creating a fiber end that is flat and perpendicular to the axis of the fiber.

coating: see buffer coating.

concentricity: the degree to which the core deviates from being in the exact center of the clad.

cone of acceptance: the cone defined by the critical angle or the numerical aperture. This is also the cone within which all of the light exits a fiber.

core: the region of an optical fiber in which most of the light energy travels.

core diameter: the diameter of the region in which most of the light energy travels, measured in micrometers.

core offset: see concentricity.

coupler: a device that allows two separate optical signals to be joined for transmission on a single fiber.

crimp ring: the device that is deformed around the backshell of a connector. The crimp ring traps the strength member of the cable, providing acceptable cable-connector strength.

critical angle: the maximum angle to the axis of a fiber at which rays of light will enter a fiber and experience total internal reflection at the core-clad boundary.

crush load, maximum recommended: the recommended maximum load that can be applied to a fiber optic cable without any permanent change in the attenuation of the cable. This can be specified as either or both of long-term or a short-term crush load.

cut-off wavelength: the wavelength below which a singlemode fiber will not experience modal dispersion. Below this wavelength, the singlemode fiber will transmit multimode light, at a higher attenuation rate than when transmitting singlemode light.

D4: a connector style.

dB: decibel, the unit of measurement of power loss. dB = 10 log (power in/power out)

dielectric: having no components that conduct electricity in the cable.

differential modal attenuation: (DMA) the mechanism by which light in a multimode fiber becomes increasingly focussed at the center of a fiber core as it moves further down a fiber.

dispersion: the spreading of pulses in fibers. There are three types of dispersion: modal, spectral, and material.

dispersion rate: the rate at which pulses spread in a singlemode fiber, expressed in units of ps/nm/km.

ESCON™: Enterprise System Connectivity, an IBM standard. Also a connector style.

expanded beam coupling: a method of transmitting light from one fiber to another with lenses.

FC: a connector style.

FC/PC: a connector style; the successor to the FC style.

FDDI: fiber data distributed data interface.

feed through: see adapter.

ferrule: that portion of the connector with which a fiber is aligned.

fiber: the structure that guides light in a fiber optic system.

filled and blocked cable: a type of cable in which all empty space is filled with compounds to prevent water from moving along the cable and from coming into contact with the fibers. Such contact would result in a degradation of both mechanical and optical properties of the fibers.

fillers: cable structures that fill otherwise empty space in a cable.

Fresnel reflection: the reflection that occurs when light travels between two media in which the speed of light (or index of refraction) differs.

FRP strength member: a fiberglass reinforced plastic or epoxy rod used as a dielectric strength member in cables. The term "plastic" may refer to either a polymer plastic or an epoxy.

fusion splice: a splice made by melting two fibers together.

gel filling compound: a compound placed inside a loose buffer tube to prevent water from contacting the fiber(s) in that buffer tube.

graded index: a type of multimode fiber in which the chemical composition of the core is not uniform.

HCS™: a hard clad silica fiber.

heat shrink tubing: tubing placed on the backshell of a connector.

HiPPI: high speed, parallel processor interface.

index of refraction (IR, η): the ratio of the speed of light in a vacuum to the speed of light in the material. This is a dimensionless number, with typical values of 1.46–1.51 in communication fibers.

inner duct: a corrugated plastic pipe in which fiber optic cables are placed.

inner jacket: any layer of jacketing plastic other than the outermost layer.

installation strength, maximum recommended: the maximum load that can be applied along the axis of a cable without any breakage of fibers and without any permanent change in attenuation.

installation temperature range: the temperature range within which a cable can be installed without damage. This range is usually determined by the plastics used in the cable.

jacket: a layer of plastic in a cable. It can be an outer jacket or an inner jacket.

jumper: a short length of a single fiber cable with connectors on both ends.

Kevlar®: an aramid yard produced by Dupont Chemical which is used to provide strength in fiber optic cables.

keying: a mechanism in connectors by which ferrules are prevented from rotating in a barrel or receptacle.

laser diode: a semiconductor that converts electrical signals to optical signals.

LED: light emitting diode. A semiconductor that converts electrical signals to optical signals.

lensed coupling: see expanded beam coupling.

loose buffer tube: a buffer tube with space between the outer diameter of the buffer coating and the inner diameter of the buffer tube.

loose tube: a cable design in which the fiber floats loosely inside an oversized tube. The tube does not contact the entire circumference of the fiber. A loose tube can contain one or more fibers.

loss: the end to end reduction in optical power as light travels through a fiber, connectors, or splices. In this book, the "loss" refers to reduction of optical power at a splice or a connector pair.

material dispersion: the spreading of pulses of light due to rays of light traveling through different regions of the core. These different regions have slightly different compositions, in each of which light travels at a slightly different speed.

maximum loss/pair: the maximum loss/pair you will experience when you install connectors correctly.

maximum recommended installation load: see installation strength, maximum recommended.

mechanical splice: a mechanism that aligns two fiber ends precisely for efficient transfer of light from one fiber to another.

MFPT: a multiple fiber per (buffer) tube cable design. This design usually has 6 or 12 fibers per loose buffer tube.

mini-BNC: a connector style.

minimum recommended long-term bend radius: see bend radius, minimum unloaded.

minimum recommended short-term bend radius: see bend radius, minimum loaded.

modal dispersion: the spreading of pulses of light due to rays traveling different paths through the core. Modal dispersion occurs in multimode fibers.

modal pulse spreading: see modal dispersion.

mode: one of the paths in which light can travel in a fiber core. Mode means roughly "path."

mode field diameter: the diameter within which the light energy field actually travels in a singlemode fiber. The mode field diameter is slightly larger than the core diameter.

monomode: a method of light transmission in which all rays of light act as though they are traveling parallel to the axis of the fiber.

multimode: a method of propagation of light in which all of the rays of light do not travel in a path parallel to the axis of the fiber. These rays travel paths that result in different transit times for different rays. This difference in transit times limits the bandwidth that any given multimode fiber can transmit.

NA: see numerical aperture.

numerical aperture: the sine of the critical angle. The NA is a measure of the solid angle within which rays of light will enter and be transmitted along the fiber.

optical amplifier: a device that increases the signal strength without an optical to electrical to optical conversion process.

optical coupling: see expanded beam coupling.

optical power budget: the maximum loss in optical power that a transmitter-receiver pair can withstand while still functioning at the specified level of accuracy.

optical return loss: see return loss.

optical rotary joint: a rotating joint that allows transmission of light from a stationary fiber to a rotating fiber.

optical switch: a switch that can direct light to more than one output path.

optical time domain reflectometer: a test device that creates a map of the loss of signal strength of a passive optical path.

optical waveguide: another term for an optical fiber.

optoelectronic device: any device that converts a signal from electrical to optical domain or vice versa.

optoelectronics: see optoelectronic device.

OTDR: optical time domain reflectometer or optical time domain reflectometry.

ovality: a measure of the degree to which a fiber deviates from perfect circularity. See clad non-circularity.

passive component: a device that manipulates light without requiring an optical signal to electrical signal conversion.

patch panel: a sheet of material which contains adapter(s).

PCS: a plastic clad silica fiber.

PD: photodiode. This device converts an optical signal to an electrical signal.

pigtail: a length of fiber or cable permanently attached to a connector or an optoelectronic device.

ping pong: a type of transmission that results from using a single LED as both a transmitter and receiver. During one cycle, the bias on the LED is reversed from its normal bias so that it will behave as a photodiode. Use of an LED reduces its potential bandwidth due to the time required to reverse the bias completely.

plug: another term for connector.

POF: plastic optical fiber. A fiber with a plastic core and a plastic clad.

polishing fixture: device to hold a connector perpendicular to the polishing surface; for use in creating a flat and perpendicular fiber surface.

premise: a tight tube cable design used most frequently in indoor applications.

pull-proof: a performance characteristic of connector styles. A connector style is pull-proof when tension on the cable attached to the connector does not produce an increase in the loss of the connector pair.

pulse dispersion: see dispersion.

pulse spreading: see dispersion.

receptacle: the device within which an active device is mounted. The receptacle is designed to mate with a specific connector style. Otherwise, a pigtail is connected to the active device.

reflectance: the ratio of reflected power to incident power for a single device, such as a connector or mechanical splice. Reflectance is measured in units of dB.

reinforced jacket: two layers of plastic separated by strength members. Strength members are usually Kevlar® or fiberglass in cables not made by AT&T. Cables made by AT&T can contain steel wires as strength members.

repeatability: the maximum change in loss between two successive measurements of two connectors.

retaining nut: device that retains connector to a receptacle or to an adapter.

return loss: see reflectance.

ribbon: a structure on which multiple fibers are precisely aligned.

SC: a connector style.

shrink tubing: plastic that covers the backshell of a connector.

simplex: a single fiber cable.

singlemode: a method of propagation of light in which all of the single wavelength energy of light arrives at the fiber at the same time. A simple, but technically inaccurate, explanation is that all rays of light behave as though they are traveling in a path parallel to the axis of the fiber.

slotted core: a cable design with a core containing helical slots.

SMA 905: a connector style.

SMA 906: a connector style.

spectral width: the measure of the width of the output power-wavelength curve at a power level equal to half the peak power.

splice: a device for permanent alignment of two fiber ends.

splice enclosure: a structure that encloses and protects splice trays and cable ends.

splice tray: a structure that encloses and protects fibers.

splitter: a device that creates multiple optical signals from a single optical signal.

spot size: the size of the area of an LED or laser diode from which light is produced.

ST®: the first of a series of connector styles designed by ATT. Other styles from ATT are the ST-II and the ST-II+.

ST®-compatible: a connector with a style compatible with the ST™ connector.

star core: see slotted core.

step index: a type of multimode fiber, in which the chemical composition of the core is uniform. This uniform composition results in a uniform index of refraction. This uniform index of refraction allows rays of light to travel in straight lines until they are reflected from the core-clad interface. This type of reflection results in a lower bandwidth-capacity than that attainable from a graded index fiber.

storage temperature range: the temperature range within which a cable can be stored without damage.

strength members: those elements of a cable design that provide strength.

style: the sum of characteristics that differentiates one connector style from another style.

TECS™: a technically enhanced clad fiber similar to hard clad silica fiber.

temperature operating range: the range of temperature within which the cable can be operated during its lifetime without degradation of either mechanical of optical properties.

tight buffer tube: a buffer tube with no space between the outer diameter of the buffer coating and the inner diameter of the buffer tube.

tight tube: see tight buffer tube.

total internal reflection: the mechanism by which optical fibers function. Internal reflection results from a difference in the speed of light in the core and clad. This difference results in reflection at the core-clad boundary if the angle of a ray of light to the axis of the fiber is less than or equal to the critical angle.

use load, maximum recommended: the maximum longitudinal load that can be applied to a cable during its entire lifetime without increase in attenuation and without breakage of fibers. This load is typically about 10 percent of the maximum recommended installation load.

water blocking compound: a compound placed in the interstices between buffer tubes in a cable or between jackets in a cable.

wavelength: a measure of the color of the light for which the performance of the fiber has been optimized. It is a length stated in nanometers (nm) or in micrometers (μm).

wavelength division multiplexer: a passive device that combines light signals with different wavelengths on different fibers onto a single fiber. The wavelength division demultiplexer performs the reverse function.

wiggle proof: a performance characteristic of connector styles. A connector style is wiggle proof if lateral pressure on the backshell does not produce an increase in the loss of the connector pair.

window: the wavelength range within which a fiber is designed to provide optimum performance. A single window product is designed for optimum performance at a single wavelength. A dual or double window product is designed for acceptable performance at two discrete windows. Typical windows for multimode fibers are 850 nm and/or 1300 nm. Typical windows for singlemode fibers are 1310 nm and/or 1550 nm.

zip-cord®: a two-fiber cable with a figure 8 cross-section. It can be either a loose tube (Belden Corporation) or a tight tube design (other manufacturers). This type of cable allows each of the two fibers to be separated in the same manner as the conductors in a standard lamp cord.

APPENDIX

2

Determining Reference Levels for Insertion Loss Tests

EIA/TIA-526-14 provides two techniques for setting a reference signal for insertion loss measurements, Method A (Figure A2–1) and Method B (Figure A2–2). Method A requires two test leads to set the reference; Method B, one lead. Both methods require the use of a second test lead during the loss measurement (Figure A2–4).

Use of Methods A or B will result in either conservative, non-conservative, or realistic loss measurements, depending on which of the two situations you are simulating during testing. Situation 1 involves measurement of a single cable segment (Figure A2–5). The cable will be directly connected to active devices at the transmitter and receiver ends. In this situation, the light

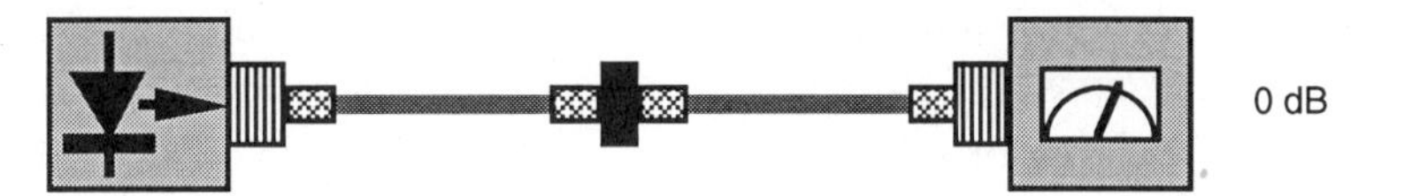

Figure A2–1 Method A, Two Lead Reference

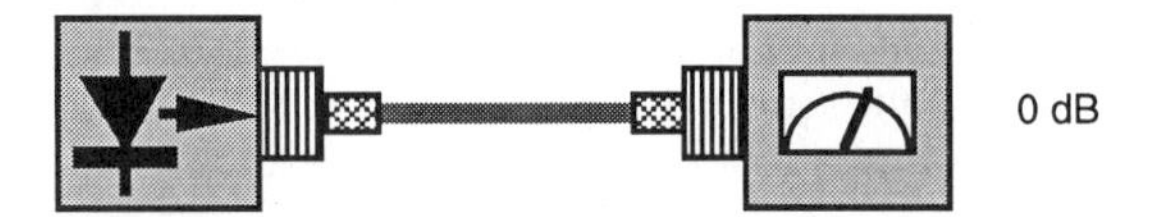

Figure A2–2 Method B, Single Lead Reference

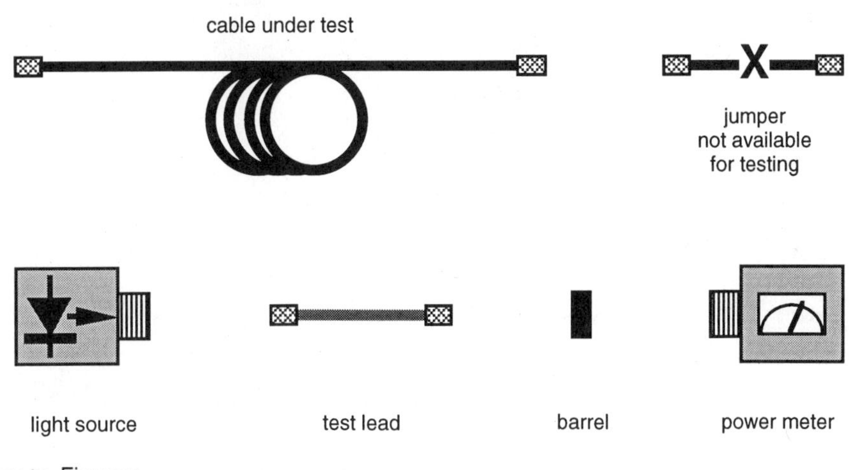

Figure A2–3 Key to Figures

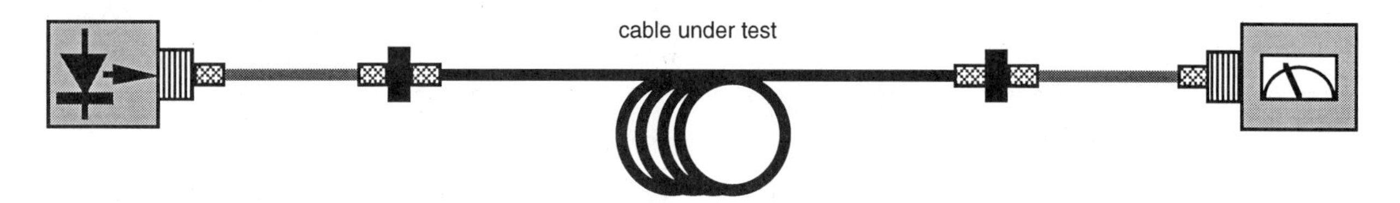

Figure A2–4 Test Set-Up for Methods A and B

travels through two connectors in the cable under test but through no connector pairs. Loss in connectors results from fiber to fiber transfer. Since the end connectors are directly connected to the active devices, these connectors cause essentially no loss of light.[1]

If the cable in Figure A2–5 is measured by Method A, the reference includes one pair of connectors (Figure A2–1) while the test set-up includes two pairs (Figure A2–4). Thus, Method A includes the loss of one pair more than the transmitter-receiver pair will experience. Thus, Method A is conservative.

If the cable in Figure A2–5 is measured by Method B, the reference includes no connector pairs (Figure A2–2), while the test set-up includes two pairs (Figure A2–4). Thus, Method B includes the loss of two pairs more than the transmitter-receiver pair will experience. Thus, Method B is conservative, and more conservative than Method A. The preceding analysis applies to any cable system of any number of segments, as long as the end connectors connect directly to the active devices in the transmitter and receiver.

Situation 2 involves a cable system with jumpers at both ends. These jumpers connect the main cable to the optoelectronics. As is often the case, these systems are installed and tested without the jumpers (Figure A2–6).

If the cable in Figure A2–6 is measured by Method A, the reference includes one pair of connectors (Figure A2–1) while the test set-up includes two pairs (Figure A2–4). Thus, the insertion loss by Method A includes the loss of one pair. Because the jumpers will be used, the transmitter receiver pair will experience the loss of two connector pairs plus two short lengths of cable.[2] Use of Method A to estimate the total of loss of the cable under test and the loss of the two jumpers will be non-conservative and will underestimate the actual loss that the transmitter-receiver pair will experience.

If the cable in Figure A2–6 is measured by Method B, the reference includes no connector pairs (Figure A2–2) while the test set up includes two pairs (Figure A2–4). Because the jumpers will be used, the transmitter-receiver pair will experience the loss of two connector pairs plus two short lengths of cable. Thus, the Method B insertion loss provides a realistic estimate of the end to end loss of the entire link.

This analysis reveals the importance of choosing the method that provides the desired data. If the cable under test is to be directly connected to the optoelectronics, Methods A and B will both provide conservative results. Method B will be more conservative than Method A.

Figure A2–5 Cable System Under Test for Situation 1

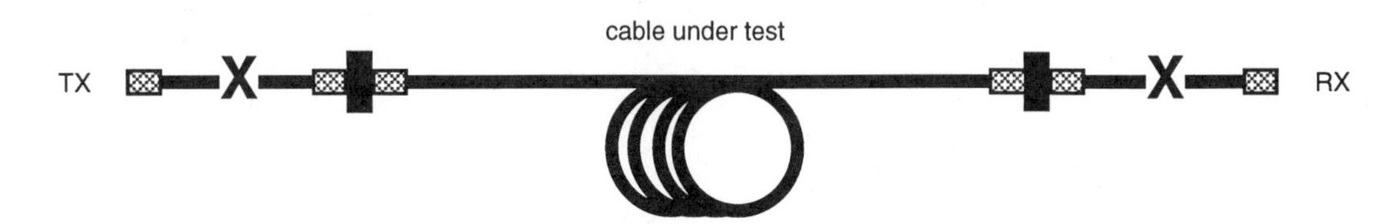

Figure A2–6 Cable System Under Test for Situation 2

If the cable under test is being tested for its loss, and if the losses[3] of the jumpers on each end will be added to the loss of the cable under test, then Method A will provide a conservative estimate of the total loss of the cable and the jumpers. However, the total loss calculated by adding three Method A losses will be higher than the loss that the transmitter-receiver pair will experience. This conservative bias will be the loss of one connector pair.

If the cable under test is being tested to provide an estimate of the total loss of that cable and the loss of the jumpers that will be attached to its ends, then Method B will provide a realistic estimate. Note that this is only an estimate. Use of this estimate requires assuming that the loss of the connectors and fiber in the jumpers will be close to the loss of the connectors in the cable under test.

APPENDIX

3

Standards

Note: Courtesy of *Fiberoptic Produce News*, Gordon Publications, Morris Plains, N.J. Used with permission. Additional standards identified from *Fiber Optic Reference Guide*, Force, Inc.,© 1995. Used with permission. Electronic Industry standards (EIA) are standards for the testing of fiber optic products. These standards can be obtained from Global Engineering Documents™, 15 Inverness Way East, Englewood, CO 80112-5776, (800) 854-7179, (303) 397-7956, (303) 397-2740 (fax).

EIA-440-A		Revision of EIA-440, Fiber Optic Terminology
EIA-445-1A	FOTP-1	Cable Flexing for Fiber Optic Interconnecting Devices
EIA-445-2A	FOTP-2	Impact Test Measurements for Fiber Optic Devices
EIA-455-3A	FOTP-3	Temperature Effects Measurement Procedure for Optical Fiber, Optical Cable, and Other Passive Components
EIA-455-4A	FOTP-4	Fiber Optic Connector/Component Temperature Life
EIA-455-5A	FOTP-5	Humidity Test Procedure for Fiber Optic Connecting Devices
EIA-455-6B	FOTP-6	Cable Retention Test Procedure for Fiber Optic Cable Interconnecting Devices
EIA-455-10	FOTP-10	Measurement of the Amount of Extractable Material in Coatings Applied to Optical Fiber
EIA-455-11A	FOTP-11	Vibration Test Procedure for Fiber Optic Connecting Devices and Cables
EIA-455-12A	FOTP-12	Fluid Immersion Test for Fiber Optic Components
EIA-455-13	FOTP-13	Visual and Mechanical Inspection of Fibers, Cables, Connectors, and Other Devices
EIA-455-14A	FOTP-14	Physical Shock (Specified Pulse)
EIA-455-15A	FOTP-15	Altitude Immersion
EIA-455-16A	FOTP-16	Salt Spray (Corrosion) Test for Fiber Optic Components
EIA-455-17A	FOTP-17	Revision, Maintenance Aging of Fiber Optic Connectors and Terminated Cable Assemblies
EIA-455-18A	FOTP-18	Acceleration Testing for Components and Assemblies
EIA-455-20	FOTP-20	Measurement of Change in Optical Transmittance
EIA-455-21A	FOTP-21	Mating Durability for Fiber Optic Interconnecting Devices
EIA-455-22A	FOTP-22	Ambient Light Susceptibility of Components
EIA-455-23A	FOTP-23	Air Leakage Testing for Fiber Optic Component Seals
EIA-455-24	FOTP-24	Water Peak Attenuation Measurement of Single-mode Fibers
EIA-455-25A	FOTP-25	Repeated Impact Testing of Fiber Optic Cables and Cable Assemblies
EIA-455-26A	FOTP-26	Crush Resistance of Fiber Optic Interconnecting Devices
EIA-455-27A	FOTP-27	Fiber Diameter Measurements
EIA-455-28B	FOTP-28	Measurement of Dynamic Tensile Strength of Optical Fiber
EIA-455-29A	FOTP-29	Refractive Index Profile (Transverse Interference Method)

EIA-455-30B	FOTP-30	Frequency Domain Measurement of Multimode Optical Fiber Information Transmission Capacity
EIA-455-31B	FOTP-31	Fiber Tensile Proof Test Method
EIA-455-32A	FOTP-32	Fiber Optic Circuit Discontinuities
EIA-455-33A	FOTP-33	Fiber Optic Cable Tensile Loading and Bending Test
EIA-455-34	FOTP-34	Interconnection Device Insertion Loss Test
EIA-455-35A	FOTP-35	Fiber Optic Component Dust (Fine Sand) Test
EIA-455-36A	FOTP-36	Twist Test for Connecting Devices
EIA-455-37	FOTP-37	Fiber Optic Cable Bend Test, Low and High Temperature
EIA-455-39A	FOTP-39	Water Wicking Test for Fiber Optic Cable
EIA-455-40	FOTP-40	Fluid Immersion, Cables
EIA-455-41	FOTP-41	Compressive Loading Resistance of Fiber Optic Cables
EIA-455-42A	FOTP-42	Optical Crosstalk in Components
EIA-455-43	FOTP-43	Output Near Field Radiation Pattern Measurement of Optical Waveguide Fibers
EIA-455-44A	FOTP-44	Refractive Index Profile (Refracted Ray Method)
EIA-455-45B	FOTP-45	Microscopic Method for Measuring Fiber Geometry of Optical Waveguide Fibers
EIA-455-46A	FOTP-46	Spectral Attenuation Measurement (Long Length Graded Index Optical Fibers)
EIA-455-47B	FOTP-47	Output Far Field Radiation Pattern Measurement
EIA-455-48B	FOTP-48	Measurement of Optical Fiber Cladding Diameter Using Laser-Based Instruments
EIA-455-49A	FOTP-49	Measurement for Gamma Irradiation Effects on Optical Fiber and Cables
EIA-455-50A	FOTP-50	Light Launch Conditions (Long Length Graded Index Fibers)
EIA-455-51A	FOTP-51	Pulse Distortion Measurement, Multimode Fiber
EIA-455-53A	FOTP-53	Attenuation by Substitution (Multimode Graded Index)
EIA-455-54A	FOTP-54	Mode Scrambler Requirements for Overfilled Launching Conditions (Multimode)
EIA-455-55B	FOTP-55	End View Methods for Measuring Coating and Buffer Geometry
EIA-455-56A	FOTP-56	Test Method for Evaluating Fungus Resistance of Optical Waveguide Fibers and Cables
EIA-455-57A	FOTP-57	Optical Fiber End Preparation and Examination
EIA-455-58A	FOTP-58	Core Diameter Measurements (Graded Index Fibers)
EIA-455-59	FOTP-59	Measurement of Fiber Point Defects Using an OTDR
EIA-455-60	FOTP-60	Measurement of Fiber or Cable Length Using an OTDR
EIA-455-61	FOTP-61	Measurement of Fiber or Cable Attenuation Using an OTDR
EIA-455-62	FOTP-62	Optical Fiber Macrobend Attenuation
EIA-455-63A	FOTP-63	Torsion Test for Optical Fiber
EIA-455-65	FOTP-65	Flexure Test for Optical Fiber
EIA-455-66	FOTP-66	Test Method for Measuring Relative Abrasion Resistance
EIA-455-68	FOTP-68	Optical Fiber Microbend Test Procedure

EIA-455-69A	FOTP-69	Evaluation of Minimum and Maximum Exposure Temperature on the Optical Performance of Optical Fiber
EIA-455-71	FOTP-71	Measurement of Temperature Shock Effects on Components
EIA-455-75	FOTP-75	Fluid Immersion Test for Optical Waveguide Fibers
EIA-455-77	FOTP-77	Procedure to Qualify a Higher-Order Mode Filter for Measurements of Single-mode Fibers
EIA-455-78A	FOTP-78	Spectral Attenuation Cutback Measurement (Single-mode)
EIA-455-80	FOTP-80	Cutoff Wavelength of Uncabled Single-mode Fiber by Transmitted Power
EIA-455-81A	FOTP-81	Compound Flow (Drip) Test for Filled Fiber Optic Cable
EIA-455-82B	FOTP-82	Fluid Penetration Test for Fluid-Blocked Cable
EIA-455-83A	FOTP-83	Cable to Interconnecting Device Axial Compressive Loading
EIA-455-84B	FOTP-84	Jacket Self-Adhesion (Blocking) Test for Cables
EIA-455-85A	FOTP-85	Fiber Optic Cable Twist Test
EIA-455-86	FOTP-86	Fiber Optic Cable Jacket Shrinkage
EIA-455-87A	FOTP-87	Fiber Optic Cable Knot Test
EIA-455-88	FOTP-88	Fiber Optic Cable Bend Test
EIA-455-89A	FOTP-89	Fiber Optic Cable Jacket Elongation and Tensile Strength Test
EIA-455-91	FOTP-91	Fiber Optic Cable Twist-Bend Test
EIA-455-92	FOTP-92	Optical Fiber Cladding Diameter and Noncircularity Measurement by Fizeau Interferometry
EIA-455-94	FOTP-94	Fiber Optic Cable Stuffing Tubing Compression
EIA-455-95	FOTP-95	Absolute Optical Power Test for Fibers and Cables
EIA-455-96	FOTP-96	Fiber Optic Cable Long-Term Storage Temperature Test for Extreme Environments
EIA-455-98A	FOTP-98	Fiber Optic Cable External Freezing Test
EIA-455-99	FOTP-99	Gas Flame Test for Special Purpose Cable
EIA-455-100C	FOTP-100	Gas Leakage Test for Gas Blocked Cable
EIA-455-101	FOTP-101	Accelerated Oxygen Test
EIA-455-102	FOTP-102	Water Pressure Cycling
EIA-455-104	FOTP-104	Fiber Optic Cable Cyclic Flexing Test
EIA-455-107	FOTP-107	Return Loss for Fiber Optic Components
EIA-455-127	FOTP-127	Spectral Characterization of Multimode Laser Diodes
EIA-455-162	FOTP-162	Fiber Optic Cable Temperature-Humidity Cycling
EIA-455-164A	FOTP-164	Measurement of Mode Field Diameter by Far-Field Scanning (Single-mode)
EIA-455-165	FOTP-165	Measurement of Mode Field Diameter by Near Field Scanning (Single-mode)
EIA-455-166	FOTP-166	Transverse Offset Method
EIA-455-167A	FOTP-167	Mode Field Diameter Measurement, Variable Aperture Method in Far-Field
EIA-455-168A	FOTP-168	Chromatic Dispersion Measurement of Multimode Graded-Index and Single-mode Optical Fiber by Phase-Shift Method
EIA-455-169	FOTP-169	Chromatic Dispersion Measurement of Optical Fibers by the Phase-Shift Method
EIA-455-170	FOTP-170	Cable Cutoff Wavelength of Single-mode Fiber by Transmitted Power

EIA-455-171	FOTP-171	Attenuation by Substitution Measurement (Short Length Multimode Graded-Index and Single-mode)
EIA-455-172	FOTP-172	Flame Resistance of Firewall Connector
EIA-455-173	FOTP-173	Coating Geometry Measurement of Optical Fiber, Side-View Method
EIA-455-174	FOTP-174	Mode Field Diameter of Single-mode Fiber by Knife-Edge Scanning in Far-Field
EIA-455-175	FOTP-175	Chromatic Dispersion Measurement of Optical Fiber by the Differential Phase-Shift
EIA-455-177A	FOTP-177	Numerical Aperture Measurement of Graded-Index Fiber
EIA-455-178	FOTP-178	Coating Strip Force Measurement
EIA-455-179	FOTP-179	Inspection of Cleaved Fiber End Faces by Interferometry
EIA-455-180	FOTP-180	Measurement of Optical Transfer Coefficients of a Passive Branching Device
EIA-455-184	FOTP-184	Coupling Proof Overload Test for Fiber Optic Interconnecting Devices
EIA-455-185	FOTP-185	Strength of Coupling Mechanism for Fiber Optic Interconnecting Devices
EIA-455-186	FOTP-186	Gauge Retention Force Measurement for Components
EIA-455-187	FOTP-187	Engagement and Separation/Force for Connector Sets
EIA-455-188	FOTP-188	Low-Temperature Testing for Components
EIA-455-189	FOTP-189	Ozone Exposure Test for Components
EIA-455-190	FOTP-190	Low Air Pressure (High Altitude) Test for Components
EIA-458-B		Standard Optical Fiber Material Classes and Preferred Sizes
EIA-472		General Specification for Fiber Optic Cable
EIA-472A		Sectional Specification for Fiber Optic Communication Cables for Outside Aerial Use
EIA-472B		Sectional Specification for Fiber Optic Communication Cables for Underground and Buried Use
EIA-472C		Sectional Specification for Fiber Optic Communication Cables for Indoor Use
EIA-472D		Sectional Specification for Fiber Optic Communication Cables for Outside Telephone Plant Use
EIA-4750000-B		Generic Specification for Fiber Optic Connectors
EIA-475C000		Sectional Specification for Type FSMA Connectors
EIA-475CA00		Blank Detail Specification for Optical Fiber and Cable Type FSMA, Environmental Category I
EIA-475CB00		Blank Detail Specification Connector Set for Optical Fiber and Cables Type FSMA, Environmental Category II
EIA-475CC00		Blank Detail Specification Connector Set for Optical Fiber and Cables Type FSMA, Environmental Category III
EIA-475E000		Sectional Specification for Fiber Optic Connectors Type BFOC/2.5
EIA-475EA00		Blank Detail Specification for Connector Set for Optical Fiber and Cables, Type BFOC/2.5, Environmental Category I
EIA-475EB00		Blank Detail Specification for Connector Set for Optical Fiber and Cables, Type BFOC/2.5, Environmental Category II

EIA-475EC00	Blank Detail Specification for Connector Set for Optical Fiber and Cables, Type BFOC/2.5, Environmental Category III
EIA-492AAAA	Detail Specification for 62.5 micron Core Diameter/125 micron Cladding Diameter Class 1A Multimode, Graded Index Optical Waveguide Fibers
EIA-5390000	Generic Specification for Field Portable Polishing Device for Preparation Optical Fiber
EIA-5460000	Generic Specification for a Field Portable Optical Inspection Device, Combined EIA-NECQ Specification
EIA-546A000	Sectional Specification for a Field Portable Optical Microscope for Inspection of Optical Waveguide and Related Devices
EIA-587	Fiber Optic Graphic Symbols
EIA-590	Standard for Physical Location and Protection of Below-Ground Fiber Optic Cable Plant
EIA-598	Color Coding of Fiber Optic Cables
ANSI Z136.2	The Safe Use of Optical Fiber Communication Systems Utilizing Laser Diode and LED Sources
IEC 693	Dimensions of Optical Fibers
IEEE STD 812	Glossary of Terms, Fiber Optics

MILITARY AND GOVERNMENT STANDARDS

Department of Defense (DOD) and military standards MIL-STD can be obtained from Global Engineering Documents™, 15 Inverness Way East, Englewood, CO 80112-5776, (800) 854-7179, (303) 397-7956, (303) 397-2740 (fax).

DOD-C-24621	Couplers, Passive, Fiber Optic
DOD-C-24621/1(SH)	Coupler, Fiber Optic Cable, Cable Splitter Passive 2 × 2, Transmission Star
DOD-C-24621/2	Coupler, Fiber Optic Cable, Cable Splitter Passive 4 × 4, Transmissive Star
DOD-C-24621/3	Coupler, Fiber Optic Cable, Cable Splitter Passive 20 × 20, Transmissive Star
DOD-C-24621/4	Coupler, Fiber Optic Cable, Cable Splitter Passive 5 × 5, Transmissive Star
DOD-C-85945C	Cables, Fiber Optics, General Specification for (Metric)
DOD-C-85045/1 A1	Single Fiber, Buffered, Unjacketed
DOD-C-85045/2B	Cables, Fiber Optics, Type 1, Class 2
DOD-C-85045/3A	Cables, Fiber Optics, Type 1, Class 2
DOD-C-85045/4A	Cables, Fiber Optics, Type 1, Class 2
DOD-C-85045/5A	Cables, Fiber Optics, Type 1, Class 2
DOD-C-85045/6C	Cables, Fiber Optics, Environment Resisting
DOD-C-85045/8(CR)	Cable, Fiber Optic, Ruggedized, Radiation Hardened, Single Jacket
DOD-C-85045/9	Fiber Optic, Breakout, Individually Jacketed Fibers
DOD-C-85045/10	Fiber Optics, Gel Filling and Flooding Compound
DOD-C-85045/11	Steel Sheath, Gel Filling and Flooding Compound
DOD-C-85045/12(NASA)	Cables, Fiber Optics, High Reliability, Single Fiber, Multimode, Graded Index, Out-Gassing Resistant
DOD-D-24620	Detectors, PIN, Fiber Optic

DOD-D-24620/1	Detectors, PIN, Fiber Optic, 820–910 nm, Glass-Pigtailed
DOD-F-49291	Fiber, Optical, General Specification
DOD-F-49291/1	Fiber, Optical 50/125, Radiation Hard
DOD-F-49291/2	Fiber, Optical, 100/140, Radiation Hard
DOD-F-49291/3	Fiber, Optical, 50/125, Non-Radiation Hard
DOD-F-49291/4	Fiber, Optical, 100/140, Non-Radiation Hard
DOD-S-24622	Sources, LED, Fiber Optic
DOD-S-24622/1	Fiber Optic Sources, LED, 820–910 nm, Glass Pigtailed
DOD-S-24622/2	Source, LED, Fiber Optic, Type A, Class 1, Style T Metric
DOD-S-24622/3(CR)	Source, LED, Type A, Class 2, Pigtail, DIP
DOD-S-24623	Splice, Fiber Optic Cable
DOD-S-24623/1	Splice, Fiber Optic Cable, Fiber Splice
DOD-S-24623/2	Splice, Fiber Optic Cable, Fiber Splice Enclosure
DOD-S-24623/3	Splice, Fiber Optic Cable, Splice, Cable/Fiber
DOD-STD-347	Product Assurance Program Requirements for Fiber Optic Components
DOD-STD-1678 Notice 1	Fiber Optic Test Methods & Instrumentation
DOD-STD-1864	Fiber Optic Symbols
FED-STD-1037B	Glossary of Telecommunication Terms (Includes Fiber Optic Terms)
MIL-A-24726	Attenuators, Fiber Optic, Shipboard, General Specifications for VSMF
MIL-A-24726/1	Attenuator, Fiber Optic, Shipboard, Fixed, Connectorized, Single-mode, Stand-alone
MIL-C-22520/10D	Crimping Tool, Terminal, Hand
MIL-C-28876	Connectors, Circular Plug and Receptacle
MIL-C-28876 B	Connectors, Fiber Optic, Environment Resisting, Circular, Screw Threads
MIL-C-28876/1B	Receptacle, Wall Mount w/o Strain Relief
MIL-C-28876/2B	Receptacle, Wall Mount, Strain Relief
MIL-C-28876/3B	Receptacle, Wall Mount, 45 degree
MIL-C-28876/4B	Receptacle, Wall Mounting, 90 degree
MIL-C-28876/5B	Receptacle, Wall Mounting, Strain Relief
MIL-C-28876/6B	Plug w/o Strain Relief
MIL-C-28876/7B	Plug, Strain Relief
MIL-C-28876/8B	Plug, Strain Relief, 45 degree
MIL-C-28876/9B	Plug, Strain Relief, 90 degree
MIL-C-28876/10B	Plug Dust Cover
MIL-C-28876/11B	Receptacle, Jam Nut w/o Strain Relief
MIL-C-28876/12B	Receptacle, Jam Nut, Strain Relief
MIL-C-28876/13B	Receptacle, Jam Nut, Strain Relief, 45 degree
MIL-C-28876/14B	Receptacle, Jam Nut, Strain Relief, 90 degree
MIL-C-28876/15B	Receptacle, Dust Cover
MIL-C-28876/16	Canceled and Superseded by MIL-T-29504/1
MIL-C-28876/17	Canceled and Superseded by MIL-T-29504/2
MIL-C-28876/18	Canceled and Superseded by MIL-T-29504/3

MIL-C-28876/26A	Receptacle, Wall Mounting, Short Backshell
MIL-C-28876/27A A1	Backshell, Straight, Strain Relief
MIL-C-28876/28A A1	Backshell, 45 degree, Strain Relief
MIL-C-28876/29A A1	Backshell, 90 degree, Strain Relief
MIL-C-28876C	Connectors, Fiber Optic, Circular, Plug and Receptacle
MIL-C-28876/1C	Receptacle, Wall Mount w/o Strain Relief
MIL-C-28876/2C	Receptacle, Wall Mount, Strain Relief
MIL-C-28876/3C	Receptacle, Wall Mount, 45 degree
MIL-C-28876/4C	Receptacle, Wall Mounting, 90 degree
MIL-C-28876/5C	Receptacle, Wall Mounting, Strain Relief
MIL-C-28876/6C	Plug w/o Strain Relief
MIL-C-28876/7C	Plug, Strain Relief
MIL-C-28876/8C	Plug, Strain Relief, 45 degree
MIL-C-28876/9C	Plug, Strain Relief, 90 degree
MIL-C-28876/10C	Dust Cover, Plug
MIL-C-28876/11C	Receptacle, Jam Nut w/o Strain Relief
MIL-C-28876/12C	Receptacle, Jam Nut, Strain Relief
MIL-C-28876/13C	Receptacle, Jam Nut, Strain Relief, 45 degree
MIL-C-28876/14C	Receptacle, Jam Nut, Strain Relief, 90 degree
MIL-C-28876/15C	Dust Cover, Receptacle
MIL-C-28876/16	Canceled and Superseded by MIL-T-29504/1
MIL-C-28876/17	Canceled and Superseded by MIL-T-29504/2
MIL-C-28876/18	Canceled and Superseded by MIL-T-29504/3
MIL-C-28876/26B	Receptacle, Wall Mounting, Short Backshell
MIL-C-28876/27B	Backshell, Straight, Strain Relief
MIL-C-28876/28B	Backshell, 45 degree, Strain Relief
MIL-C-28876/29B	Backshell, 90 degree, Strain Relief
MIL-C-28876/1D	Connectors, Circular, Plug and Receptacle Style, Screw Threads, Multiple Removable Termini, General Specifications
MIL-C-49292(CR)	Cable Assembly, Nonpressurized
MIL-C-49292/4(CR)	Cable Assembly, Dual Channel
MIL-C-49292/7(CR)	Cable Assembly, Test (Not Pressure Proof)
MIL-C-83522D	Connector, Single Terminus, General Specifications
MIL-C-83522/1E	Connector, Plug, Single Terminus, Threaded (Step-Down Nose Interface). Lensless, Epoxy, for 50/125 µm, 62.5/125µm, and 100/140 µm
MIL-C-83522/2F	Connector, Plug, Single Terminus,Threaded (Straight Nose Interface). Lensless, Epoxy, for 50/125 µm, 62.5/125µm, and 100/140 µm
MIL-C-83522/3F	Connector, Plug-Receptacle-Adapter Style, Fixed Single Terminus, Threaded, Adapter Bulkhead Mount
MIL-C-83522/4E	Connector, Plug-Receptacle-Adapter Style, Fixed Single Terminus, Threaded, Receptacle, Low Profile Parallel PC Mount
MIL-C-83522/5D	Connector, Plug-Receptacle-Adapter Style, Fixed Single Terminus, Threaded, Receptacle, Bulkhead Mount

MIL-C-83522/6D	Connector, Plug-Receptacle-Adapter Style, Fixed Single Terminus, Threaded, Plug, Expanded Beam Lens, Epoxy, for 50/125 μm, and 100/140 μm fiber
MIL-C-83522/7B	Connector, Plug-Receptacle-Adapter Style, Fixed Single Terminus, Threaded, Receptacle, Parallel PC Mount
MIL-C-83522/8B	Connector, Plug-Receptacle-Adapter Style, Fixed Single Terminus, Threaded, Receptacle, Hex Mount
MIL-C-83522/9	Connector, Receptacle, PC Mount, Active
MIL-C-83522/10	Connector, Receptacle, Flange Mount, Active (Used with MIL-T-29504/4 Pin Terminus, Size 16)
MIL-C-83522/11	Connector, Adapter, In Line Cable Panel Mount (Used with MIL-T-29504/4 Pin Terminus, Size 16)
MIL-C-83522B	Connector, Fiber Optic, Single Fiber, General
MIL-C-83522/1D	Connector, Fiber Optic, Single Fiber
MIL-C-83522/2D	Connector, Fiber Optic, Single Fiber
MIL-C-83522/2E(NA)	Connector, Fiber Optic, Single Fiber
MIL-C-83522/3E	Connector, Fiber Optic, Single Fiber
MIL-C-83522/4E	Connector, Fiber Optic, Single Fiber
MIL-C-83522/5D	Connector, Fiber Optic, Single Fiber
MIL-C-83522/6C	Connector, Fiber Optic, Expanded Beam, Plug
MIL-C-83522/7B	Connector, Fiber Optic, Receptacle, Parallel P.C. Mount
MIL-C-83522/8B	Connector, Fiber Optic, Receptacle, Hex Mount Bulkhead
MIL-C-83522/12A	2.5 mm Bayonet, Epoxy, for 50/125 μm, 62.5/125 μm, 100/140 μm
MIL-C-83522/13A	Connector, Adapter, 2.5 mm Bayonet, Bulkhead Panel Mount, Coupling Receptacle
MIL-C-83522/14A	Connector, Active Device Receptacle, 2.5 mm Bayonet, PC Mount
MIL-C-83526A	Connectors, Fiber Optic, Hermaphroditic, Circular, Environment Resistant
MIL-C-83526/1A	Connectors, Fiber Optic, Hermaphroditic, Circular, Inline Mounting, 2 and 4 Positions
MIL-C-83526/2A	Connectors, Fiber Optic, Hermaphroditic, Circular, Inline Mounting, 6 Positions
MIL-C-83526/3A	Connectors, Fiber Optic, Hermaphroditic, Circular, Jam-Nut Mounting, 2 and 4 Positions
MIL-C-83526/4A	Connectors, Fiber Optic, Hermaphroditic, Circular, Jam-Nut Mounting, 6 Positions
MIL-C-83526/5A	Connectors, Fiber Optic, Hermaphroditic, Circular, Wall Mounting, 6 Positions
MIL-C-83526/6A	Dust Cover, Connectors, Fiber Optic, Hermaphroditic, Circular, Jam-Nut Mounting, 2 and 4 Positions
MIL-C-83526/7A	Dust Cover, Connectors, Fiber Optic, Hermaphroditic, Circular, Jam-Nut Mounting, 4 and 6 Positions
MIL-C-83526/8A	Dust Cover, Connectors, Fiber Optic, Hermaphroditic, Circular, Jam-Nut Mounting, 6 Positions
MIL-C-83526/9A	Dummy Connector, Fiber Optic, Hermaphroditic, Circular, 2 and 4 Positions, Test Plug

MIL-C-83526/10A	Dummy Connector, Fiber Optic, Hermaphroditic, Circular, 6 Positions, Test Plug
MIL-C-83526/11A	Connectors, Receptacle, Hermaphroditic, Circular, Jam Nut Mounting, Without Strain Relief, 2 and 4 Positions
MIL-D-24620A	Detectors, PIN, and APD, Fiber Optic, General Specification
MIL-D-24620/1	Fiber Optic Detectors, PIN, 820–910 nm, Glass Pigtailed Type
MIL-D-24620/2	Detector, PIN Type A and B, Class 1, Style 3
MIL-D-24620/3	Detector, APD, Type A and B, Class 1, Style 3
MIL-D-24620/4A	Detector, PINFET, 1100–1600 nm, Glass Pigtailed, Hermetically Sealed, Dual In-Line Package (DIP) Couplers, Passive
MIL-C-24621/1	Couplers, Cable Splitter, Passive Glass Connectorized Output, Type I, 820–910 nm 2 × 2, Transmission Star, 50/125 μm Fiber, Graded-Index, MM, Bidirectional
DOD-C-24621/2	Couplers, Cable Splitter, Passive Glass Pigtailed Output, Type II, 820–910 nm 4 × 4, Transmission Star, 50/125 μm Fiber, Graded-Index, MM, Bidirectional
DOD-C-24621/3	Couplers, Cable Splitter, Passive Glass Pigtailed Output, Type II, 820–910 nm 20 × 20, Transmission Star, 100/140 μm 50/125 μm Fiber, Graded-Index
DOD-C-24621/4	Couplers, Cable Splitter, Passive Glass Pigtailed Output, Type II, 820–910 nm 5 × 5, Transmission Star, 100/140 μm 50/125 μm Fiber, Graded-Index
MIL-C-24621/5	Couplers, Cable Splitter, Passive Glass Pigtailed Output, Type II, 1280–1360 nm, 2 × 2, Transmission Star, 62.5/125 μm Fiber, Graded-Index, Bidirectional
MIL-C-24733	Controller Interface Unit, Fiber Optic, General Specification for VSMF
MIL-C-24733/1	Controller Interface Unit, Fiber Optic, 2 Fiber Channels, Multimode Fiber
MIL-C-28688	Cable, Fiber Optic, Packaging of
MIL-H-24626(NAVY)	Harness Assemblies, Cable, Pressure Proof
MIL-H-24626/1(NAVY)	Harness, Cable, Pressure Proof, Fiber Optic
MIL-HDBK-277(NAVY)	Fiber Optic Checkout Procedure for Military Applications
MIL-HDBK-278(NAVY)	Systems Design Guide for Applying Fiber Optic Technology to Shipboard Systems
MIL-HDBK-282(NAVY)	Fiber Optic Cable Installation Procedure
MIL-HDBK-415	Design Handbook for Fiber Optic Communication Systems
MIL-I-24728A	Interconnection Box, Fiber Optic, General Specification for VSMF
MIL-I-24728A/1	Interconnection Box, Fiber Optic, Shipboard, Submersible
MIL-I-24728/2A	Interconnection Box, Fiber Optic, Submersible
MIL-I-81969/46A	Installing and Removal Tools, Type I, Class 2, Composition C, Size 16 Termini
MIL-I-81969/47A	Installing and Removal Tools, Type II, Class 2, Composition C, Size 16 Termini
MIL-I-81969/48A	Installing and Removal Tools, Type III, Alignment, Sleeve, Class 1, Composition C, Size 16 Termini

MIL-I-81969/49B	Installing and Removal Tools, Hand, Right Angle, Type I, Alignment, Sleeve, Class 2, Composition A, Size 16 Termini
MIL-K- 83525	Kit, Portable Optical Microscope, for Fiber Optic Inspection
MIL-M-24731	Multiplexers, Demultiplexers, Frequency-Division, Fiber Optic Interfaceable, Shipboard, Metric, General Specification for VSMF
MIL-M-24731/1	Multiplexers, Demultiplexers, Frequency-Division, Fiber Optic Interfaceable, Shipboard, 4:1 Channels, 30 MHz (Double Sideband)
MIL-M-24731/2	Multiplexers, Demultiplexers, Frequency-Division, Fiber Optic Interfaceable, Shipboard, 4:1 Channels, 30 MHz (Double Sideband)
MIL-M-24736	Multiplexers, Demultiplexers, Frequency-Division, Fiber Optic Interfaceable, Shipboard, General Specification for VSMF
MIL-M-24736/1	Multiplexers, Demultiplexers, Frequency-Division, Fiber Optic Interfaceable, Shipboard, 8:1 Channels, 311 Mbps per Input Channel
MIL-M-24736/2	Multiplexers, Demultiplexers, Frequency-Division, Fiber Optic Interfaceable, Shipboard, 1:8 Channels, 311 Mbps per Output Channel
MIL-P-24627(NAVY)	Penetrators, Bulkhead, Connectorized, Fiber Optic
MIL-P-24627/1(NAVY)	Penetrators, Bulkhead, Connectorized, Fiber Optic, 32-Channel
MIL-P-24628(NAVY)	Penetrators, Hull, Connectorized, Connectors, Pressure-Proof, Fiber Optic, Submarine
MIL-P-24628/1(NAVY)	Penetrators, Hull, Connectorized, Connectors, Pressure-Proof, Fiber Optic, Submarine Plug Connector
MIL-R-24720	Receivers, Digital, Fiber Optic Shipboard, General Specification for VSMF
MIL-R-24720/1	Receivers, Digital, Fiber Optic Shipboard, 0.5 to 16 Mbps, Manchester Encoded
MIL-R-24721	Transmitters, Digital, Fiber Optic Shipboard, General Specification for VSMF
MIL-R-24721/1	Transmitters, Digital, Fiber Optic Shipboard, DC to 16 Mbps, Manchester Encoded
MIL-R-24727	Rotary Joints, Fiber Optic, Shipboard, Multiple Fibers, On-Axis
MIL-R-24727/1	Rotary Joints, Fiber Optic, Shipboard, Single Fiber, On-Axis, Multimode Cable Pigtail
MIL-R-24727/2	Rotary Joints, Fiber Optic, Shipboard, Multiple Fibers, On-Axis
MIL-R-24737	Receivers, Light Signal, Analog, Fiber Optic, Ship, General Specification for VSMF
MIL-R-24737/1	Receivers, Light Signal, Analog, Fiber Optic, Ship, 0.5 to 60 MHz (0.5 dB Passband)
MIL-S-24622A	Sources, LED Type General Specification
MIL-S-24622/2	LED, Type A, Class 1, Style 3
MIL-S-24622/3A	LED, 1290 nm, Pigtailed, Hermetically Sealed, Dual In-line Package
MIL-S-24623B	Splice, Fiber Optic Cable, Fiber Splice
MIL-S-24623/3	Splice, Fiber Optic, Cable/Fiber
MIL-S-24623/4A	Splice, Fiber Optic, Housing, Fiber
MIL-S-24623/5A	Splice, Fiber Optic, Housing, Cable
MIL-S-24725	Switches, Fiber Optic, Shipboard, General Specification for VSMF

MIL-S-24725/1	Switches, Fiber Optic, Shipboard, Electrical, Nonlatching, Bypass, Multimode Cable, Stand-alone
MIL-STD-188-111	Subsystem Design and Engineering Standards for Common Long Haul and Tactical Fiber Optic Communications
MIL-STD-188-111A	Interoperability and Performance Standards for Fiber Optic Communication Systems
MIL-STD-1773	Fiber Optic Mechanization of an Aircraft Internal Time Division Command/Response Multiplex Data Bus
MIL-STD-1863A, Notice 1	Interface Designs and Dimensions for Fiber Optic Interconnection Devices
MIL-STD-1864	Fiber Optic Symbols
MIL-STD-2163C	Insert Arrangements for MIL-C-28876 Connectors, Circular, Plug and Receptacle Style, Multiple Removable Termini
MIL-T-24735	Transmitters, Light Signal, Analog, Fiber Optic, Ship, General Specification for VSMF
MIL-T-24735/1	Transmitters, Analog, Fiber Optic, Ship, 0.5-60 MHz, 90.5 dB Passband)
MIL-T-29504A	Termini, Connector, Removable, General Specification
MIL-T-29504/1A	Termini, Connector, Removable, Environmental, Class 2, Type II, Style A, Pin Terminus (for MIL-C-28876 and MIL-C-83526 Connectors)
MIL-T-29504/2A	Termini, Connector, Removable, Environmental, Class 2, Type II, Style A, Socket Terminus (for MIL-C-28876 and MIL-C-83526 Connectors)
MIL-T-29504/3A	Termini, Connector, Removable, Environmental, Class 2, Type II, Style A, Dummy Terminus (for MIL-C-28876 and MIL-C-83526 Connectors)
MIL-T-29504/4B	Termini, Connector, Removable, Environmental, Pin Terminus, Size 16 Rear Release—MIL-C-38999, Series I, III, and IV
MIL-T-29504/5B	Termini, Connector, Removable, Environmental, Socket Terminus, Size 16 Rear Release—MIL-C-38999
MIL-T-29504/6	Termini, Removable, Environmental, Class 5, Type II, Style A, Pin Terminus, Size 16, Rear Release—DOD-C-83527
MIL-T-29504/7	Termini, Environmental, Class 5, Type II, Style A, Pin Terminus, Size 16, Rear Release—DOD-C-83527
MIL-T-29504/8	Termini, Environmental, Class 5, Type II, Style A, Socket Terminus, Size 16, Front Release—MIL-C-28840
MIL-T-29504/9	Termini, Environmental, Class 5, Type II, Style A, Socket Terminus, Size 16, Front Release—MIL-C-28840
MIL-T-29504/10	Termini, Environmental, Class 5, Type II, Style A, Pin Terminus, Size 16, Rear Release—MIL-C-26482, Series II
MIL-T-29504/11	Termini, Environmental, Class 5, Type II, Style A, Socket Terminus, Size 16, Rear Release—MIL-C-83723, Series III, MIL-C-26482, Series II
MIL-T- 83523	Tools, Fiber Optic
MIL-T- 83523/1	Hand Tool, Fiber Optic Termini
MIL-T- 83523/2A	Tools, Fiber Optic, Polishing Assembly, Type IV
MIL-T- 83523/3A	Tools, Fiber Optic, Scribe Carbide, Type I
MIL-T- 83523/4A	Tools, Fiber Optic Scribe Diamond/Sapphire
MIL-T- 83523/5	Retaining Band Strain Relief
MIL-T- 83523/6	Polishing, Fixture Assembly, Type IV

MIL-T- 83523/7	Wrench and Adapter, Torque, Strain Relief
MIL-T- 83523/8	Wrench, Spanner, Type IV
MIL-T- 83523/9	Scribe, Conical and Wedge, Type I
MIL-T- 83523/10	Tool, Carbide, Hand, Type VII, Carbide/Diamond, Composition C
MIL-T- 83523/16A	Tools, Fiber Optic Stripping, (50/125 μm), (100/140 μm), (400/430 μm), (600/630 μm), (1000/1030 μm), Type X, Composition C
MIL-C-28876/2D	Connectors, Circular, Receptacle Style, Screw Threads, Multiple Removable Termini, Wall Mounting, With Strain Relief, Environmental
MIL-C-28876/3D	Connectors, Circular, Receptacle Style, Screw Threads, Multiple Removable Termini, Wall Mounting, With 45° Strain Relief, Environmental
MIL-C-28876/4D	Connectors, Circular, Receptacle Style, Screw Threads, Multiple Removable Termini, Wall Mounting, With 90° Strain Relief, Environmental
MIL-C-28876/5D	Connectors, Circular, Receptacle Style, Screw Threads, Multiple Removable Termini, With Straight Strain Relief, Environmental
MIL-C-28876/6D	Connectors, Circular, Plug Style, Screw Threads, Multiple Removable Termini, Without Strain Relief, Environmental
MIL-C-28876/7D	Connectors, Circular, Plug Style, Screw Threads, Multiple Removable Termini, Straight Strain Relief, Environmental
MIL-C-28876/8D	Connectors, Circular, Plug Style, Screw Threads, Multiple Removable Termini, With 45° Strain Relief, Environmental
MIL-C-28876/9D	Connectors, Circular, Plug Style, Screw Threads, Multiple Removable Termini, 90° Strain Relief, Environmental
MIL-C-28876/10D	Connectors, Circular, Plug Style, Multiple Removable Termini, Dust Cover, Screw Threads, Environmental
MIL-C-28876/11D	Connectors, Circular, Receptacle Style, Multiple Removable Termini, Jam-Nut Mounting, Screw Threads, Without Strain Relief, Environmental
MIL-C-28876/12D	Connectors, Circular, Receptacle Style, Multiple Removable Termini, Jam-Nut Mounting, Screw Threads, With Straight Strain Relief, Environmental
MIL-C-28876/13D	Connectors, Circular, Receptacle Style, Multiple Removable Termini, Jam-Nut Mounting, Screw Threads, With 45° Strain Relief, Environmental
MIL-C-28876/14D	Connectors, Circular, Receptacle Style, Multiple Removable Termini, Jam-Nut Mounting, Screw Threads, With 90° Strain Relief, Environmental
MIL-C-28876/15D	Connectors, Circular, Receptacle Style, Multiple Removable Termini, With Receptacle Dust Cover, With 45° Strain Relief, Environmental
MIL-C-28876/19	Connectors, Environmental, Insertion Tool
MIL-C-28876/20	Connectors, Environmental, Removal Tool
MIL-C-28876/21	Connectors, Environmental, Alignment Sleeve
MIL-C-28876/26C	Connectors, Circular, Receptacle Style, Multiple Removable Termini, Screw Threads, Wall Mounting, With Short Straight Back Shell
MIL-C-28876/27C	Connectors, Circular, Plug and Receptacle Style, Multiple Removable Termini, Screw Threads, Straight Back Shell With Straight Strain Relief, Environmental

MIL-C-28876/28C	Connectors, Circular, Plug and Receptacle Style, Multiple Removable Termini, 45° Back Shell, Screw Threads, With Straight Strain Relief, Environmental
MIL-C-28876/29D	Connectors, Circular, Plug and Receptacle Style, Multiple Removable Termini, 90° Back Shell, Screw Threads, With Strain Relief, Environmental Resisting
MIL-T-29504/12	Termini, Environmental, Class 5, Type II, Style A, Pin Terminus, Front Release, Stainless Steel (for MIL-C-28876 Connectors)
MIL-T-29504/13	Termini, Environmental, Class 5, Type II, Style A, Socket Terminus, Front Release, Stainless Steel (for MIL-C-28876 Connectors)
MIL-T-29504/14	Termini, Connector, Removable, Environmental, Class 5, Type II, Style A, Pin Terminus, Front Release, Ceramic Guide Bushing (for MIL-C-28876 Connectors)
MIL-T-29504/15	Termini, Connector, Removable, Environmental, Class 5, Type II, Style A, Socket Terminus, Front Release, Ceramic Guide Bushing (for MIL-C-28876 Connectors)
MIL-F-49291B	Optical Fiber, General Specifications
MIL-F-49291/1A	Optical Fiber, 50/125 μm, Radiation Hardened
MIL-F-49291/2A	Optical Fiber, 100/140 μm, Radiation Hardened
MIL-F-49291/3A	Optical Fiber, 50/125 μm
MIL-F-49291/4A	Optical Fiber, 100/140 μm
MIL-F-49291/6A	Optical Fiber, 62.5/125 μm, Radiation Hardened
MIL-F-49291/7A	Optical Fiber, Single-mode, Dispersion Unshifted, Radiation Hardened
MIL-F-49291/08	Optical Fiber, 400/430 μm, Avionic Rated
MIL-F-49291/09	Optical Fiber, 400/430 μm, Radiation Hardened, Avionic Rated
MIL-C-49292A	Cable Assemblies, Non-Pressure Proof, Fiber Optic, Metric, General Specification
MIL-C-49292/1	Cable Assemblies, Non-Pressure Proof, Fiber Optic, Metric, Branched, MIL-C-83522, MIL-C-83526, Connectors, and DOD-C-85045 Cables
MIL-C-49292/2	Cable Assemblies, Non-Pressure Proof, Fiber Optic, Metric, Single Bundle, 6-Position, MIL-C-83526 Connectors, and DOD-C-85045 Cables
MIL-C-49292/3	Cable Assemblies, Non-Pressure Proof, Fiber Optic, Metric, Connector, 2-Position, Hermaphroditic, Jam-Nut, MIL-C-83526 Connectors, and DOD-C-85045 Cables
MIL-C-49292/4A	Cable Assemblies, Non-Pressure Proof, Fiber Optic, Metric
MIL-C-49292/6	Cable Assemblies, Non-Pressure Proof, Fiber Optic, Metric, Single Fiber, Glass, MIL-C-83522 Connectors, and DOD-C-85045 Cables
MIL-C-49292/7A	Cable Assemblies, Non-Pressure Proof, Fiber Optic, Test, Metric
MIL-F-50533	Fiber, Acrylic, Fibrillated
MIL-F-50809	Fiber, Acrylic, Fibrillatable
MIL-C-83255/15	Connector, Plug-Receptacle-Adapter Style, Fixed Single Terminus, Environmental, Crimp and Cleave
MIL-C-83522/16A	Connector, Single Terminus, Adapter Style, 2.5 mm, Bayonet Coupling, Bulkhead Panel Mount

MIL-C-83522/17A	Connector, Single Terminus Plug, Adapter Style, 2.5 mm, Bayonet Coupling, Epoxy
MIL-C-83522/18A	Connector, Single Terminus, Adapter 2.5 mm, Bayonet Coupling PC Mount
MIL-T-83523	Tools, Fiber Optic, General Specification
MIL-T-83523/1	Tools, Fiber Optic, Hand Terminating, Type II
MIL-T-83523/2A	Tools, Fiber Optic, Polishing Bushing Assembly, Type IV
MIL-T-83523/3A	Tools, Fiber Optic, Scribe, Carbide, Type I
MIL-T-83523/4A	Tools, Fiber Optic, Scribe, Diamond or Sapphire, Type I
MIL-T-83523/5	Tools, Fiber Optic, Scribe, Retaining Band Strain Relief, Type VIII
MIL-T-83523/6	Tools, Fiber Optic, Polishing Fixture Assembly, Type IV
MIL-T-83523/7	Tools, Fiber Optic, Wrench, Spanner, Type V
MIL-T-83523/8	Tools, Fiber Optic, Wrench, Spanner, Type V
MIL-T-83523/9	Tools, Fiber Optic, Scribe, Conical and Wedge, Type I
MIL-T-83523/MIL-T-83523/	
MIL-T-83523/16A	
MIL-M-83524	Microscope, Optical, Monocular, Hand-Held, Portable, Militarized, 200x Magnification
MIL-C-83526/12	Connectors, Circular, Hermaphroditic, In-Line Mounting, 2 Positions
MIL-C-83526/13	Connectors, Circular, Hermaphroditic, Bulkhead Mounting, 2 and 4 Positions
MIL-C-83526/14	Connectors, Circular, Hermaphroditic, 2 Positions, Test Plugs
MIL-C-83526/15	Dust Cover, Connectors, Circular, Hermaphroditic, 2 Positions
MIL-C-85045E	Fiber Optic Cables (Metric), General Specification
DOD-C-85045/1	Type 1, Class 2, Composition A, Attenuation Range A, Single Fiber
DOD-C-85045/2B	Cable, Fiber Optic, 1,2,4 and 6 Fiber, Heavy Duty, Metric
DOD-C-85045/3A	Cable, Fiber Optic, Heavy Duty, Metric, With Gel Filling and Flooding Compound
DOD-C-85045/4A	Cable, Fiber Optic, Heavy Duty, Ruggedized with Steel Sheathing Rodent Protection, Gel Filling and Flooding Compound
DOD-C-85045/5A	Cable, Fiber Optic, Heavy Duty, Ruggedized with Non-Metallic Sheathing Rodent Protection, Gel Filling and Flooding Compound
DOD-C-85045/6C	Cable, Fiber Optic, Environmental, Type II
DOD-C-85045/8	Cable, Fiber Optic, Ruggedized, Radiation Hardened
DOD-C-85045/9	Cable, Fiber Optic, Break Out Individually Jacketed Fibers
DOD-C-85045/10	Cable, Fiber Optic, with Gel Filling and Flooding Compound
DOD-C-85045/11	Cable, Fiber Optic, Steel Sheathing Rodent Protection, Gel Filling and Flooding Compound
MIL-C-85045/13A	Cable, Cross-Linked, Eight Fibers, Cable Configuration Type 2, Loose Tube, Cable Class SM and MM
MIL-C-85045/14A	Cable, Cross-Linked, One Fiber, Cable Configuration Type 2, Application B, Cable Class SM and MM
MIL-C-85045/15	Cable, Cross-Linked, Four Fibers, Cable Configuration Type 2, Application B, Cable Class SM and MM
MIL-C-85045/16	Cable, Cross-Linked, One Fiber, Cable Configuration Type 2, Tight Buffered, Cable Class SM and MM

MIL-C-85045/17	Cable, Cross-Linked, Eight Fibers, Cable Configuration Type 2, Application B, Cable Class SM and MM
MIL-C-85045/18	Cable, Cross-Linked, Four Fibers, Cable Configuration Type 2, Application B, Cable Class SM and MM

AStanP-4, Volume 6060, Chapter 1

Section 2.1.1	Connector, Single Fiber, Fiber Optic
Section 2.1.2	Connector, Single Fiber, Fiber Optic
Section 2.1.3	Connector, Single Fiber, Fiber Optic
Section 2.1.4	Connector, Single Fiber, Fiber Optic
Section 2.1.5	Connector, Single Fiber, Fiber Optic

The following Engineering Practices (EP) Studies are available from Defense Electronics Supply Center (DESC-EMD), Dayton, Ohio 45444-5282, telephone (513) 296-8199, (AV) (513) 986-8199.

EP Study Connector Requirements
EP Study Modem Devices
EP Study Fiber Optic Switches
EP Study Hazardous Material Study
Bellcore Standards

Bellcore standards can be obtained from Bellcore at (800) 521-2673. Bellcore standards are of two types: technical advisories (TA) and technical requirements (TR).

TA-NWT-000020	Generic Requirement for Optical Fiber and Optical Fiber Cable, Issue 8 Dec. 1991
TR-TSY-000020	Generic Requirement for Optical Fiber and Optical Fiber Cable, Issue 4 Mar. 1989
TA-TSY-000038	Digital Fiber Optic Systems, Requirements and Objectives
TR-TSY-000196	Generic Criteria for Optical Time Domain Reflectometers, Issue 2 Sept. 1989
TR-TSY-000198	Generic Criteria for Optical Loss Test Sets, Issue 2, Mar. 1990
TR-NWT-000253	Synchronous Optical Network (SONET) Transport Systems: Common Generic Criteria (A module of TSGR, FR-NWT-000440), Issue 2 Dec. 1991
TR-TSY-000264	Optical Fiber Cleaving Tools, Issue 1 Dec. 1986
TR-TSY-000266	Optical Patch Panels, Issue 1, Oct. 1985
TR-NWT-000326	General Requirements for Optical Fiber Connectors and Connectorized Jumper Cables, Issue 2, Mar, 1991
TR-NWT-000332	Reliability Prediction Procedure for Electronic Equipment (a module of RQGR, FR-NWT-000796), Issue 3, Sept. 1990
TR-TSY-000409	General Requirements for Intrabuilding Optical Fiber Cable, Issue 2, Sept. 1990
TA-NWT-000418	General Reliability Assurance Requirements for Fiber Optic Transport Systems (A module of RQGR, FR-NWT-000796), Issue 1, May 1988
TR-TSY-000441	Submarine Splice Closures for Fiber Optic Cable
TR-OPT-000449	General Requirements and Design Considerations for Fiber Distributing Frames, Issue 1, Dec. 1991
TR-NWT-000468	Reliability Assurance Practices for Optoelectronic Devices in Central Office Applications, Issue 1, Dec. 1991
TR-TSY-000761	General Criteria for Chromatic Dispersion Test Sets, Issue 1, Dec. 1991
TR-NWT-000764	General Criteria for Optical Fiber Identifiers, Issue 1, Aug. 1990
TR-TSY-000765	Splicing Systems for Single-mode Optical Fibers, Issue 1, Dec. 1989

TR-TSY-000769	Splice Organizer Assemblies for Optical Fibers, Issue 1, Mar. 1988
TR-NWT-000771	General Requirement for Universal Splice Closures for Fiber Optic Cable, Issue 2, April 1991
TR-TSY-000786	Optical Source Module General Requirements for Subscriber Loop Distribution, Issue 1, Dec. 1989
TR-TSY-000901	General Requirements for WDM (Wavelength Division Multiplexing) Components, Issue 1, Aug. 1989
TR-NWT-000909	General Requirements and Objectives for Fiber in the Loop Systems, Issue 1, Dec. 1991
TA-NWT-000910	General Requirements for Fiber Optic Attenuators, Issue 1, July 1989
TR-TSY-000944	General Requirements for Optical Distribution Cable, Issue 1, July 1990
TR-TSY-000949	General Requirements for Service Terminal Closures with Optical Cable, Issue 1, June 1990
TR-NWT-000950	General Requirements for Distribution/Service Closures Used with Optical Cable, Issue 1, Sept. 1990
TR-NWT-000955	General Requirements for Optical Fiber Stripping (Coating Removal) Tools, Issue 1, Dec. 1990
TR-NWT-001009	General Requirements for Fiber Contacting Devices, Issue 1, Sept. 1990
TR-NWT-001042	General Requirements for Operations Interfaces Using OSI Tools: Synchronous Optical Network (SONET) Transport Information Model, Issue 1, Mar. 1992
TR-NWT-001095	General Requirements for Multifiber Splicing Systems for Singlemode Optical Fiber, Issue 1, June 1991
TR-NWT-001121	General Requirements for Self-Supporting Optical Fiber Cable, Issue 1, Oct. 1991
TR-NWT-001137	General Requirements for Handheld Optical Power Meters, Issue 1, Dec. 1991
SR-TSY-000686	A Methodology for Cost Comparison of 45 Mb/s Video Over Fiber with 139 Mb/s Video Over Fiber and Other Technologies, Issue 1, Aug. 1987
SR-NWT-000821	Field Reliability Performance Study Handbook (a module of RQGR, FR-NWT-000796), Issue 3, Dec. 1990
SR-TSY-001171	Methods and Procedures for System Reliability Analysis (a module of RQGR, FR-NWT-000796), Issue 1, Jan. 1989
ST-TEC-000051	Telecommunications Transmission Engineering Textbook—Volume 1: Principles, Third Edition, Issue 1, May 1989
ST-TEC-000052	Telecommunications Transmission Engineering Textbook—Volume 2: Facilities, Third Edition, Issue 1, May 1989
ST-TEC-000053	Telecommunications Transmission Engineering Textbook—Volume 3: Networks and Services, Third Edition, Issue 1, May 1989
ST-CSP-000054	Trademarks, Acronyms, and Abbreviations Commonly Used in the Telecommunications Industry, Issue 1, Oct. 1987
TR-73536	Technical Requirements for Optical Connectors, Oct. 1989
TR-73539	BellSouth General Requirements for Fiber Optic Splice Closures, Issue A, Nov. 1989
TR-73540	BellSouth Technical Requirements for Fiber Optic Splice Closures, Issue A, Nov. 1989
TR-73541	BellSouth Technical Requirements for Fiber Optic Mechanical Splice Systems, Issue A, Nov. 1990
TR-73542	BellSouth Technical Requirements for Fiber Terminating Equipment for Remote Cabinets and Customer Premise Locations, Issue A, Nov. 1989

APPENDIX

4

Additional References

Allard, Frederick C. *Fiber Optics Handbook for Engineers and Scientists.* New York: McGraw-Hill Publishing Company, 1990.

Hays, Jim (ed.). *Fiber Optics Technology Manual.* Albany, NY: Delmar Publishers, 1996.

Hecht, Jeff. *Understanding Fiber Optics.* Indianapolis: Howard W. Sams & Company, 1989.

Hentschel, Christian. *Fiber Optics Handbook.* Federal Republic of Germany: Hewlett Packard Company, 1988.

Hoss, Robert J. *Fiber Optic Communications Design Handbook.* Englewood Cliffs, NJ: Prentice Hall, 1990.

Sterling, Jr., Donald J. *Technician's Guide to Fiber Optics*, 2nd ed. Albany, NY: Delmar Publishers, 1993.

Weik, Martin H. *Fiber Optics Standard Dictionary*, 2nd ed. New York: Van Nostrand Reinhold, 1989.

APPENDIX

5

The Complete Tool Kit

PRODUCTS TO BE INSTALLED

cable
cable end boxes/enclosures
connectors
mechanical splices
splice trays
splice enclosures
fusion splice covers
splice trays with furcation tubing
pigtails to be spliced to main cable
inner duct
cable hangers

NECESSARY TOOLS: CABLE INSTALLATION

cable puller or Kellems grip with break away swivel
pulleys or slings
pulley supports
pay off stands
cones

NECESSARY TOOLS: SPLICING

fusion splicer
fiber cleaver

NECESSARY TOOLS: CONNECTOR INSTALLATION

safety glasses
tube cutter or jacket remover for tight tube cable
buffer coating/buffer tube stripper
crimper
bottle for bare fiber
polishing tool
polishing plate, pad, or machine
wedge scriber
curing oven
electronic thermometer for checking oven temperature

OPTIONAL TOOLS

fiber phone or walkie talkies
connector shrouds
connector cleaver

TEST EQUIPMENT

stabilized light source with receptacles to accept test leads
calibrated light power meter with receptacles to accept test leads
test leads (at least four for each combination of core diameter, NA, and connector style)
four barrels for each connector style to be tested
OTDR or mini-OTDR
two lead out cables for OTDR for each connector style to be tested
200x[1] connector inspection microscope with adapter for each connector style
high intensity light for white light testing

SUPPLIES: CABLE INSTALLATION

sign stating "NO EATING, DRINKING, OR SMOKING"
yellow 3-inch wide caution tape
cable ties
cable lubricant

SUPPLIES: CABLE END PREPARATION

gel remover for gel-filled cable
95 percent–100 percent isopropyl alcohol with no additives
dry fiber lubricant
clean rags or paper towels

SUPPLIES: SPLICING

lens, or optical, grade compressed air
lens, or optical, grade tissues or Alco pads
95–100 percent isopropyl alcohol with no additives

SUPPLIES: CONNECTOR INSTALLATION

premeasured two part epoxy
syringes and needles
anaerobic adhesive (two parts) with needles
polishing films (three grades)
lens, or optical, grade compressed air
Scotch™ tape
lens, or optical, grade tissues or Alco pads
95 percent–100 percent isopropyl alcohol with no additives
distilled or deionized water
pipe cleaners
toothpicks
batteries for microscope
bulbs for microscope
bulbs for high intensity light source

SUPPLIES: TESTING

test data forms with maximum and expected values
tags to be placed on connectors that look bad
disks for OTDR traces
batteries for light source and power meter

APPENDIX 6

Sources of Products and Supplies

PRODUCTS TO BE INSTALLED

This listing is a partial list, which includes products with which the author has knowledge and/or experience. For a more complete listing, consult one of the fiber optic buyers guides, such as those issued by *Fiberoptic Products News* (Gordon Publications, Morris Plains, NJ) or Kessler Marketing Intelligence (Newport, RI).

Cable
- Alcoa-Fujikura Ltd.
- ATT
- Belden Wire and Cable
- Chromatic Technologies Inc.
- Northern Lights Cable Inc.
- Siecor

Cable End Boxes/Enclosures
- Anixter
- Fiber Instrument Sales Inc.
- Fiber Optic Center Inc.
- Graybar

Connectors
- Anixter
- Automatic Tool and Connector Co. Inc.
- Fiber Instrument Sales Inc.
- Fiber Optic Center Inc.
- Graybar
- 3M

Mechanical Splices
- AMP Inc.
- Advanced Custom Applications Inc.
- Norland Products
- 3M

Splice Trays
- Anixter
- Fiber Instrument Sales Inc.
- Fiber Optic Center Inc.
- Graybar

Splice Enclosures
- Anixter
- Fiber Instrument Sales Inc.
- Fiber Optic Center Inc.
- Graybar

Fusion Splice Covers
- ACP International Inc.
- Anixter
- Fiber Instrument Sales Inc.
- Fiber Optic Center Inc.
- Graybar

Furcation Tubing
- Fiber Instrument Sales Inc.
- Northern Lights Cable Inc.

Pigtails
- Fiber Instrument Sales Inc.

Inner Duct
- Anixter
- Fiber Instrument Sales Inc.
- Fiber Optic Center Inc.
- Graybar

NECESSARY TOOLS: CABLE INSTALLATION

Condux International, Inc.
Greenlee Textron

NECESSARY TOOLS: SPLICING

Fusion Splicer
- Alcoa-Fujikura Ltd.
- Anixter
- Preformed Line Products Inc.
- Fiber Instrument Sales Inc.
- Siecor

Fiber Cleaver
- Alcoa-Fujikura Ltd.
- Fiber Instrument Sales Inc.
- Fitel
- Thomas and Betts

NECESSARY TOOLS: CONNECTOR INSTALLATION

Anixter
Fiber Instrument Sales Inc.
Fiber Optic Center Inc.
Graybar

TEST EQUIPMENT

Anixter
Arcade Electronics Inc.
EXFO
Fiber Instrument Sales Inc.
Fiber Optic Center Inc.
Fotec Inc.
Graybar
Laser Precision Corp.
Leica
Tektronix Inc.

SUPPLIES: CABLE INSTALLATION AND CABLE END PREPARATION

American Polywater Corporation
Anixter
Fiber Instrument Sales Inc.
Fiber Optic Center Inc.
Graybar
P-T Technologies Inc.
3M

SUPPLIES: SPLICING

ACP International
Anixter
Fiber Instrument Sales Inc.
Fiber Optic Center Inc.
Graybar

SUPPLIES: CONNECTOR INSTALLATION

Anixter
Fiber Instrument Sales Inc.
Fiber Optic Center Inc.
Graybar

APPENDIX

7

Company Contact Information

3M PRIVATE NETWORK PRODUCTS
6801 River Place Blvd.
Austin TX 78726
(512) 984-1800

ACP INTERNATIONAL
1010 Oakmead
Arlington, TX 76011
(817) 640-0992
(817) 633-3131 (fax)

ALCOA FUJIKURA LTD.
Box 3127
Spartanburg, NC 29304
(800) 235-3423
(864) 433-5353 (fax)

AMERICAN POLYWATER CORP.
11222 North 60th St., Box 53
Stillwater, MN 55082
(612) 430-2270
(800) 328-9384
(612) 430-3634 (fax)

AMP INC.
P.O. Box 3608
Harrisburg, PA 17105-3608
(717) 986-5279
(717) 986-7321 (fax)

AMPHENOL/FIBER OPTIC PRODUCTS
1925A Ohio St.
Lisle, IL 60532
(800) 678-0141
(708) 810-5640 (fax)

AT&T NETWORK CABLE SYSTEMS
2000 NE Expressway
Norcross, GA 30007
(800) 824-1931

AUGAT COMMUNICATIONS GRP.
23315 66th Ave. S.
Kent, WA 98032
(206) 854-9802
(206) 813-1001 (fax)

AUTOMATIC TOOL AND CONNECTOR
COMPANY INC.
1 Gary Rd.
Union, NJ 07083
(800) 524-2857
(908) 686-7710 (fax)

BELDEN WIRE & CABLE
P.O. Box 1980
Richmond, IN 47375
(317) 983-5200
(317) 983-5257 (fax)

CHROMATIC TECHNOLOGIES INC.
9 Forge Park
Franklin, MA 02038
(508) 520-1200
(508) 528-9950

CLAUSS CUTLERY CO.
Electronic Products Division
223 N. Prospect St.
Fremont, OH 43420
(800) 225-2877
(419) 332-7344
(419) 332-8077 (fax)

CONDUX INTERNATIONAL, INC.
145 Kingswood Rd.
Mankato, MN 56002
(507) 387-6576
(507) 387-1442 (fax)

EXFO AMERICA
903 North Bowser, Suite 360
Richardson, TX 75801
(214) 907-1505
(800) 663-EXFO
(214) 907-2297 (fax)

FIBER INSTRUMENT SALES
161 Clear Rd.
Oriskany, NY 13424
(315) 736-2206
(800) 5000-FIS
(315) 736-2285 (fax)

FIBER OPTIC CENTER INC.
488 Pleasant St.
New Bedford, MA 02740
(800) IS FIBER
(508) 992-6464
(508) 991-8876 (fax)

FOTEC INC.
151 Mystic Ave.
Medford, MA 02155-4615
(800) 537-8254
(617) 396-6155
(617) 396-6395 (fax)

GN NETTEST LASER PRODUCTION DIV.
109 N. Genesee St.
Utica, NY 13502-2596
(315) 797-4449
(315) 789-4038 (fax)

GREENLEE TEXTRON
4455 Boeing Dr.
Rockford, IL 61109
(815) 397-7070
(815) 397-8289 (fax)

KLEIN TOOLS INC.
7200 McCormick Blvd.
P.O. Box 599033
Chicago, IL 60649-9033
(708) 677-9500

LEETEC/WALTECH INC.
96 Vreeland Ave.
S. Hackensack, NJ 07606
(201) 641-7300
(201) 641-7229 (fax)

LEICA INC.
111 Deer Lake Rd.
Deerfield, IL 60015
(708) 405-0123
(708) 405-0147 (fax)

MESON FIBER OPTICS
4 Valley St.
Binghamton, NY 13905
(800) 45MESON
(607) 722-3776 (fax)

METHODE ELECTRONICS INC.
7444 W. Wilson Ave.
Chicago, IL 60656
(708) 867-9600
(708) 867-9130 (fax)

MOLEX/FIBER OPTIC INTERCONNECT
TECHNOLOGIES INC.
2111 Oxford Rd.
Des Plaines, IL 60018
(708) 803-3600
(800) A1-FIBER
(708) 803-3608 (fax)

NORLAND PRODUCTS INC.
P.O. Box 145
North Brunswick, NJ 08902
(908) 545-7828
(908) 545-9542 (fax)

NORTHERN LIGHTS CABLE INC.
BCIC Bldg. Box 758
North Bennington, VT 05257
(802) 442-5411
(802) 447-3845 (fax)

PREFORMED LINE PRODUCTS
660 Beta Dr.
Cleveland, OH 44143
(216) 461-5200
(216) 442-8816 (fax)

P-T TECHNOLOGIES, INC.
108 4th Ave. So.
Safety Harbor, FL 34695
(800) 441-7874
(813) 725-9544 (fax)

SIECOR CORP.
P.O. Box 13625
Research Triangle Park, NC 27709
(800) SIECOR4

SPECTRAN SPECIALTY
150 Fisher Dr., Box 1260
Avon, CT 06001
(203) 678-0371
(203) 674-8818 (fax)

TEKTRONIX, INC.
P.O. Box 500
Beaverton, OR 97077
(800) 426-2200
(413) 448-8033 (fax)

TRA-CON INC.
55 North St.
Medford, MA 02155
(800) TRA-CON1
(617) 391-5550
(617) 391-7380 (fax)

Chapter Notes

CHAPTER 1

1. The wavelength is an indicator of the color of light.
2. For clarity, we will use the term "attenuation" to refer to fibers and cables. We will use the term "loss" to refer to connections.
3. In digital systems, this mechanism results in pulse spreading. In analog systems, the mechanism results in signal smearing. Both spreading and "smearing" result in a reduction in the accuracy of the signal. When considering pulses of light, we are assuming a digital transmission system. However, fiber optics is used for both digital and analog transmission.
4. For clarity, we use the word "type" to refer to the fiber, the word "design" to refer to cable, and the word "style" to refer to connectors.
5. The specified measurement conditions are those of no chromatic dispersion. However, almost all fiber communication systems have operating conditions that result in chromatic dispersion. Thus, most systems differ significantly from the precisely defined test conditions. Because of this difference, you cannot divide the bandwidth-distance product by the distance of transmission to calculate the effective bandwidth, or maximum real bandwidth, that a link can transmit.

CHAPTER 4

1. Connectors other than that specified in this procedure may require a different crimper. Follow the instructions of the connector manufacturer.
2. A shroud can be made by drilling out a metal ST-compatible barrel. Some manufacturers sell shrouds.
3. The oven temperature and cure time depend on the epoxy, the core diameter, the clad diameter, and the ferrule material. If not using the combination in this book, follow the instructions of the connector manufacturer.
4. Note that some cables will not work with this connector. Check with Automatic Tool and Connector Co. to determine whether the cable you have will work with this connector installation technique.
5. While condition of the core is important to the optical performance of the connector, the condition of the clad provides information concerning the installation procedure. For example, a clad that is not round will not produce high loss, but indicates the use of excessive force in some part of the installation procedure. You should identify the excessive force.
6. Occasionally, a connector will have a round, clear, featureless core and high loss. This situation indicates a problem inside the connector or in the cable. For instance, a fiber broken inside the backshell will result in high loss, but the core can be round, clear, and featureless.

CHAPTER 5

1. It is common for cables to have optical performance specifications at two different wavelengths. These two wavelengths allow for increase in capacity by two methods: changing the wavelength of operation to a wavelength at which the fiber has an increased capacity (Table 1–5); and using multiple wavelengths on the same fiber (wavelength division multiplexing), each of which carries a separate data stream.
2. 0 dBm is 1 milliwatt of optical power. 0 dBμ is 1 microwatt of optical power. -10 dBm is 10 dB below 1 milliwatt. A 3.01 dB loss is the same as a 50 percent loss. A 6.02 dB loss is the same as a 75 percent loss.
3. With the exception of one-fiber cables and two-fiber zip-cord duplex cables, the fiber is longer than the cable. This excess fiber length is due to the spiral path of the fiber, either in a buffer tube or of a buffer tube around a central strength member. Typical excess lengths are on the order of 0.15–0.2 percent.

CHAPTER 7

1. Under some conditions, an OTDR can be used to determine lack of continuity. An OTDR can be used to determine lack of continuity as long as the discontinuity does not occur in the optical dead zone, in an event zone, or near the end of the cable.
2. This correction factor depends on the cable type and cable manufacturer. However, it ranges from 0.15–0.2 percent.

APPENDIX 2

1. The loss that occurs is included in the optical power budget.
2. The attenuation in short jumpers can be ignored at all wavelengths. For glass fibers, the loss will be less than 0.004 dB/m at 850 nm.
3. These losses must also be measured by Method A.

APPENDIX 5

1. 400x for singlemode fibers.

Index